Group Theory and its Applications in Physics–1980

(Latin American School of Physics, Mexico City)

AIP Conference Proceedings
Series Editor: Hugh C. Wolfe
Number 71

Group Theory and its Applications in Physics–1980

(Latin American School of Physics, Mexico City)

Editor
Thomas H. Seligman
Instituto de Fisica, University of Mexico

American Institute of Physics
New York 1981

L.C. Catalog Card No. 81-66132
ISBN 0–88318–170–3
DOE CONF- 8006126

FOREWORD

The Latin American School of Physics 1980 was held in Mexico City from June 2nd to 20th. It's subject was "Group Theory and its Applications" with emphasis on application in particle and nuclear physics. This subject related the school closely to the International Colloquium on the same subject that was held the week after the school in Cocoyoc near Mexico City.

The subject was covered in ten more extended and three shorter series of lectures; eight of these lectures are reproduced in the present volume. The other lectures are not reproduced because some covered introductory textbook material and others because the authors could not prepare a manuscript. Furthermore the students of the school had opportunity to present their own work in a series of seminars not reproduced here.

The scientific proceedings of the school may be summarized by looking briefly at the subjects discussed. In representation theory new points of view and extensions of classical material on non-compact groups where presented (K.B. Wolf, J. Plebański) while the subject of compact groups was covered in an introductory course (E. Chacón). This clearly indicates the direction in which the more mathematical aspects of the game are pointing these days. In nuclear physics the trend clearly points towards an algebraic understanding of coherent and collective phenomena (D. Rowe, V. Vanagas, W. Zahn) as opposed to the more conventional methods related to the shell model. Concerning particle physics the trend leans toward gauge theories (O 'Raifeartaigh, J. Boya-Balet, L. Jacobs) but very interesting alternatives where shown by A.O. Barut, in the context of non-perturbative treatments. The role of group theory in modern problems of mathematical physics was emphasized in courses relating to the representation of canonical transformations (A. Grossmann, M. Moshinsky) and topics in non-linear systems (F. Pirani).

The number of participants (Students and Lecturers) totaled 75 from 21 countries breaking up into 29 from Mexico 20 from other Latin American countries and 26 from countries outside Latin America. The school was financially supported by the organization of American States, the Consejo Nacional de Ciencia y Tecnología (México) and the Universidad Nacional Autónoma de México. Furthermore the visit of individual lectures was subsidized by the Ministry of Education, the Republic of France and the Federal Republic of Germany.

The school was held at the Institute of Physics whose director Jorge Flores and whose staff I wish to thank for their continuous support essential for the outcome of this school.

I also wish to thank A. Giachetti from the Organization of American States for his support and understanding as well as L. Jacobs, P.A. Mello, M. Moshinsky and K.B. Wolf, for their advice and help. Last but not least I am grateful to A. Angeles who took care of the administrative part of the organization.

T.H. Seligman
Organizer LASP 1980

LECTURERS

A. O. BARUT	Department of Physics & Astronomy University of Colorado. USA.
E. CHACON	Instituto de Física, UNAM, Mexico
L. J. BOYA	Departamento de Física Teórica Universidad de Salamanca, España.
A. GROSSMAN	CNRS, Marseille, France.
L. JACOBS	Instituto de Física, UNAM, Mexico.
M. MOSHINSKY	Instituto de Física, UNAM, Mexico.
F. PIRANI	Centro de Investigación y Estudios Avanzados, IPN. Mexico y/o King's College, London. G.B.
J. F. PLEBAÑSKI	Centro de Investigación y Estudios Avanzados, IPN. Mexico.
L. O'RAIFEARTAIGH	Dublin Institute for Advanced Studies Dublin, Ireland.
D. J. ROWE	Department of Physics University of Toronto. Canada.
V. VANAGAS	Institute of Physics, Academy of Sciences of Lithuanian SSR. USSR.
K.B. WOLF	IIMAS, UNAM, Mexico.
W. ZAHN	Institut für Theoretische Physik, Universität Erlangen, W. Germany.

ORGANIZER

T. H. SELIGMAN

Instituto de Física, UNAM, México.

PARTICIPANTS

R. ABLAMOWICZ	University of Carbondale, Carbondale, Illinois, USA.
B. ANGERMANN	Institut für Theoretische Physik der Technischen Universität Clausthal, Clausthal, W. Germany.
A. ANTILLON	Instituto de Física, UNAM, México.
J. BROECKHOVE	Universiteit Antwerpen, Antwerpen, Belgium.
P. H. BUTLER	Physics Department, University of Canterbury, Christchurch, New Zealand.
A. S. CAMACHO	Universidad Industrial de Santander, Bucaramanga, Colombia.
J. CARIÑENA	Departamento de Física Teórica, Universidad de Zaragoza, España.
J. A. CASTILHO A.	Instituto de Fisica Teorica, Sao Paulo (S.P.) Brasil.
D. B. CLINE	Department of Physics, University of Wisconsin-Madison, Madison, Wi. USA.
W. CROMBIE	Brown University, Providence, Rhode Island. USA.
P. CHRISTILLIN	European Organization for Nuclear Research, Geneve, Suisse.
A. DAIGNEAULT	Department of Mathematics and Statistics, Université de Montréal, Montréal Quebec, Canada.
J. DUKELSKY	Comisión Nacional de Energía Atómica, Buenos Aires, Argentina.

E. DEUMENS	Dienst Theoretische en Wiskundige Natuurkunde, Universiteit Antwerpen, Antwerpen, Belgium.
E. ESTEBAN	Notre Dame University, Notre Dame, In. USA.
B. FERNANDEZ	Departamento de Física, Universidad de Panamá, Panamá.
J. L. GALAN ROJAS	Universidad Industrial de Santander, Bucaramanga, Colombia.
J. C. GARRIDO GARCIA	Escuela de Física, Universidad de Panamá, Panamá.
I. GARZON SOSA	Instituto de Física, UNAM, México.
G. GERMAN VELARDE	Instituto de Física, UNAM, México.
R. J. GLEISER	Universidad Nacional de Córdoba, Córdoba, Argentina.
A. GONGORA TREVIÑO	Instituto de Física, UNAM, México.
E. HENESTROZA	Instituto de Física, UNAM, México.
A. HENRIQUEZ	Institute of Physics, Universiteteit I Oslo, Oslo, Norway.
E. HERNANDEZ S.	Instituto de Física, UNAM, México.
P. HESS	Institut für Theoretische Physik der J. W. Goethe, Universitat Frankfurt, Frankfurt, West Germany.
R. HUERTA Q.	CIEA -IPN, México.
M. JARAMILLO	Departamento de Física, Universidad de los Andes, Bogotá, Colombia.
R. B. JUAREZ	Instituto de Física, UNAM, México.
M. KIBLER	Institut de Physique Nucléaire, Villeurbanne, France.

A. K. KWASNIEWSKI	Universytet Wrocławski, Wrocław, Polonia.
S. A. LABIANO	Facultad de Ciencias, UNAM, México.
L. M. LEON ROSSANO	Instituto de Física, UNAM, México.
J. LEON TELLEZ	Universidad Industrial de Santander, Bucaramanga, Colombia.
C. LIZARRAGA	Instituto de Física, UNAM, México.
J. C. LOPEZ	Escuela de Altos Estudios, Universidad de Sonora, México.
G. LOPEZ VELAZQUEZ	Instituto de Física, UNAM, México.
H. MARTINEZ	Instituto de Física, UNAM, México.;
C. P. MASSOLO	Facultad de Ciencias, Exactas, La Plata, Argentina.
C. MEDRANO	Instituto de Física, UNAM, México.
J. MILLAN HENAO	Departamento de Física, Universidad del Valle, Calí, Colombia.
L. MOCHAN	Instituto de Física, UNAM, México.
R. MONTEMAYOR	Facultad de Ciencias, UNAM, México.
A L. MONTES B.	Departamento de Física, Universidad de Antioquía, Medellín, Colombia.
L. E. MORALES	CIEA - IPN, México.
M. MORENO YNTRIAGO	Instituto de Física, UNAM, México.
T. MURPHY	University of Wisconsin - Madison, Department of Physics, Madison, Wisconsin. USA.
J. H. PACHECO	Instituto de Física, UNAM, México.

J. E. PAEZ	Escuela de Física, Universidad de Costa Rica, San José, Costa Rica.
G. A. PONCE F.	Universidad del Valle de Guatemala, Guatemala, Guatemala.
F. QUEVEDO R.	Universidad de San Carlos de Guatemala, Guatemala.
J. F. RECAMIER	Instituto de Física, UNAM, México.
S. REINECKE	Instituto de Física, Universidade Federal do Rio de Janeiro, Rio de Janeiro, Brasil.
I. RODITI	Centro Brasileiro de Pesquisas Fisicas, Rio de Janeiro, Brasil.
R. W. J. ROEL	University of Amsterdam, Amsterdam, The Netherlands.
A. RUEDA	Universidad de los Andes, Departamento de Fisica, Bogotá, Colombia.
Y. SAINT-AUBIN	Centre de Recherches de Mathematiques Appliquées, Université de Montréal, Montréal, Quebec, Canada.
M. SARACENO	Comisión Nacional de Energía Atómica, Buenos Aires, Argentina.
D. SEPULVEDA	Instituto de Física, UNAM, México.
J. L. SORIA	CEN -UNAM, Mexico.
C. VILLAREAL	Facultad de Ciencias, UNAM, México.
C. J. WOTZASEK	Universidade Federal do Rio de Janeiro, Instituto de Fisica, Rio de Janeiro, Brasil.

TABLE OF CONTENTS

TOPICS IN NONCOMPACT GROUP REPRESENTATIONS

Kurt Bernardo Wolf
Instituto de Investigaciones en Matemáticas
y en Sistemas.
Universidad Nacional Autónoma de México
Apartado Postal 20-726
México 20, D. F. México

CONTENTS

ISSN: 0094-243X/81/710001-72$1.50

INTRODUCTION

The theory of Lie algebras, Lie groups and their representations is a subject which hardly needs of a lengthy intoduction to highlight its intrinsec mathematical interest and manifold applications in physics. It is also a subject whose exposition in a course for physicists may be done from a wide variety of starting points, directions, speeds and objectives. We assume that the audience of physicists is reasonably confident in the use of quantum mechanics, angular momentum theory and the standard SU(3) methods of the old quark model, so we shall proceed presenting some of the Lie algebraic and group theoretical concepts, offering alternatives and generalizations, with the aim that the subject acquire a certain systematic structure. This should enable the student to work more at ease when applying the methods of Lie algebras and groups to the problems in his field.

The territory we will try to make more accessible is that of representations of noncompact algebras and groups. Symmetry algebras and groups à la Wigner-Raccah are probably standard in any physics curriculum, but dynamical algebras and the associated groups are less so. The similarities between the two are sufficient to use the concepts of the former in order to gain familiarity with the latter; but the differences in formalism and applications are sometimes dramatic. Unitary representations of noncompact groups are infinite dimensional, and generally describe covariance instead of invariance of a system.

Chapter 1 deals with the Heisenberg-Weyl algebra of quantum mechanics, and Chapter 2 with the corresponding group. This is an easy case, as the group is nilpotent and close to the abelian case. Chapter 3 explores a semisimple noncompact case, the 2+1 Lorentz algebra and group, as well as its covering groups.

We do not claim to present everything you always wanted to know in this field, as have other authors in References (1) and (2), so we hardly need to start with a disclaimer. We do wish to point out, however, that many of the prime areas of research in the last years have been left out, notably higher dimensional cases and their classification, similarity methods in differential equations, induced representations, pseudogroups, superalgebras and gauge theories.

It is hoped that this material will prove useful for mathematically inclined physicists. We mut apologize both to the purer mathematical analysts for glossing over defining concepts or presenting proofs; also to the applied physicists, for not presenting the energy levels' match of any molecule, atom, nucleus or resonance. We hope our list of references will partially atone for the sins of omission. Those of comission, we also hope, will entice the reader to explore the source literature.

CHAPTER 1: A LIE ALGEBRA

We start this set of lecture notes with an example -a rather simple and presumably well-know one- of a Lie algebra. We shall then examine its representations, its irreducible representations, and finally, its unitary irreducible representations.

1.1 A Lie algebra and its bases.

We consider the Lie algebra w with elements Q, P and I, over the field of complex numbers, defined by its Lie bracket operation

$$[Q,P]=iI,\ [Q,I]=0,\ \ [P,I]=0. \tag{1.1}$$

The symbols used for Q and P suggest, of course, that these are the quantum-mechanical operators of position (multiplication by the coordinate q), momentum ($-i\hbar$ times differentiation with respect to q) while I should be a multiple $\hbar$ of the unit operator 1, on the space of quantum mechanical wavefunction completed with respect to an inner product so as to form a Hilbert space, normally $L^2(R)$. We shall later obtain these as a realization of w, but meanwhile they are to be taken only as formal symbols.

The three elements Q, P and I constitute a *vector* basis for the algebra, in the sense that any element of w can be written as

$$E = xQ + yP + zI \quad , x,y,z \in C \tag{1.2}$$

An *algebraic* basis for w is provided by Q and P alone, as I is produced through the Lie bracket in the first of Eqs. (1.1).

Particular elements in w which we will refer to are

$$R = \frac{1}{\sqrt{2}}\,(Q-iP)\,, \qquad L = \frac{1}{\sqrt{2}}\,(Q+iP)\,. \tag{1.3a}$$

These are to be related, later, with the raising and lowering operators for the harmonic oscillator wavefunctions. They satisfy

$$[L,R] = I\ ,\ \ [L,I] = 0,\ [R,I] = 0, \tag{1.3b}$$

and may thus also serve to define w over the field C.

1.2 Representations.

A *representation* ρ of a Lie algebra a on a vector space V is a *homomorphism* (i. e. a mapping which preserves the operators of the algebra: Linear combination and Lie brackets)

$$\rho : a \longrightarrow g\ell\,(V)\ , \tag{1.4}$$

from a *into* the algebra gl(V) of linear operators on V, where the Lie bracket is the commutator.

If $\{X_j\}_{j=1}^{D}$ is a D-dimensional vector basis for a, abstractly defined through the set of structure constants c^{ℓ}_{jk} as

$$[X_j, X_k] = i \sum_{\ell=1}^{D} c^{\ell}_{jk}\, x_\ell , \tag{1.5}$$

and $\mathrm{X}_j = \rho(x_j) \in$ gl(V), then the homomorphism requirements are

$$\rho(c_j X_j + c_k X_k) = c_j\, \rho(X_j) + c_k\, \rho(X_k) = c_j \mathrm{X}_j + c_k \mathrm{X}_k \tag{1.6a}$$

$$\rho(\,[X_j, X_k]\,) = \mathrm{X}_j \mathrm{X}_k - \mathrm{X}_k \mathrm{X}_j = i \sum_{\ell=1}^{D} c^{\ell}_{jk}\, \mathrm{X}_\ell . \tag{1.6b}$$

When V is an N-dimensional vector space, gl(V) is the set of $N\times N$ matrices. Amongst these, for w, we should be able to find three matrices $\mathbf{Q}$, $\mathbf{P}$ and $\mathbf{I}$ such that (1.1) holds replacing each symbol for its boldface homonym, and the bracket meaning commutation. Abstractly, you will recall that the Lie bracket is only required to be skew-symmetric $[A,B] = -[B,A]$, bilinear $[aA+bB,C] = a[A,C]+b[B,C]$, and to satisfy the Jacobi identity $[A,[B,C]]+[B,[C,A]]+[C,[A,B]]$. Poisson (3, Sect III-D and V-A) and Moyal (4,5) brackets, as well as commutators, satisfy these three requirements.

When V is a function- or other infinite-dimensional space, gl(V) is the set of all linear operators with domain and range in that space. Amongst these we should be able to find, for w, three linear operators $\mathbb{Q}$, $\mathbb{P}$ and $\mathbb{I}$ such that (1.1) holds for them. Clearly here we run headlong into trouble, since we can easily find that some proposed operator does not '*quite*' have V for its domain, or that it may send elements of V out of V. We may require that it do the job only in a dense subspace of V, and that will take us into Hilbert spaces.

1.3 The adjoint representation.

For any D-dimensional Lie algebra a [with a D-dimensional vector basis $\{X_j\}_{j=1}^{D}$, defined by the structure constants $c^{\ell}_{jk} = -c^{\ell}_{kj}$ in the Lie Bracket (1.5)] one can always produce a $D\times D$ matrix representation, called the *adjoint* representation ρ^A of a, through

$$\mathrm{X}^A_k = \rho^A(X_k), \qquad (\mathrm{X}^A_k)_{mn} = i\, c^{n}_{m\,k} = -\,i\, c^{n}_{k\,m} . \tag{1.7a}$$

To prove this, replace (1.7a) into (1.5), X_k's replacing X_k's, thus finding for the m-n element of this equality

$$\sum_s [\, c^{s}_{mj}\, c^{n}_{ks} + c^{s}_{km}\, c^{n}_{js} + c^{s}_{jk}\, c^{n}_{ms} \,] = 0, \tag{1.7b}$$

after a dummy-index change. Equation (1.6) is equivalent to the statement of the Jacobi identity for the algebra.

The adjoint representation of w may be obtained from (1.7), numbering Q, P and I in (1.1) by 1, 2, 3. As only $c^3_{21} = -1$ and $c^3_{12} = 1$ are nonzero,

$$Q^A = \begin{pmatrix} 0 & 0 & 0 \\ 0 & 0 & -i \\ 0 & 0 & 0 \end{pmatrix}, \quad P^A = \begin{pmatrix} 0 & 0 & i \\ 0 & 0 & 0 \\ 0 & 0 & 0 \end{pmatrix}, \quad I^A = 0 . \tag{1.8}$$

This is a representation, but it is not *faithful*, as the algebra element I is mapped on the zero matrix. This is a general feature of algebras with *a centre*, i. e. element(s) which commute with every element of the algebra.

1.4 A faithful 3×3 matrix representation.

For the case of w we have to use other arguments: If Q is represented by a matrix with a single off-diagonal nonzero element in the (a,b) position, $a \neq b$, and P by another such matrix with a nonzero (c,a) element, $c \neq a$, the commutator representing I will have non-zero elements in the (b,c) position and, if $b=c$, in the (a,a) position. The latter will commute with the former two when the (a,a) element is zero, i. e., when $b \neq c$. Adjusting signs and i's so that (1.1) be obtained, we may thus provide a simple 3×3 faithful representation through $a=1$, $b=2$, $c=3$:

$$Q^{\delta} = \begin{pmatrix} 0 & 1 & 0 \\ 0 & 0 & 0 \\ 0 & 0 & 0 \end{pmatrix}, \quad P^{\delta} = \begin{pmatrix} 0 & 0 & 0 \\ 0 & 0 & 0 \\ 1 & 0 & 0 \end{pmatrix}, \quad I^{\delta} = \begin{pmatrix} 0 & 0 & 0 \\ 0 & 0 & 0 \\ 0 & i & 0 \end{pmatrix}. \tag{1.9a}$$

As a representation is a linear mapping, every element (1.2) of w may be represented through

$$E^{\delta} = x\, Q^{\delta} + y P^{\delta} + z I^{\delta} = \begin{pmatrix} 0 & x & 0 \\ 0 & 0 & 0 \\ y & iz & 0 \end{pmatrix}. \tag{1.9b}$$

1.5 Equivalent representations and isomorphisms of w.

We may perform any similarity transformation on representation matrices (as $M^{\delta} \to M^{\delta e} = A M^{\delta} A^{-1}$ for $M^{\delta} = Q^{\delta}$, P^{δ} and I^{δ}) and obtain another *equivalent* representation. For w we may do more, however: We can perform linear combinations of its vector basis elements obtaining new basis elements with the *same* commutation relations:

$$\begin{pmatrix} \bar{Q} \\ \bar{P} \\ \bar{I} \end{pmatrix} = \begin{pmatrix} a & b & u \\ c & d & v \\ 0 & 0 & ad-bc \end{pmatrix} \begin{pmatrix} Q \\ P \\ I \end{pmatrix}, \qquad a, b, c, d, \; u, v, \in C , \tag{1.10}$$

i. e. (1.1) for Q, P and I implies (1.1) also holds for Q, P and

I implies (1.1) also holds for Q, P and I. The linear transformation (1.10) is an isomorphism of w. This is a rather exceptional case, as usually the isomorphism group of a Lie algebra is only the adjoint action of the Lie group associated to it. For w it is $\mathrm{I}^2\mathrm{GL}(2,\mathrm{C})$.

The above paragraph binds the 'simplest' representation (1.9) to the more popular representation [3, Eq' (2.4); 6, Eqs. (2.1)] given by

$$Q^p = \begin{pmatrix} 0 & -i & -i \\ i & 0 & 0 \\ -i & 0 & 0 \end{pmatrix}, \quad P^p = \begin{pmatrix} 0 & 1 & 1 \\ 1 & 0 & 0 \\ -1 & 0 & 0 \end{pmatrix}, \quad I^p = \begin{pmatrix} 0 & 0 & 0 \\ 0 & 2 & 2 \\ 0 & -2 & -2 \end{pmatrix}. \tag{1.11}$$

It is more popular since it may be generalized straighforwardly as an (N+2) × (N+2) representation of the (2N+1)-dimensional Heisenberg-Weyl algebra used in N-dimensional quantum mechanics, through 'vectorizing' the nonzero elements of Q^p and P^p. As matrices, the link between (1.11) and (1.9) is

$$A^{-1}\ Q^p\ A = \alpha Q^{\delta} + \beta P^{\delta} + \gamma I^{\delta} \tag{1.12a}$$

$$A^{-1}\ P^p\ A = -i\alpha Q^{\delta} + i\beta P^{\delta} + i\gamma I^{\delta} \tag{1.12b}$$

$$A^{-1}\ I^p\ A = -2i\alpha\beta I^{\delta} \tag{1.12c}$$

where in this case

$$A = \begin{pmatrix} \beta & i\gamma & 0 \\ 0 & \alpha\beta & 1 \\ 0 & 0 & -1 \end{pmatrix}. \tag{1.12d}$$

1.6 The Fock realization.

The Heisenberg-Weyl algebra can be also furnished with infinite-dimensional representations. To this end we prefer to work with the basis R, L and I as given by (1.3). We may propose to associate to each of these elements the formal Fock operators

$$R \longrightarrow R^F = z, \quad L \longrightarrow L^F = \frac{d}{dz}, \quad I \longrightarrow I^F = I, \tag{1.13}$$

i. e. L^F is the differentiation operator, $R^F f(z) = z f(z)$ is the multiplication-by-argument operator, and I^F, the unit operator. Acting on any differentiable function $f(z)$ we can see that (1.13) follow the commutation relations (1.3b). We prefer to speak of (1.13) as a *realization* of the algebra w, since a representation is defined only when a proper vector space V in (1.4) is fully specified. The Fock operators are purely formal up to now, with no mention as to their domain.

1.7 The power-function basis.

Clearly, we need some functions. Consider the infinite sequence of *power* functions

$$p_n^{\nu}(z) = z^{\nu+n}, \quad n \in Z = \{0,\pm 1,\pm 2,\ldots\}, \mathrm{Re}\ \nu \in [0,1). \tag{1.14}$$

These provide a proper basis for a function space V as they are differentiable, and (1.13) belong to $\mathrm{gl}(V)$:

$$\mathrm{R}^F\, p_n^{\nu}(z) = p_{n+1}^{\nu}(z), \tag{1.15a}$$

$$\mathrm{L}^F\, p_n^{\nu}(z) = (n+\nu)\, p_{n-1}^{\nu}(z), \tag{1.15b}$$

$$1^F\, p_n^{\nu}(z) = p_n^{\nu}(z). \tag{1.15c}$$

The action of (1.13) can be extended to the whole of V through linear combiation. Note carefully that the vector space axioms speak only of finite linear combinations, and hence V is *not* the space of z^{ν} times analytic functions in some annullus around the origin (having a convergent Laurent expansion), but only the subspace of finite sums of power functions. The limit points are the *closure* of that space, of which we shall have a right to talking only when we introduce a *norm* into that space. Meanwhile, though, we do have a representation through infinite matrices with rows and columns $n,m \in Z$, the set of integers:

$$\mathrm{R}^{F(\nu)} = \begin{array}{c} n\backslash m \\ \downarrow \\ 2 \\ \downarrow \\ 1 \\ \downarrow \\ 0 \\ \downarrow \\ -1 \\ \downarrow \\ -2 \\ \downarrow \\ \vdots \end{array} \begin{array}{c} \rightarrow 2 \rightarrow 1 \rightarrow 0 \rightarrow -1 \rightarrow -2 \rightarrow \cdots \\ \left(\begin{array}{cc|ccc} 0 & 1 & & & \\ & 0 & 1 & & \\ \hline & & 0 & 1 & \\ & & & 0 & 1 \\ & & & & 0 \\ & & & & \\ \end{array}\right) \end{array} \tag{1.16a}$$

$$
L^{F(\nu)} \qquad
\begin{array}{c|ccccc}
n\backslash m & 2 & 1 & 0 & -1 & -2 \\
\hline
2 & 0 & & & & \\
1 & \nu+2 & 0 & & & \\
0 & & \nu+1 & 0 & & \\
-1 & & & \nu & 0 & \\
-2 & & & & \nu+1 & 0 \\
\vdots & & & & &
\end{array}
\tag{1.16b}
$$

$$
I^{F(\nu)} = 1, \tag{1.16c}
$$

acting on infinite-component vectors, where $p_n^{\nu}(z)$ is represented by a column with a single nonzero entry 1 in the nth position. This entry is raised by one place by the raising matrix (1.16a), lowered and multiplied by $n + \nu$ by the lowering matrix (1.16b), and left invariant by (1.16c).

Clearly also $R^{F(\nu)}$, $L^{F(\nu)}$ and $I^{F(\nu)}$ given by (1.16) satisfy the commutation relations (1.3b): $L^{F(\nu)}R^{F(\nu)}$ and $R^{F(\nu)}L^{F(\nu)}$ are diagonal matrices with entries $\nu + n + 1$ and $\nu + n$ along the diagonal.

1.8 Irreducible and indecomposable representations.

When ν is not zero, every $p_n^{\nu}(z)$ may be moved and, through repeated application of $R^{F(\nu)}$ or $L^{F(\nu)}$, taken to a function proportional to any other $p_m^{\nu}(z)$, $m \in Z$. No subspace being invariant, under the three algebra generators; the whole space spanned by $\{p_n(z)\}_{n\in Z}$ is thus required as a basis for V, which is thus an *irreducible* representation basis.

When $\nu=0$ then $L^{F(\nu)}$ has a zero in the 0th column, and something dramatic happens: If $n < 0$, $p_n^0(z)$ may be lowered or raised to any other $p_m^0(z)$, $m \in Z$, but if $n \geqslant 0$ it may be raised, but not lowered below $n=0$. The space spanned by $\{p_n^0(z),\ n \in Z\}$ may be divided thus into two disjoint subbases: $\{p_n^0(z),\ n \geqslant 0\}$ which still transforms irreducibly under the algebra, and the subspace generated by $\{p_n^0(z),\ n < 0\}$ which does not: It '*spills over*' into the first through repeated action or the raising operator. Every element of w in the Fock representation (1.16) for $\nu = 0$ and its powers, are represented by a block-triangular infinite matrix acting on the basis vectors as

$$\left(\begin{array}{c|c} \frac{A}{O} & \frac{B}{C} \end{array}\right)\left(\begin{array}{c} P^{0}_{(\geq 0)}(z) \\ \hline P^{0}_{(<0)}(z) \end{array}\right) = \left(\begin{array}{c} A\,P^{0}_{(\geq 0)}(z) + B\,P^{0}_{(<0)}(z) \\ \hline C\,P^{0}_{(<0)}(z) \end{array}\right) \quad (1.17)$$

where we have separated by horizontal and vertical lines on 0th and $(-1)^{th}$ rows and columns. The representation of w provided by A is still irreducible, but (1.16) is reducible, though *indecomposable* since, as can be verified, it can *not* be decomposed fully into a block-diagonal form signifying two (or more) irreducible parts. Triangular matrices cannot be diagonalized.

1.9 Self-adjoint representations.

In Lie group theory one is generally interested in hermitian or self-adjoint irreducible representations of Lie algebras (1.5); thence the apparently unconfortable i in (1.5)-(1.6b), which in-sures that we may revert to the algebra a over the real field and still retain hermiticity for every element of the algebra. For our Lie algebra w, this is needed by the current axioms of Quantum Mechanics as given by Dirac and von Neumann (7, 8 Sect. 2.1, 9) which require that the position and momentum operators $\mathbb{Q}$ and $\mathbb{P}$ representing Q and P in w be self-adjoint -or rather, have self-adjoint extensions- in $L^2(R)$. This turns into a statement of intention to reduce our attention to representations (1.4) where V is a *Hilbert* space endowed with and Cauchy-sequence closed under an appropiate positive definite inner product $(\cdot,\cdot)$: $V \times V \to C$. Axioms require that it be linear in one argument (physicists prefer the second argument) antilinear in the first, $(\underset{\sim}{f},\underset{\sim}{f}) \geq 0$ and $(\underset{\sim}{f},\underset{\sim}{f}) = 0$ iff $\underset{\sim}{f} = \underset{\sim}{0}$. For some purposes a *norm* rather than an inner product is needed, so V is taken to be only a Banach space, but we shall forego this weaker scenario.

1.10 Hermiticity imposed.

Returning to our now self-adjoint representation for Q, P and I in w, the reader will no doubt recognize the Schrödinger realization as possessing a self-adjoint extension in $L^2(R)$. In accord with our emphasis on alternate approaches, however, let us insist in staying with the Fock realization (1.13) which from (1.3a) must be such that $\mathbb{L}^{F\dagger} = \mathbb{R}^F$, $\mathbb{R}^{F\dagger} = \mathbb{L}^F$ and of course $\mathbb{I}^{F\dagger} = \mathbb{I}^F$. What we need is a Hilbert space B such that these adjunction properties hold, i. e., an inner product $(\cdot,\cdot)_B$ where the formal Fock operators (1.13) satisfy

$$(\mathbb{L}^F \underset{\sim}{f}, \underset{\sim}{g})_B = (\underset{\sim}{f}, \mathbb{R}^F \underset{\sim}{g})_B \quad \text{for all } \underset{\sim}{f}, \underset{\sim}{g} \in B, \qquad (1.18)$$

plus some function-analytic minutae to determine the geometry of the space of functions which belong to $\mathcal{B}$ so that the limit points of Cauchy sequences of these are in the space.

It has been the work of Bargmann (10) to turn (1.18) into a set of two coupled partial differential equations for the weight function of a measure on the complex plane, and to determine precisely the geometry of the space -thereafter called *Bargmann's* space. Generalizations of this procedure have been performed by Barut and Girardello (11) and the author (12,13,14 Sect. 9.2). Let us follow a different argumentation here, based on the matrix representation (1.16) which we have already, and which we want to turn into a hermitian matrix representation.

An orthonormal denumerable basis for the putative Hilbert space $\mathcal{B}$, $\{g_n^\nu(z),\ n \in ?\}$ should have the property of providing a matrix representation of ω through

$$(\mathbb{L}^{F'(\nu)})_{mn} = (\underset{\sim}{q}_m^\nu, \mathbb{L}^F \underset{\sim}{q}_n^\nu)_{\mathcal{B}} = (\mathbb{R}^F \underset{\sim}{q}_m^\nu, \underset{\sim}{q}_n^\nu)_{\mathcal{B}} = (\underset{\sim}{q}_n^\nu, \mathbb{R}^F \underset{\sim}{q}_m^\nu)^*_{\mathcal{B}} = (\mathbb{R}^{F'(\nu)})^*_{nm}, \tag{1.19}$$

and $(\mathbb{I}^{F'(\nu)})_{mn} = (\underset{\sim}{q}_m^\nu, \underset{\sim}{q}_n^\nu)_{\mathcal{B}} = \delta_{mn}$. The representation (1.15)-(1.16) is not far off the mark The nonzero elements of $\mathbb{L}^{F(\nu)}$ are indeed in the position of the nonzero elements of $\mathbb{R}^{F(\nu)\dagger}$ (the dagger meaning transposition and conjugation), only the normalization is not quite right. If we were to set

$$q_n^\nu(z) = a_n^\nu\, p_n^\nu(z) = a_n^\nu\, z^{n+\nu}\ , \qquad a_n^\nu \in C,\ n \in Z, \tag{1.20a}$$

then (1.15) and (1.19) lead to $|a_n^\nu|^2(n+\nu) = |a_{n-1}^\nu|^2$. As absolute values are positive real numbers, this implies first that ν must be real. As for n, two recurrence relations for $|a_n^\nu|^2$ in terms of $|a_0^\nu|^2$ may be set up, depending on whether $n > 0$ or $n < 0$:

$$|a_n^\nu|^2 = |a_0^\nu|^2\ \Gamma(\nu + 1)\ /\ \Gamma(n + \nu + 1), \tag{1.20b}$$

$$|a_{-n}^\nu|^2 = |a_0^\nu|^2\ \Gamma(\nu + 1)\ /\ \Gamma(\nu - n + 1),\ n \geqslant 0. \tag{1.20c}$$

Equation (1.20b) yields a recursion relation, while (1.20c) is a limitant for $\nu \neq 0$ as $\Gamma(\nu - n + 1)$ alternates in sign for $n > 1$, yet absolute values must be positive. It follows that only $\nu = 0$ in (1.20b) will produce submatrices in (1.17) following the adjuntion property (1.19), necessary for a self-adjoint representation of ω. There, an orthonormal basis is $a_n\, p_n^0(z)$, with

$$a_n = (n!)^{-1/2}\, a_0,\quad n \geqslant 0,\quad a_{-n} = 0,\ n > 0. \tag{1.20d}$$

1.11 A measure for a Hilbert space.

What is the Hilbert space? In that space, an orthonormal basis should be provided by $q_n(z) = a_n p_n^0(z) = (n!)^{-1/2} a_0 z^n$, $n \geq 0$ (we have set the phase of a_n independent of n). What is a possible inner product which would satisfy this requirement? We may follow Galbraith and Louck (15) in proposing, for $f(z)$ and $g(z)$ analytic in a common open circle containing the origin,

$$(\underset{\sim}{f},\underset{\sim}{g})_{GL} = |a_0|^{-2}\, f\,(d/dz)^*\, g(z)\,\Big|_{z=0}. \tag{1.21a}$$

In this way, one ensures that

$$(\underset{\sim}{q}_n,\underset{\sim}{q}_{n'})_{GL} = (n!\, n'!)^{-1/2}\, \frac{d^n}{dz^n}\, z^{n'}\,\Big|_{z=0} = \delta_{n,n'}. \tag{1.21b}$$

The inner product (1.21) is very handy for computations involving quantum creation and annihilation operators, since (1.21a) is in shell-model language just $\langle 0|f(a)g(a^\dagger)|0\rangle$ while f and g are usually polynomials. Mathematicians -and many physicists- prefer inner products defined through integrals since Hilbert space theory is usually cast in that way. If we momentarily assume that g in (1.21a) satisfies also the conditions of the Fourier integral theorem -although the $q_n(z)$ clearly do not- we may write

$$\begin{aligned}
(f,g)_{GL} &= |a_0|^{-2}\, f(d/dz)^*\, \frac{1}{2\pi} \int_{-\infty}^{\infty} dp \int_{-\infty}^{\infty} dz'\, e^{ip(z-z')} g(z')\Big|_{z=0} \\
&= |a_0|^{-2}\, \frac{1}{2\pi} \int_{-\infty}^{\infty} dp \int_{-\infty}^{\infty} dz'\, f^*(ip)\, g(z') e^{-ipz'} \\
&= |a_0|^{-2}\, \frac{i}{2\pi} \int_{-\infty}^{\infty} dz \int_{-i\infty}^{i\infty} dz^*\, f^*(z^*) g(z) e^{-z^* z} \\
&= |a_0|^{-2}\, \frac{1}{\pi} \int_C d^2z\, f(z)^* g(z)\, e^{-|z|^2} = (\underset{\sim}{f},\underset{\sim}{g})_B.
\end{aligned} \tag{1.22}$$

After taking $z=0$ we have changed variables to the independent $z=z'$, $z^* = ip$, then to variables $x = \text{Re}\, z = \frac{1}{2}(z + z^*)$ and $y = \text{Im}\, z = -i\,\frac{1}{2}(z - z^*)$ which are readily interpreted as the real and imaginary parts of a complex variable z, of which both f and g are analytic functions, the integration ranges of z and z^* are convertible into an integration over the complex $z = x + iy$ plane C, with measure $d^2z = dxdy = d\,\text{Re}\,z\; d\,\text{Im}\,z = |z|d|z|\; d\arg z$.

If we set $a_0 = 1$, we have exactly *Bargmann's* inner product $(\cdot,\cdot)_B$ (10, Eq. (1.6)), where (1.21b) can be easily verified, in spite of the fact that $q_n(z)$ does not abide the Fourier transform conditions. We may abandon the definition of $(\tilde{f},\tilde{g})_{GL}$ altogether and regard now only $(f,g)_B$ which poses its existence requirements quite openly: $f(z)$ must be *entire* analytic functions (i. e. no poles or branch cuts are allowed on the finite z-plane) and the growth at infinity should be overcome by the weight function $e^{-|z|^2}$. In fact, the functions may grow in some direction in C -as analytic functions do-, but may not grow faster than $\exp(\frac{1}{2}\phi\, z^2)$ for some phase ϕ as otherwise they would overwhelm the weight functions in some direction of the complex plane.

The space of entire analytic functions of growth (2,1/2) constitute -as proved by *Bargmann* (10, Sect. 1)- a separable Hilbert space under $(\cdot,\cdot)_B$. In Bargmann's space, the Fock operators (1.13) constitute thus a self-adjoint irreducible representation of the Heisenberg-Weyl algebra (1.13).

1.12 The enveloping algebra.

The construction we have just completed is straightforward, except, we should say, in our original choice of basis functions (1.14). Had we chosen a set of functions other than power functions, for example trigonometric ones, the construction of the representations of the algebra and the argument about their irreducibility and self-adjointness would have been considerably more involved. We chose power functions for the reason that we knew that $\{p_n^{\nu}(z)\}$ would map among themselves up and down the ladder under multiplication and differentiation: They are eigenfunctions of the operator $\mathbb{R}^F\mathbb{L}^F = z\,d/dz$ with eigenvalue $n+\nu$. The product $R\,L$ between algebra elements is undefined within the framework of the algebra. It may be incorporated through the simple device of defining a (noncommutative) product operation for the algebra elements and placing this product in a new set called the *enveloping algebra* $\bar{a}$ of the original algebra a.

If we want $\bar{a}$ to be an algebra itself, the product in $\bar{a}$ must be bidistributive with respect to the sum,

$$(c_1X_i + c_2X_j)\, X_k = c_1X_iX_k = c_1X_iX_k + c_2X_jX_k, \tag{1.23a}$$

$$X_i(c_3X_j + c_4X_k) = c_3X_iX_j + c_4X_iX_k,\ c_\ell \in C, \tag{1.23b}$$

and with respect to the Lie bracket it must satisfy the '*Leibnitz rule*'

$$[X_i, X_jX_k] = [X_i, X_j]X_k + X_j[X_i, X_k], \tag{1.23c}$$

which is an identity if the Lie bracket is the commutador. The enveloping algebra $\bar{a}$ of a has the structure of a ring (i.e. a non-commutative bilinear product is defined, but inverses under the product are not, and there is generally no identity under this product -although it may be defined for algebras with a centre when we take equivalence classes modulo the centre).

1.13 The Schrödinger realization.

The enveloping algebra is a very useful concept when it comes to find ways of building irreducible representations and classifying them, as the Casimir operators of semisimple algebras lie there. The properties of $N = RL = \frac{1}{2}(P^2 + Q^2 - I)$ under commutation with the elements of w are $[N,R] = R$, $[N,L] = -L$ and $[N,I] = 0$ so that if in some representation we manage to construct an eigenvector Ψ of $\mathbb{N}$ with eigenvalue μ, then by a well-known argument, $\mathbb{R}^n \Psi$ will also be an eigenvector of $\mathbb{N}$ with eigenvalue $\mu + n$, and so will $\mathbb{L}^n \Psi$ with eigenvalue $\mu - n$, unless such a vector is zero.

We can put these concepts to work on the Schrödinger realization of the Heisenberg-Weyl algebra:

$$Q \to \mathbb{Q}^S = q, \quad P \to \mathbb{P}^S = -i\frac{d}{dq}, \quad I \to \mathbb{I}^S = 1, \tag{1.24a}$$

whereby

$$R \to \mathbb{R}^S = \frac{1}{\sqrt{2}}\left(q - \frac{d}{dq}\right), \quad L \to \mathbb{L}^S = \frac{1}{\sqrt{2}}\left(q + \frac{d}{dq}\right), \quad N \to \mathbb{N}^S = \frac{1}{2}\left(-\frac{d^2}{dq^2} + q^2 - 1\right). \tag{1.24b}$$

From the arguments of the previous section, we look for the solutions $\psi^\nu(q)$ of

$$\mathbb{N}^S \psi_n^\nu(q) = (n+\nu)\psi_n^\nu(q), \quad n \in Z, \ \mathrm{Re}\ \nu \in [0,1). \tag{1.25a}$$

These are any linear combination of

$$\Psi_n^\nu(q) = c_n^\nu \ U(-[n+\nu] - \tfrac{1}{2}, \sqrt{2}\, q),$$
$$\Upsilon_n^\nu(q) = d_n^\nu \ V(-[n+\nu] - \tfrac{1}{2}, \sqrt{2}\, q), \tag{1.25b}$$

where c_n^ν and d_n^ν are arbitrary constants, and U and V are the Parabolic Cylinder functions classified in attention to their asymptotic behaviour (see 16, Sect. 19.3). They are related to the more familiar Whittaker D and U funtions through

$$U(\sigma,y)=D_{-\sigma-1/2}(y)=2^{-\sigma/2}y^{-1/2}\,W_{-\sigma/2,-1/4}(y^2/2)=$$

$$=2^{-1/4-\sigma/2}\,e^{-y^2/4}\,U(\frac{\sigma}{2}+\frac{1}{4},\frac{1}{2},\frac{x^2}{2}), \tag{1.26a}$$

$$V(\sigma,y)=\pi^{-1}\Gamma(\sigma+\frac{1}{2})\{\sin\pi\sigma\,D_{-\sigma-1/2}(y)+D_{-\sigma-1/2}(-y)\}. \tag{1.26b}$$

It is an easy matter to verify that $\mathbb{R}^S$ and $\mathbb{L}^S$ indeed raise and lower the values of n in $\{\Psi_n^\nu, \mathrm{T}_n^\nu\}$ by units (16, Sect. 19.6):

$$\mathbb{R}^S\,\Psi_n^\nu(q)=(c_n^\nu/c_{n+1}^\nu)\,\Psi_{n+1}^\nu(q), \tag{1.27a}$$

$$\mathbb{L}^S\,\Psi_n^\nu(q)=(n+\nu)(c_n^\nu/c_{n-1}^\nu)\,\Psi_n^\nu(q), \tag{1.27b}$$

$$\mathbb{R}^S\,\mathrm{T}_n^\nu(q)=(n+\nu+1)(d_n^\nu/d_{n+1}^\nu)\,\mathrm{T}_{n+1}^\nu(q), \tag{1.28a}$$

$$\mathbb{L}^S\,\mathrm{T}_n^\nu(q)=(d_n^\nu/d_{n-1}^\nu)\,\mathrm{T}_{n-1}^\nu(q). \tag{1.28b}$$

With (1.27) we are at the same point of the program for the Schrödinger realization as we were for the Fock realization, when we wrote (1.15a) for $c_n^\nu = 1$. The irreducible representations of W we obtain is thus identical to that of Eqs. (1.16) for $\mathrm{R}^{F(\nu)}$ and $\mathbb{L}^{F(\nu)}$, but now for $\mathrm{R}^{S(\nu)}_{(U)}$ and $\mathrm{L}^{S(\nu)}_{(U)}$. We have now two sets of basis vectors, $\{\Psi_n^\nu(q)\}_{n\in Z}$ and $\{\mathrm{T}_n^\nu(q)\}_{n\in Z}$, instead of the apparently single one $\{p_n^\nu(z)\}_{n\in Z}$ in (1.14). As far as the representation of the algebra is concerned, we relate them as $\mathbb{R}^{S(\nu)}_{(U)}=\mathrm{L}^{F(\nu)\dagger}_{(U)}$ and $\mathbb{L}^{S(\nu)}_{(V)}=\mathrm{R}^{F(\nu)\dagger}_{(V)}$. This provides the 'two' representations for the Fock case through $p_n^\nu \leftrightarrow p_{-n-1}^\nu$.

1.14 The special $\nu = 0$ cases.

Again, for $\nu=0$ we obtain from $\{\underset{\sim}{\Psi}_n^0\}_{n\in Z}$ the upper - triangular, and from $\{\underset{\sim}{\mathrm{T}}_n^0\}_{n\in Z}$ the lower-triangular reducible indecomposable representations (1.17). The irreducible parts are provided by $\{\underset{\sim}{\Psi}_n^0\}_{n\geqslant 0}$ and $\{\underset{\sim}{\mathrm{T}}_n^0\}_{n<0}$. Specifically, the $\underset{\sim}{\mathrm{T}}_n^0$-functions are the following:

$$\Psi_n^0(q)=c_n^0\,D_n(\sqrt{2}\,q)=c_n^0\,2^{-n/2}e^{-q^2/2}\,H_n(q),\; n=0,1,2,\ldots \tag{1.29a}$$

$$\Psi^0_{-n}(q) = c^0_{-n}\; 2^{n/2-1}\sqrt{\pi}\, e^{q^2/2}\, i^{n-1}\mathrm{erfc}\, q, \quad n=1,2,\ldots, \tag{1.29b}$$

where $H_n(q)$ are the Hermite polynomials and $i^m\mathrm{erfc}\, q$ the repeated integrals $\int_q^\infty dq'\, i^{m-1}\mathrm{erfc}\, q'$ of the complementary error function $i^0\mathrm{erfc}\, q = \mathrm{erfc}\, q = 2\pi^{-1/2}\int_q^\infty dq' e^{-q'^2}$. The T^0_n-functions, on the other hand, are related for $n \geq 0$ to the repeated integrals of Dawson's integral $F(q) = e^{-q^2}\int_0^q dq'\, e^{q'^2}$ as

$$T^0_n(q) = d^0_n\, e^{-q^2/2}\{\sqrt{2}\,(d^0_{n-1})^{-1}\int_0^q dq' e^{q'^2/2} T^0_{n-1}(q') - \frac{\sin(n\pi/2)}{2^{n/2}\Gamma(n/2+1)}\},$$

$$n=1,2,\ldots, \tag{1.30a}$$

$$T^0_0(q) = d^0_0\, 2\pi^{-1/2}\, e^{q^2/2}\, F(q), \tag{1.30b}$$

$$T^0_{-n}(q) = d^0_{-n}\, 2^{1-n/2}\,\pi^{-1/2}\, e^{q^2/2}(-i)^{n-1}\, H_{n-1}(iq), \quad n=1,2,\ldots \tag{1.30c}$$

For negative n's, the $T^0_{-n}(q)$ are thus simply proportional to $\Psi_n(iq)$.

We have gone into some detail in spelling out the eigenfunctions of the Schrödinger operator $\mathbb{N}^S = \mathbb{R}^S\mathbb{L}^S$. The reader will have recognized that (1.29a) are proportional to the well-known quantum harmonic oscilator wavefunctions, and $\mathbb{N}^S$ the corresponding *number* operator related to the Hamiltonian as $\mathbb{H}^{ho} = \mathbb{N}^S + \frac{1}{2}\mathbb{I}$.

1.15 Self-adjoint extensions of the number operator.

When looking for self-adjoint representations of ω, one can follow the same argument which lead us to (1.18) and conclude that we must stay with the subset of Ψ^0_n and T^0_{-n-1}, $n \geq 0$. In that direction, the search for an appropiate Hilbert space will lead us to L^2-spaces, where the Ψ^0_n, $n \geq 0$ are orthogonal and dense.

It is at least as instructive, however, to adopt a different approach which will yield results useful in other contexts as well. Since it was the number operator $\mathbb{N}^S = \mathbb{H}^{ho} - \frac{1}{2}\mathbb{I}$ which provided the basis functions (1.25) for the representations of ω, let us examine the L^2-spaces where $\mathbb{N}^S$ may be self-adjoint. This is a consequence of but not a requisite for $\mathbb{Q}^S$, $\mathbb{P}^S$ and $\mathbb{I}^S$ to have this property. We recall some facts: In the $L^2(a,b)$ Hilbert space defined through an inner product $(\underset{\sim}{f},\underset{\sim}{g})_{(a,b)} = \int_a^b dq \; f(q)^* \, g(q)$, we may show that $(\underset{\sim}{f},\mathbb{N}^S \underset{\sim}{g})_{(a,b)} = -\tilde{W}(\underset{\sim}{f},\underset{\sim}{g}) \Big|_a^b + (\mathbb{N}^S \underset{\sim}{f}, \underset{\sim}{g})_{(a,b)}$ holds, provided the norms of $\underset{\sim}{f}$, $\underset{\sim}{g}$, $\mathbb{N}^S \underset{\sim}{f}$ and $\mathbb{N}^S \underset{\sim}{g}$ are finite. The Wronskian $W(\underset{\sim}{f},\underset{\sim}{g}) = \underset{\sim}{f}\underset{\sim}{g}' - \underset{\sim}{g}\underset{\sim}{f}'$ valued at a and b, will be zero and $\mathbb{N}^S$ Hermitean, for spaces of functions with fixed logarithmic derivatives at those points, i.e. $h'(a) = p_a \, h(a)$ and $h'(b) = p_b \, h(b)$. Each pair of values p_a and p_b thus determines, together with certain further technical requirements related with the domain of $\mathbb{N}^S$ and its adjoint, a *self-adjoint* extension of $\mathbb{N}^S$ is determined. If $-a$ and/or b become infinity, the condition of asymptotic decrease -so that the norm of $\underset{\sim}{h}$ remain finite- takes precedence. We shall examine first the possibilities for the cas $(a,b) = (-\infty,\infty)$, then $(0,\infty)$ and last briefly, (a,b) finite.

1.16 $L^2(R)$.

The Ψ- and T-functions in (1.25b) were chosen for their asymptotic properties (16, Sect. 19.8):

$$\Psi_n^{\nu}(q) \underset{q\to\infty}{\sim} (c_n^{\nu} \; 2^{(n+\nu)/2}) \; q^{n+\nu} \; e^{-q^2/2} \quad , \tag{1.31a}$$

$$T_n^{\nu}(q) \underset{q\to\infty}{\sim} (d_n^{\nu} \; 2^{-(n+\nu)/2} \; \pi^{-1/2}) \; q^{-n-\nu-1} e^{q^2/2} \quad . \tag{1.31b}$$

The growing Gaussian behavior of T_n^{ν} -for all n and ν is sufficient reason to discard these as elements of the $L^2(R)$ space where $\mathbb{N}^S$ is to be self-adjoint. The next condition, at $q\to -\infty$, may be obtained through the special function relation

$$\Psi_n^{\nu}(-q) = \cos[\pi(n+\nu)]\,\Psi_n^{\nu}(q) + [\pi c_n^{\nu}/d_n^{\nu} \; \Gamma(-n-\nu)]\, T_n^{\nu}(q), \tag{1.31c}$$

and a similar one for $T_n^{\nu}(-q)$ which we omit as it is now unnecessary. From (1.31a) and (1.31c) it follows that for $q\to\infty$, only when the second summand in (1.31c) is zero will $\Psi_n^{\nu}(-q)$ be in $L^2(R)$. The finitude of $\Psi_n^{\nu}(q)$ is proven as an immediate consequence of (1.25a), since $\mathbb{N}^S$ has no singularities for finite q. This means that $-n-\nu = 0, -1, -2, \ldots$ i. e. that $\nu=0$ and n is a non-negative integer. The end result: Only $\{\Psi_n^0(q)\}_{n\geq 0}$ as given in (1.29a) may serve as basis functions for a self-adjoint representation of ω on $L^2(R)$. This set is closed under the algebra (c. f. Eqs. (1.27a), (1.27b)) and functional analysis tells us that the closure of the linear hull of $\{\underset{\sim}{\Psi}_n^0\}_{n\geq 0}$ is $L^2(R)$. This *quantizes* the problem through restricting the spectrum of $\mathbb{N}^S$ to be the set of nonnegative integers. We thus reconstitute the representation afforded by the Fock realization in Bargmann's Hilbert space. The unitary equivalence of the two will be implemented in Sect. 1.18.

1.17 $L^2(R^+)$.

Let us now turn to $L^2(0,\infty)$ spaces where the dense subset of once-differentiable functions have fixed logarithmic derivative at the origin, i. e. where $h'(0) = p\, h(0)$. The integrability conditions demanded by asymptotic behaviour still demand that only the $\Psi_n^\nu(q)$ appear. We find from special-function tables that

$$\Psi_n^\nu(0) = c_n^\nu\, 2^{(n+\nu)/2}\, \pi^{1/2} / \Gamma(\tfrac{1}{2}[1-n-\nu]), \tag{1.32a}$$

$$d\Psi_n{}^\nu(q)/dq|_{q=0} = -\, c_n^\nu\, 2^{(n+\nu)/2+1} \pi^{1/2}\, \Gamma(\tfrac{1}{2}[-n-\nu]). \tag{1.32b}$$

The basis of the $p=0$ space of functions whose derivative vanishes at the origin are $\Psi_n{}^\nu(q)$ such that (1.32b) is zero. The Gamma function provides this behaviour through its poles at $\frac{1}{2}[-n-\nu] = 0, -1, -2, \ldots$ This means $\nu=0$ and $n=2N$ where $N=0,1,2,\cdots$. The even-n harmonic oscillator wavefunctions in (1.29a) are thus obtained. Next, the basis of the $p=\infty$ space of functions which vanish at the origin require that (1.32a) be zero. Again, this is provided by the Gamma function for $\frac{1}{2}[1-n-\nu] = 0, -1, -2, \ldots$; i. e. $\nu=0$ and $n=2N+1$ where $N=0,1,2,\cdots$. These are the odd-n harmonic oscillator wavefunctions. Lastly, for fixed, finite p, $\Psi_n^{\nu\prime}(0) = p\, \Psi_n{}^\nu(0)$ implies the equality

$$\Gamma\,(\tfrac{1}{2}[1-n-\nu]) = -\,(2p)^{-1}\;\Gamma(\tfrac{1}{2}[-n-\nu]) \;. \tag{1.33}$$

This is a transcendental equation whose set of solutions for $n+\nu$ gives the spectrum of the self-adjoint extension of $\mathbb{N}^S$ determined by p. It is not difficult to see that for p positive, the solutions $n+\nu$ are all positive and that only one solution exists between any two consecutive integers. Similarly, we check that if some $n+\nu$ is a solution of (1.33), no $(n+M)+\nu$ with M integer may be a solution to to the same equation. For p negative, the spectrum has both positive and negative values, but the spacing property is the same.

Consequence: *Only* the $p=0$ and $p=\infty$ self-adjoint extensions of $\mathbb{N}^S$ have spectra with equally-spaced eigenvalues. Enter (1.27). The elements of w take us between functions whose $n+\nu$ differs by units or, as $\mathrm{Re}\,\nu \in [0,1)$, whose n differs by units. When applied in spaces determined by self-adjoint extensions of $\mathbb{N}^S$, the elements of w will *not* respect the constant p which therefore *cannot* be used to classify irreducible representations. Repeated application of $\mathbb{R}^S$ on the $p=0$ $\Psi_0^0(q)$ function will take us, at each step, between $p=0$ and $p=\infty$ functions. A self-adjoint representation of w requires thus the union of these two self-adjoint extension spaces of $\mathbb{N}^S$. In fact, when the basis functions are extended to all of R as even and odd functions, we obtain the previous $L^2(R)$ space. There, in addition, $\mathbb{Q}^S$ and $\mathbb{P}^S$ are not only hermitean, but self-adjoint. The hermiticity properties of $\mathbb{N}^S$ in the union of the $p=0$ and $p = \infty$ spaces is a consequence of the presence of functions which vanish at the origin, so that the crossed-p Wronskian continues to be zero. For any other self-adjoint extension space (determined by fixed, finite p) this is not so: The repeated application of the elements of w to some function in that space will never take us back to that same space.

We thus conclude that in order that an algebra of operators have a self-adjoint representation, it is necessary but not sufficient that the operators chosen to classify the basis functions be self-adjoint. For semisimple groups these are usually the Casimir operators.

The case worked out above makes it unnecessary to further analyze $L^2(a,b)$ for a and b finite: The spectrum of $\mathbb{N}^S$ is never equally spaced, as moreover, it asymptotically resembles the n^2 - spectrum of an impenetrable box. This argument will also eliminate from consideration any other self-adjoint extensions for the elements of the Lorentz algebra built out of $\mathbb{N}^S$, $(\mathbb{R}^S)^2$ and $(\mathbb{L}^S)^2$, which are elements in the universal covering algebra $\bar{w}$ of w.

1.18 The Bargmann transform.

The last point in this chapter will be to relate the Fock realization in Bargmann Hilbert space, Eqs. (1.13) and (1.22), with the Schrödinger realization in the $L^2(R)$ Hilbert space. Both of these are separable and hence should be unitarily equivalent. The integral kernel $A(z,q)$ which relates them as

$$f^B(z) = \int_R dq\, A(z,q)\, f(q) \in \mathcal{B}, \tag{1.34a}$$

$$f(q) = \int_C e^{-|z|^2}\, d^2z\, A(z,q)^*\, f^B(z) \in L^2(R), \tag{1.34b}$$

may be found (10, Introduction) through requiring that if $f(q) \in C^2 \cap L^2(R)$ is mapped on $f^B(z)$, then $\mathbb{R}^S f(q)$ and $\mathbb{L}^S f(q)$ be mapped on $\mathbb{R}^F f^B(z)$ and $\mathbb{L}^F f^B(z)$ respectively. This leads to a set of two coupled first-order differential equations whose solution was given by Bargmann (10). An equivalent solution may be found as a generating function built out of two dense orthonormal bases: Bargmann-normalized power functions $\{q_n^0\}_{n\geq 0}$ in (1.20a)-(1.20d) with $a_0 = 1$ for $\mathcal{B}$, and the harmonic oscillator wavefunctions $\{\Psi_n^0\}_{n\geq 0}$ in (1.29a) normalized with $c_n^0 = (\pi^{1/2}\, n!)^{-1/2}$ for $L^2(R)$:

$$\begin{aligned} A(z,q) &= \sum_{n=0}^{\infty} q_n^0(z)\, \Psi_n^0(q)^* \\ &= e^{-q^2/2}\, \pi^{-1/4} \sum_{n=0}^{\infty} (n!\, 2^{n/2})^{-1}\, z^n\, H_n(q) \qquad (1.35) \\ &= \pi^{-1/4} \exp\left[-\tfrac{1}{2}(z^2+q^2)+\sqrt{2}\, z\, q\right], \end{aligned}$$

where we have made use of the Hermite polynomial generating function.

CHAPTER 2: A LIE GROUP.

Out of the Heisenberg-Weyl Lie algebra of last Chapter we shall develop the Heisenberg-Weyl Lie group through the exponential map. The basic concepts of harmonic analysis on the group and coset manifolds will follow. Out of these we shall find various equivalent unitary irreducible representations through infinite matrices and integral kernels.

2.1 Ado's theorem and the exponential map.

A theorem by Ado (17) states that every Lie algebra $\mathcal{a}$ over C is isomorphic to some matrix algebra. That is, if we have a finite dimensional algebra we can find a finite dimensional faithful $N \times N$ matrix representation which will be a subalgebra of gl(N,C). For the Heisenberg-Weyl algebra w defined in the first chapter through (1.1), a faithful representation and subalgebra of gl(3,C) is given by (1.9); w is *not* contained in gl(2,C).

We can use this representation in order to define the *exponential map* of $\mathcal{a}$ into a Lie group G, which will be a subgroup of GL(N,C), the group of $N \times N$ nonsingular, complex matrices. If $\mathcal{a}$ has a vector basis $\{X_k\}_{k=1}^{D}$ faithfully represented by $N \times N$ matrices $\{\mathsf{X}_k\}_{k=1}^{D}$, then to every element $X=\Sigma_{k=1}^{D} x_k X_k$, $x_k \in C$ represented by $\mathsf{X}= \Sigma_{k=1}^{D} x_k \mathsf{X}_k$, we associate the matrix $\mathsf{G}(x_1,\ldots,x_D)= exp\, \mathsf{X}$, element of GL($N$,$C$) and faithful analytic representation of the Lie group G. The parameters $\{x_k\}_{k=1}^{D}$ constitute the *canonical coordinate* system of G. The exponential map exists since -it is easy to show- the exponential of an arbitrary $N \times N$ matrix with finite elements is an absolutely convergent series which yields another such matrix which is, moreover, invertible, (as $det\, exp\, \mathsf{X} = exp\, tr\, \mathsf{X}$) and sends the zero element in $\mathcal{a}$ to the unit element in G. Lastly, the matrix elements of $exp\, \mathsf{X}$ are analytic functions of the canonical coordinates. They thus satisfy one last requirement of representations of topological groups, namely that these be continuous maps of the abstract group into a matrix subgroup of GL(N,C).

2.2. The Heisenberg-Weyl group.

The algebra w with general element represented by X^{σ} in (1.9) is easy to exponentiate since the matrices are nilpotent. We have the elements of W represented by 3×3 matrices as

$$\mathsf{G}(x,y,z) = exp\; i(x\mathsf{Q}^{\sigma} + y\mathsf{P}^{\sigma} + z\mathsf{I}^{\sigma}) = exp\; i\begin{pmatrix} 0 & x & 0 \\ 0 & 0 & 0 \\ y & iz & 0 \end{pmatrix} \qquad (2.1)$$

$$= I + i\begin{pmatrix} 0 & x & 0 \\ 0 & 0 & 0 \\ y & iz & 0 \end{pmatrix} - \frac{1}{2!}\begin{pmatrix} 0 & 0 & 0 \\ 0 & 0 & 0 \\ 0 & xy & 0 \end{pmatrix} = \begin{pmatrix} 1 & ix & 0 \\ 0 & 1 & 0 \\ iy & -z-\frac{xy}{2} & 1 \end{pmatrix}.$$

$$= exp\ (ix\ Q^{\delta})\ exp\ (iy\ P^{\delta})\ exp(\ i\ [z + xy/2]\ I^{\delta}) \qquad \text{(2.1 cont.)}$$

$$= exp\ (iy\ P^{\delta})\ exp\ (ix\ Q^{\delta})\ exp(\ i\ [z - xy/2]\ I^{\delta}).$$

The last two lines go under the name of the *Weyl commutation relations* (18). Through introduction of the i in the exponential map we are assuring that self-adjoint representations of w exponentiate to unitary representations of W.

The matrix representation (2.1) yields the composition law of the abstract group elements $g(x,y,z) \in W$ as

$$g(x_1,y_1,z_1)g(x_2,y_2,z_2) = g(x_1+x_2, y_1+y_2, z_1+z_2+\tfrac{1}{2}[y_1x_2-x_1y_2]), \qquad \text{(2.2a)}$$

$$e = g(0,0,0)\ ,\ g(x,y,z)^{-1} = g(-x,-y,-z). \qquad \text{(2.2b)}$$

Associativity clearly holds. Finally, w is the Lie algebra of W, since

$$\partial G(x,y,z)/\partial x\ \Big|_{g=e} = i\ Q,\ \partial G(x,y,z)/\partial y\ \Big|_{g=e} = i\ P,\ \partial G(x,y,z)/\partial z\ \Big|_{g=e} = i\ I. \qquad \text{(2.3)}$$

All parameters range over R, the group manifold is thus isomorphic to R^3, non-compact (Sect. 2.19) and simply connected. The centre of W generated by I is the subgroup of elements $g(0,0,z)$.

2.3 Functions on the group and on coset spaces.

We consider now complex-valued functions $f(g) = f(x,y,z)$ on R^3 identified as the W manifold. We may act on this space of functions with $g' \in W$ through an action from the *right*

$$f(g) \xrightarrow{g'(R)} f^R_{g'}(g) = f(g\,g'), \qquad \text{(2.4a)}$$

and an action from the *left*

$$f(g) \xrightarrow{g'(L)} f^L_{g'}(g) = f(g'^{-1}g). \qquad \text{(2.4b)}$$

In this rather trivial way, W becomes a *Lie transformation group* on the differentiable manifold $M = W$.

The general definition of a Lie transformation group is such that to each pair (p,g'), $p \in M$ (a differentiable manifold), $g' \in G$ (a Lie group), there is associated an element $p_{g'} \in M$ (which may be denoted and mean pg' or g'^{-1} according to convenience), such that

a) the map is differentiable
b) $p\,e = p$ or $e\,p = p$
c) $(p\,g_1)g_2 = p(g_1\,g_2)$ or $g_2^{-1}(g_1^{-1}\,p) = (g_1 g_2)^{-1}\,p$.

These axioms are the result of the group axioms when M is G itself, as above. But they also hold when M is a *coset space*: if $H \subset G$, we consider the points of M to be the sets $p^L = Hg$ or $p^R = gH$. There is a standard argument to show that any two of these sets are either disjoint or they coincide, and that they *partition* G. They are called $M^L = H\backslash G$ (left cosets) or $M^R = G/H$ (right cosets) respectively.

Left cosets map into each other under the *right* action of the group, i. e. $p^L = Hg \xrightarrow{g'(R)} (Hg)g' = H(gg') = p^{L'}$. *Right* cosets map under the *left* action of the group $p^R = gH \xrightarrow{g'(L)} g'^{-1}(gH) = (g'^{-1}g)H = p^{R'}$.

Consider our example $G=W$ and $H_{\mathcal{Q}} = \{g(x,0,0)\}_{x\in R}$. Since

$$g(x,y,z - \tfrac{1}{2}\,xy) = g(x,0,0)\;g(0,y,z), \tag{2.5a}$$

we may partition W into cosets by $H_{\mathcal{Q}}$ letting $g(x,0,0)$ range over $H_{\mathcal{Q}}$ (i. e. x over R):

$$c_{\mathcal{Q}}^{L}\;(y,z) = \{g(x,0,0)\}_{x\in R}\;g(0,y,z) = \{g(x,y,z - \tfrac{1}{2}\,xy)\}_{x\in R}. \tag{2.5b}$$

The manifold $M = H_{\mathcal{Q}}\backslash W$ is then isomorphic to R^2 and its representative elements may be labelled by $g(0,y,z)$. The space of left cosets $M = H_{\mathcal{Q}}\backslash W$ may be subject to transformations through the right action of W as

$$\begin{aligned} c_{\mathcal{Q}}^{L}(y,z) \xrightarrow{g(x',y',z')\ (R)}\ & c_{\mathcal{Q}}^{L}(y,z)\,g(x',y',z') \\ &= \{g(x,0,0)\}_{x\in R}\;g(0,y,z)\;g(x',y',z') \\ &= \{g(x,0,0)\}_{x\in R}\;g(x',y+y',\,z+z'+\tfrac{1}{2}\,x'y) \\ &= \{g(x,0,0)\}_{x\in R}\;g(x',0,0)\,g(0,y+y',z+z'+\tfrac{1}{2}x'y'+x'y) \\ &= c_{\mathcal{Q}}^{L}\;(y+y',\;z+z'+\tfrac{1}{2}\,x'y'+x'y). \end{aligned} \tag{2.5c}$$

2.4 Transitive and effective action.

The action of W on the right on $M = H_Q\backslash W$ given by (2.5c) is transitive and effective. We recall these concepts. A group G acts *transitively* on a manifold M when for any two points $p,\bar{p}\in M$ there exists a $g \in G$ which maps p on $\bar{p}$, i. e. $\bar{p} = p_g$. In our example above, given two cosets $c_Q^L(y,z)$ and $c_Q^L(\bar{y},\bar{z})$, the $g \in W$ which does the job is any $g(x', \bar{y}-y, \bar{z}-z-\frac{1}{2}x'(\bar{y}+y)), x' \in \mathbb{R}$.

A differentiable manifold where a Lie group acts transitively is a *homogeneous* space for the group. Every coset space of a Lie group is homogeneous for it, and in fact, every homogeneous space is a coset space for these groups (19, Chapter II, Theorem 3.2).

The *isotropy group* (also called *stability* or *little* group) of any fixed point $p \in M$ is the set of elements $h \in I_p \subset G$ such that $p_h = p$. In the previous paragraph, setting $\bar{y} = y$ and $\bar{z} = z$, the isotropy group of $c_Q^L(y,z) \in H_Q\backslash W$ is $I_{c(y,z)} = \{g(x',0,x'y)\}_{x'} \in \mathbb{R}$. The isotorpy group, quite clearly, may depend on the coordinates of p. In fact, if G is transitive over M, the isotropy groups of any two points in M are isomorphic and conjugate: $I_{pg} = g^{-1}\, I_p g$.

The action of G on M is said to be *effective* if no transformation in G except the identity leaves all of the points in M fixed. That is $\bigcap_{p\in M} I_p = \{e\}$ and is the case in the example(2.5c). It would *not* have been the case had we chosen $H_I = \{g(0,0,z)\}_{z \in \mathbb{R}}$ to define the coset space $H_I\backslash W = W/H_I$ with cosets $c_I(x,y)$. There the non-abelianity of W would have been completely lost as H_I would be the isotropy group for every coset $c_I(x,y)$. It is straightforward to show that a group G acts effectively on a coset space $H\backslash G$ if and only if H does not contain a normal subgroup N of G. (A normal subgroup $N \lhd G$ we recall, is a subgroup of G such that $gn\,g^{-1}=n' \in N$ for all $n \in N$; in our example H_I is normal in W as well as its centre.) The proof goes as follows : Every $n \in N\subset H \subset G$ applied to a coset $H\,g$ yields $Hgn=Hn'g = Hg$ independently of the coset and hence is the isotropy group for the whole coset space. Conversely, if some n exist such that $Hgn= Hg$ for all g, they form a subgroup $N \subset G$; since $Hg\bar{g}$ is a coset $Hg\bar{g}n\bar{g}^{-1} = Hg$ it follows that $N \lhd G$.

2.5 Multiplier representations.

Normal or central subgroups cannot be used to divide the group into coset spaces without loosing the effectiveness of the group action on the coset manifold, but we may use them to obtain *multiplier* representations. This is a subject closely related to induced representations (20, Chapters 16 and 17; 21, Chapter 9) which we will apply to the Heinsenberg-Weyl case. The point in that, the group N being normal in G , there is a natural action of G on N given by $n \xrightarrow{g} gng^{-1} = n_g \in N$. If we have a manifold $M=H\backslash G$ where the action of G is effective, we may build $N\backslash M = N\cdot H\backslash G$ where the action of G is no longer effective, and supplement it with a more intimate knowledge of the group N , its representations in particular.

We may apply this principle to our example in considering functions over the space of cosets $c_Q^L(y,z)$ of the special form

$$F^{\lambda}\ (c^{L}_{Q}(y,z)) = f^{\lambda}(y)\ e^{i\lambda z}, \quad \lambda \in C. \tag{2.6}$$

The coset space is effectively acted upon by W as shown in (2,5c), but the factorized form (2.6) and the number λ in it are respected, as the points (y,z) move under $g(x',y',z')$ to $(y+y', z+z'+\frac{1}{2}x'y'+x'y)$. This induces a transformation of the functions on y (functions on $H_I\ H_Q\backslash W$)

$$f^{\lambda}(y) \xrightarrow{g(x',y',z')} f^{\lambda}_{g'}(y) = f^{\lambda}(y+y')\exp\left[i\lambda(z'+\tfrac{1}{2}x'y'+x'y)\right]. \tag{2.7}$$

The factor $\mu(y,g) = e^{i\lambda\cdots}$ is called a *multiplier* factor. It is a result of function theory that there is no proper subspace of $L^2(R)$ which is invariant under the action of all translation and multiplication-by-exponential operators. The action of (2.7) is thus irreducible on $L^2(R)$, since it cannot be broken into invariant proper subspaces.

2.6 Infinitessimal generators in 3, 2 and 1 variable.

We have seen the action of W on three-variable functions on W in (2.4), on two-variable functions on $H_Q\backslash W$ in (2.5), and on one-variable functions with multiplier in (2.6)-(2.7). We shall show what this means in terms of the Lie algebra generator realization. We consider transformations $g(\delta x', \delta y', \delta z')$ near to the group indentity $g(0,0,0)$ and collect Taylor expansion terms to first order in δ, so that

$$f(p) \xrightarrow{\delta g'} f(p_{\delta g'}) = (1+i\,[\,\delta x'\mathbb{Q}^{M}+\delta y'\,\mathbb{P}^{M}+\delta z'\,\mathbb{I}^{M}]+o(\delta^2))\,f(p). \tag{2.8}$$

The coordinates of $p \in M$ may be three, two or one, so that $\mathbb{Q}^{M}$, $\mathbb{P}^{M}$ and $\mathbb{I}^{M}$ will be the generator of the group of transformations on that manifold.

Setting (2.2) into the action from the right in the 3-variable case we obtain

$$\mathbb{Q}^{W(R)} = -i(\partial_x + \tfrac{1}{2}\,y\partial_z),\quad \mathbb{P}^{W(R)} = -i(\partial_y - \tfrac{1}{2}x\partial_z),\quad \mathbb{I}^{W(R)} = -\,i\partial_z. \tag{2.9a}$$

A similar procedure for the action from the left (2.4b) yields

$$\mathbb{Q}^{W(L)} = i(\partial_x - \tfrac{1}{2}\,y\partial_z),\quad \mathbb{P}^{W(L)} = i\,(\partial_y + \tfrac{1}{2}x\partial_z),\quad \mathbb{I}^{W(L)} = i\partial_z. \tag{2.9b}$$

The generators (2.9a) commute with those in (2.9b). From (2.5c), for $H_Q\backslash W$ we find

$$\mathbb{Q}^{H_Q\backslash W} = -iy\partial_z,\ \mathbb{P}^{H_Q\backslash W} = -i\partial_y,\ \mathbb{I}^{H_Q\backslash W} = -i\partial_z. \tag{2.9c}$$

Clearly, we can repeat this procedure for $H_p\backslash W$, W/H_Q and W/H_p. Finally, for (2.7)

$$\mathbb{Q}^{(\lambda)} = \lambda y,\ \mathbb{P}^{(\lambda)} = -i\partial_y\ ,\ \mathbb{I}^{(\lambda)} = \lambda\ . \tag{2.9d}$$

The last set of generators of W constitute the Schrödinger realization of W given in (1.24a) for $\lambda y = q$. We can identify λ with $\hbar$, which is assigned a particular value by Nature.

Perhaps the most striking feature of the generators (2.9d) stemming from the multiplier action (2.7) is that Sophus Lie would not have recognized them as generators of a Lie algebra (22). He considered Lie transformation groups, where groups act effectively on manifolds, so his generators are all and only first-order differential operators (with coefficients which are functions of the coordinates). Generators with zeroth order operators, as $\mathbb{Q}^\lambda$ and $\mathbb{I}^\lambda$ above, stem from multiplier group action (23).

Lie algebras of operators of order higher that first are interesting in physics. The associated Lie groups are generally groups of integral transforms (12, 13, 14 Part 4), rather than manifold mappings of the type (2.4).

2.7 On representations of groups on homogeneous spaces.

A representation ρ of a group G on a vector space V is defined in a very similar way as those of an algebra, namely, as a homomorphism $\rho : G \longrightarrow GL(V)$ from the group G into the group $GL(V)$ of linear operators on V. The homomophism means that $\rho(g_1)\rho(g_2) = \rho(g_1g_2)$ and $\rho(e) = 1$. One is generally interested in the case when V is a separable Hilbert space and when the representation is strongly continuous, i. e. $\|\rho(g)f - \rho(g_0)f\| \to 0$ as $g \to g_0$.

The physicists' experience with compact groups, where the V are finite-dimensional, leads one to search for matrix representations which may be of infinite dimension, or for their generalizations as integral kernels. A finite-dimensional representation on $V=R^3$ is already provided by (2.1), although it is non-unitary. An appropiate Hilbert space may be constructed for functions of three, two or one variable on which we defined the action of W.

Infinite-dimensional matrix representations may be obtained if we give a complete denumerable basis for the separable Hilbert space, $\{\Psi_n(y)\}_{n=0}^\infty$, and build the matrix $\rho(g) = \|\mathcal{D}_{nn'}(g)\|$ with elements $\mathcal{D}_{nn'}(g) = (\Psi_n, \Psi_{n'}{}^g)$, the inner product being that of $L^2(R)$ and the group action given by (2.7). Similarly, integral-kernel representations may be obtained through working with a Dirac-orthonormal basis for the space, $\{\chi_\nu(y)\}_{\nu\in\mathcal{S}}$, where $\mathcal{S}$ is some index interval. There, the integral kernels are $\underset{\sim}{\rho}(g) = \|\mathcal{D}_{\nu\nu'}(g)\|$ with $\mathcal{D}_{\nu\nu'}(g) = (\chi_\nu, \chi_{\nu'}{}^g)$.

The one-variable $L^2(R)$ space the is manifestly mapped onto itself by W. Its choice is dictated over any other $L^2(a,b)$ by the fact that the group action (2.7) includes translations in the argument y, so that only spaces of periodic functions of period $b-a$ may be contemplated, but the exponential factor $e^{i\lambda x'y}$ does not respect this periodicity over any finite interval.

2.8 The Q-subgroup basis.

Let us start first with the integral-kernel representations afforded by the normalized Dirac eigenbasis of the 'position' operator $\mathbb{Q}^{(\lambda)}$,

$$\mathbb{Q}^{(\lambda)} \chi_q^\lambda(y) = q\, \chi_q^\lambda(y) \;,\quad q,\ \lambda \in R, \tag{2.10a}$$

$$\chi_q^\lambda(y) = |\lambda|^{1/2}\, \delta(q - \lambda y\,). \tag{2.10b}$$

Then, the representation of W labelled by $\lambda \in R$ in the $\mathbb{Q}$-eigenbasis is

$$\begin{aligned} D_{qq'}^{\lambda\,(Q)}(g(x',y',z')) &= (\chi_q^\lambda,\ \chi_{q'g}^\lambda) \\ &= \int_{-\infty}^{\infty} dy\ \chi_q^\lambda(y)^*\ \chi_{q'}^\lambda(y+y')\exp[i\lambda(z'+\tfrac{1}{2}\,x'y'+\,x'y)] \\ &= \delta\,(q-q'+\lambda y')\ \exp\, i[\lambda(z'+\tfrac{1}{2}\,x'y')+\,x'q\,]. \end{aligned} \tag{2.11}$$

The representation property holds:

$$\int_{-\infty}^{\infty} dq'\ D_{qq'}^{\lambda(Q)}(g_1)\ D_{q'q''}^{\lambda(Q)}(g_2) = D_{q\,q''}^{\lambda(Q)}(g_1g_2), \tag{2.12}$$

and $D_{qq'}^{\lambda(Q)}(e) = \delta(q-q')$ is the unit operator. The strong continuity of (2.11), meaning the strong continuity of (2.7) is easily ascertained. The subgroup $g(x',0,0)$ is represented by a 'diagonal' integral kernel (i. e. one with a factor of $\delta(q-q')$). This representation diagonalizes the subgroup generated by Q.

A representation of the operators in the Lie algebra w and W through integral kernels may be obtained subjecting the integral kernel (2.11) to the limiting procedure in (2.8):

$$Q_{q\,q'}^{\lambda(Q)} = -\,i\partial_{x'}\, D_{qq'}^{\lambda(Q)}\ (g(x',y',z'))|_{g=e} = q\ \delta(q-q'), \tag{2.13a}$$

$$P_{q\,q'}^{\lambda(Q)} = -\,i\partial_{y'}\, D_{qq'}^{\lambda(Q)}\ (g(x',y',y'))|_{g=e} = -i\lambda\delta'(q-q'), \tag{2.13b}$$

$$I^{\lambda(Q)}_{q\,q'} = -i\,\partial_{z'}\,D^{\lambda(Q)}_{q\,q'}\,(g(x',y',z'))\Big|_{g=e} = \lambda\,\delta(q-q'). \qquad (2.13c)$$

These integral kernels represent the differential operators of the Schrödinger realization on the space of differentiable functions in $L^2(R)$, a space dense in the latter.

2.9 The P -subgroup basis.

A second generalized Dirac basis of $L^2(R)$ is provided by the generalized eigenfuctions of $\mathbb{P}^{(\lambda)}$ in (2.9c):

$$\mathbb{P}^{(\lambda)}\tilde{\chi}^{\lambda}_{p}\,(y) = p\,\tilde{\chi}^{\lambda}_{p}\,(y), \quad p,\ \lambda \in R, \qquad (2.14a)$$

$$\tilde{\chi}^{\lambda}_{p}\,(y) = (2\pi)^{-1/2}\,e^{ipy}. \qquad (2.14b)$$

The representation thus obtained is

$$D^{\lambda(p)}_{p\,p'}\,(g(x',y',z')) = (\tilde{\chi}^{\lambda}_{p}, \tilde{\chi}^{\lambda}_{p'g})$$

$$= \delta\,(p-p'-\lambda x')\,\exp i[\lambda(z'+\tfrac{1}{2}x'y') + p'y'] \qquad (2.15)$$

$$= \int_{-\infty}^{\infty} dq \int_{\infty}^{\infty} dq'\ (\tilde{\chi}^{\lambda}_{p}, \chi^{\lambda}_{q})\ D^{\lambda(Q)}_{qq'}\,(g)\,(\chi^{\lambda}_{q'}, \tilde{\chi}^{\lambda}_{p'}).$$

It satisfies the representation property (2.12), and diagonalizes the subgroup $g(0,y,0)$ generated by P. The last line in Eq. (2.15) shows that the $D^{\lambda(P)}_{pp'}\,(g)$ are *equivalent* to the $D^{\lambda(Q)}_{qq'}\,(g)$ in (2.11), since they are basically the double Fourier transforms of the $D^{\lambda(Q)}_{qq'}\,(g)$ in (2.11), as $(\chi^{\lambda}_{q}, \tilde{\chi}^{\lambda}_{p}) = (2\pi|\lambda|)^{-1/2}\,e^{ipq/\lambda}$ is the Fourier transform kernel. The *overlap* coefficients between the two eigenbases (2.10) and (2.14) yield the tranformation kernel between the two representations (2.11) and (2.15). We shall return to this point below when we comment on the unitarity and completeness of the representation set. Meanwhile, let us produce further representations equivalent to the above ones.

The Lie algebra representation obtained from (2.15) may be written as

$$Q^{\lambda(P)}_{p\,p'} = i\,\lambda\delta'(p-p'), \quad P^{\lambda(P)}_{pp'} = p\delta\,(p-p'), \quad I^{\lambda(P)}_{pp'} = \lambda\delta(p-p') \qquad (2.16)$$

which, again, is only the Fourier transform of the former.

2.10 The quantum free-fall nonsubgroup basis.

Further generalized bases of $L^2(R)$, unrelated to subgroups of W may be produced as eigenbases of self-adjoint operators in the enveloping algebra of w. Quantum Mechanics has a large supply of such operators. Let us start with the free-fall Schrödinger Hamiltonian and its Dirac-orthonormal eigenfunctions:

$$\mathbb{H}^{\ell}\Lambda_{\alpha}^{\lambda}(y) = [\,\tfrac{1}{2}\,\mathbb{P}^{(\lambda)2} + \mathbb{Q}^{(\lambda)}]\,\Lambda_{\alpha}^{\lambda}(y) = \alpha\Lambda_{\alpha}^{\lambda}\,(y)\,,\ \alpha \in R \tag{2.17a}$$

$$\Lambda_{\alpha}^{\lambda}(y) = (2|\lambda|^{-1/2})^{1/3}\quad \mathrm{Ai}([\,2\lambda]^{1/3}\,(y-\alpha/\lambda)), \tag{2.17b}$$

where $\mathrm{Ai}(z)$ is the Airy function of the first kind (14, subsect. 9.5.3). Then, a calculation aided by the Fourier transform shows that

$$\mathcal{D}_{\alpha\alpha'}^{\lambda(\ell)}\,(g(x',y',z')) = e^{-i\pi/4}(2\pi x')^{1/2}\,|\lambda|^{-1}\,\times$$
$$\times\, exp\ i[\,\lambda z' + \tfrac{1}{2}\,x'(\alpha+\alpha') + \frac{1}{2x'}(y' + \frac{\alpha-\alpha'}{\lambda})^{2} - \frac{1}{24}\,\lambda^{2}x'^{3}\,]. \tag{2.18}$$

Rather tediously, we can verify that the analogue of the representation property (2.12) holds. Using

$$\underset{\varepsilon\to 0^{+}}{\mathrm{l.i.m.}}\ e^{-i\pi/4}(2\pi\varepsilon)^{-1/2}\ exp\ i(s^{2}/2a^{2}\varepsilon) = |a|\,\delta(s), \tag{2.19}$$

we can show that (2.18) is indeed $\delta(\alpha-\alpha')$ for $g=e$, as well as some further properties which will be seen later. (If the powers of λ throughout produce any distress, the reader may check that "units" of λ are properly given if the arguments of transcendental functions are to be dimensionless, while wavefunctions have units of $(\ell enght)^{-1/2}$ and integral kernels of $(\ell enght)^{-1}$.

The task of finding the Lie algebra of integral kernels now takes us to find through (2.19) and its derivatives with respect to s, as in (2.13), the integral kernel representations:

$$Q_{\alpha\alpha'}^{\lambda(\ell)} = \tfrac{1}{2}\,\lambda^{2}\delta''(\alpha-\alpha') + \alpha\delta(\alpha-\alpha'), \tag{2.20a}$$

$$P_{\alpha\alpha'}^{\lambda(\ell)} = -i\,\lambda\,\delta'\,(\alpha-\alpha'), \tag{2.20b}$$

$$I_{\alpha\alpha'}^{\lambda(\ell)} = \lambda\,\delta\,(\alpha-\alpha'). \tag{2.20c}$$

Note that, indeed,

$$\left(\frac{1}{2}P^2+Q\right)^{\lambda(\ell)}_{\alpha\alpha'} = \alpha\,\delta(\alpha-\alpha') \tag{2.20d}$$

is a diagonal *number* integral kernel, as it ought to be from (2.17a).

2.11 The quantum harmonic oscillator basis.

A most convenient denumerable orthonormal basis is provided by the eigenfunctions of the quantum harmonic oscillator Hamiltonian

$$\mathbb{H}^n\,\Psi^\lambda_n(y) = \frac{1}{2}\left[\,\mathbb{P}^{(\lambda)^2}+\mathbb{Q}^{(\lambda)^2}\right]\Psi^\lambda_n(y) = |\lambda|\left(n+\frac{1}{2}\right)\Psi^\lambda_n(y), \tag{2.21a}$$

$$\Psi^\lambda_n(y) = \left(2^n n!\,[\pi/|\lambda|]^{1/2}\right)^{-1/2} e^{-|\lambda|y^2/2}\,H_n(|\lambda|^{1/2}y) \tag{2.21b}$$

$$n = 0,\ 1,\ 2,\ldots\ .$$

In this basis we must calculate

$$D^{\lambda(h)}_{nn'}(g(x',y',z')) = \int_{-\infty}^{\infty} dy\,\Psi^\lambda_n(y)^*\,\Psi^\lambda_{n'}(y+y')\,\exp[i\lambda(z'+\tfrac{1}{2}x'y'+yx')], \tag{2.22a}$$

which may be done through multiplying (2.19a) by $2^{(n+n')/2}(n!n!)^{-1/2}\times$ $\times\,s^n t^{n'}$, summing over n and n' so as to use the known generating functions for the Hermite polynomials (16, Eq. 22.9.17), integrating, and finally using the generating function (3, Eq. (2.42))

$$\exp(ab+ac-bd) = \sum_{n=0}^{\infty}\sum_{n=0}^{\infty} L^{(n-n')}_{n'}(cd)\,a^n b^{n'} c^{n-n'}/n!\ ,$$

in order to separate powers of s and t. Setting $a=\sqrt{2s}$, $b=\sqrt{2t}$ and $c=(|\lambda|/2)^{1/2}(-y'+ix') = d^*$ we obtain the result

$$D^{\lambda(h)}_{nn'}(g(x',y',z')) = \exp\lambda\left(iz+\frac{1}{4}[x'^2+y'^2]\right)(n'!/n!)^{1/2}\times$$

$$\times\left([\,|\lambda|/2]^{1/2}[-y'+ix']\right)^{n-n'} L^{(n-n')}_{n'}\left(\frac{1}{2}|\lambda|[x'^2+y'^2]\right), \tag{2.22b}$$

valid for $n \geqslant n'$. Similarly, for $a = \sqrt{2t}$, $b = \sqrt{2s}$ and $c = (|\lambda|/2)^{1/2}(y + ix) = d^*$ we obtain

$$D^{\lambda(h)}_{nn'}\ (g(x',y',z')) = \exp \lambda(iz + \frac{1}{4}[x'^2+y'^2])(n!/n'!)^{1/2}\ \times$$

$$\times\ ([|\lambda|/2]^{1/2}[y'+ix'])^{n'-n}\ L_n^{(n'-n)}\ (\tfrac{1}{2}|\lambda|[x'^2+y'^2]),$$

(2.22c)

valid for $n \leqslant n'$.

The representation properties of composition $D(g_1)\ D(g_2) = D(g_1 g_2)$ may be shown rather tediously to be valid - in fact, it is rather to be used as a proof of the addition theorem

$$\sum_{n'=0}^{\infty} D^{\lambda}_{nn'}(g_1)\ D^{\lambda}_{n'n''}(g_2) = D^{\lambda}_{nn''}(g_1 g_2) \qquad (2.23)$$

involving Laguerre polynomials (24, Sects. 1.4 and 5.2). The Lie algebra may be found obtaining derivatives with respect to the group parameters and valuating at the group identity as in (2.13). Again we find the algebra, but given by the half-infinite matrices (1.16) in the linear combinations (1.3a). The defining number operator (1.24b)-(2.21) is in this representation also diagonal, with values $|\lambda|(n + 1/2)$ on the diagonal.

2.12 The bilateral Mellin eigenbasis.

Not all representation bases must come from subgroup of Hamiltonian-type ($\frac{1}{2}\mathbb{P}^2 + V(\mathbb{Q})$) self-adjoint operators; the representing matrices/kernels may have both continuous and discrete rows and columns. Consider as a non-standard example the operator

$$\mathbb{D}\ \mu^{\lambda}_{\sigma\rho}\ (y) = \frac{1}{2}\{\mathbb{Q}^{(\lambda)}\mathbb{P}^{(\lambda)} + \mathbb{P}^{(\lambda)}\mathbb{Q}^{(\lambda)}\}\ \mu^{\lambda}_{\sigma\rho}(y) =$$

$$= -i\lambda(y\frac{d}{dy} + \frac{1}{2})\ \mu^{\lambda}_{\sigma\rho}(y) = \rho\ \mu^{\lambda}_{\sigma\rho}\ (y),$$

(2.24a)

$$\mu^{\lambda}_{\sigma\rho}\ (y) = (2\pi)^{-1/2}\ y_{\sigma}^{-\frac{1}{2} + i\rho/\lambda} \qquad \rho \in R, \quad \sigma = \pm, \qquad (2.24b)$$

where we have used the 'cut' power-functions

$$y_+ = \begin{cases} y\ ,\ y \geqslant 0 \\ 0\ ,\ y < 0 \end{cases} \qquad\qquad y_- = \begin{cases} 0\ ,\ y > 0 \\ -y\ ,\ y \leqslant 0. \end{cases} \qquad (2.24c)$$

The representation kernels are now 2 × 2 matrices (rows and colums labelled by $\sigma = \pm$) with integral kernel elements. We must consider the four pieces ($\sigma = \pm, \sigma' = \pm$) for $y' \geqslant 0$ and $y' \leqslant 0$ separately. In constructing the kernels through inner products of (2.24b) we obtain integral representations for the confluent hypergeometric functions $M(a,c,z)$ and $U(a,c,z)$ (16, Eqs. 13.2.1 and 13.2.5):

$$\mathcal{D}^{\lambda(\mathcal{D})}_{\sigma\rho,\sigma'\rho'}(g(z',y',z')) = (2\pi)^{-1}\int_{-\infty}^{\infty} dy\ \mu_{\sigma\rho}(y)^*\mu_{\sigma'\rho'}(y+y') \times$$

$$\times \exp\left[i\lambda(z'+x'\ y'/2 + x'y)\right]$$

$$= (2\pi)^{-1}\exp\left[i\lambda(z'+x'y'/2\right]\ K^{\sigma\sigma'}_{\rho\rho'}(x',y'), \qquad (2.25a)$$

$$K^{++}_{\rho\rho'}(x',\ y' > 0) = \Gamma(1/2 - i\rho/\lambda)\ y'^{\,i(\rho'-\rho)/\lambda} \times$$

$$\times\ U(1/2 - i\ \rho/\lambda, 1 + i\ (\rho'-\rho)/\lambda,\ -i\lambda x'y'). \qquad (2.25b)$$

$$K^{--}_{\rho\rho'}(x',y' > 0) = -\ e^{\pi(\rho-\rho')/\lambda}\ e^{-i\lambda x'y'}K^{++}_{-\rho',-\rho}(-x',\ y' > 0), \qquad (2.25c)$$

$$K^{\pm\pm}_{\rho\rho'}(x',y' < 0) = e^{-i\lambda x'y'}\ K^{\pm\pm}_{-\rho',-\rho}(x',\ -y' > 0), \qquad (2.25d)$$

$$K^{-+}_{\rho\rho'}(x',\ y' > 0) = -\ e^{\pi(\rho-\rho')/\lambda}\ \Gamma(1/2 + i\ \rho'/\lambda)\ \Gamma(1/2 - i\ \rho/\lambda)\ \times$$

$$\times \left[\ \Gamma(1 + i\ (\rho'-\rho)\ /\lambda)\right]^{-1}\ y'^{\,i(\rho'-\rho)/\lambda} \times$$

$$\times\ M(1/2 - i\ \rho/\lambda,\ 1 + i\ (\rho'-\rho)/\lambda, -i\lambda x'y'), \qquad (2.25e)$$

$$K^{+-}_{\rho\rho'}(x', y' < 0) = e^{-\lambda x' y'} K^{-+}_{\rho\rho'}(x', -y' > 0), \tag{2.25f}$$

$$K^{\pm\pm}_{\rho\rho'}(x', y' \gtrless 0) = 0. \tag{2.25g}$$

The 2×2 matrix is lower-triangular for $y' > 0$, and upper-triangular for $y' < 0$ the first and second rows and columns are labelled by $\sigma = +$ and $\sigma = -$ respectively). It should be interesting to verify that the group representation composition and indentity properties hold for the integral kernel (2.25). Differentiating with respect to the group parameters should yield the integral kernels for $Q^{\lambda(\mathcal{D})}_{\sigma\rho,}$, $P^{\lambda(\mathcal{D})}_{\sigma\rho,\sigma'\rho'}$, and $I^{\lambda(\mathcal{D})}_{\sigma\rho,\sigma'\rho'} = \lambda\delta_{\sigma\sigma'}\delta(\rho-\rho')$. These will constitute yet another form for the Heisenberg-Weyl algebra such that $1/2(QP+PQ)$ is the number operator $\delta_{\sigma\sigma'}\rho\delta(\rho-\rho')$. Any self-adjoint operator in $L^2(R)$ can be seen thus to have associated to it an eigenbasis which determines a corresponding representation. The first two cases we gave had Q and P for number operators. These determine that the subgroups $g(x,0,0)$ and $g(0,y,0)$, respectively, be represented by purely diagonal integral kernels (c.f. Eqs. (2.11) and (2.15), these are $\delta(q-q')\,e^{ix'q}$ and $\delta(p-p')e^{ipy}$, and hence the corresponding representations are said to be *reduced* or *classified* according to a subgroup.

The other three cases seen involve nonsubgroup bases, as the corresponding number operator was given by two Schrödinger Hamiltonians (ℓ and h cases) or by the dilatation generator $\mathbb{D}$. In these cases, the representation is given by matrices/kernels which are not diagonal in any subgroup, as suggested by the absence of Dirac δ's in (2.18) and (2.25). They are truly integral kernel representations of the group.

2.13 Unitary irreducible representations.

We now define and verify the unitarity and irreducibility of the representations obtained in (2.11) (Q), (2.14)(P), (2.18)(ℓ), (2.22) (h) and (2.25) ($\mathcal{D}$). In each of these cases we chose a basis for $L^2(R)$ through a self-adjoint operator and the bases are therefore orthonormal. The Heisenberg-Weyl algebra generators are self-adjoint in their Schrödinger realization in $L^2(R)$, and their exponentiated group operators representing (2.1) *unitary*. A unitary operator ($\mathbb{U}(g)f = f_g$ such that $(\mathbb{U}(g)f, (\mathbb{U}(g)h) = (f,h)$ in an orthonormal basis is represented by a unitary matrix or integral kernel:

$$\mathcal{D}^{\lambda}_{\mu\mu'}(g) = [\,\mathcal{D}^{\lambda}_{\mu'\mu}(g^{-1})]^{*}, \tag{2.26}$$

where $\mu \in R$ for cases Q, P and ℓ, $\mu \in \{0,1,2,\ldots\}$ for case h and $\mu = (\sigma,\rho)$, $\sigma = \pm$, $\rho \in R$ for case $\mathcal{D}$. For λ real, this is indeed the case, manifestly, for cases P, Q and ℓ. In the latter care has to be taken to define the phase of the parameters in $g^{-1}(x',y'z') = g(-x',-y',-z')$. The matrix for the h-case also satisfies (2.26) with both (2.24b) and (2.24c) required. The last case, $\mathcal{D}$, can also be seen from (2.25) using the pairs (b,d), (c,d) and (f,g) which correspond under the adjunction of the 2×2 matrix. In all of these matrices no block-decomposition occurs and the representations are in fact irreducible. This is to be expected, as $L^2(R)$ itself is irreducible under (2.7).

2.14 Equivalence of representations.

By equivalence of two representations $\underset{\sim}{\mathcal{D}}^{(1)}(g)$ and $\underset{\sim}{\mathcal{D}}^{(2)}(g)$ of a group G, associated to given number operators $\mathbb{H}^{(1)}$ and $\mathbb{H}^{(2)}$, we mean that there exist (fixed) invertible transformations $\underset{\sim}{C}$ of $L^2(R)$ such that the two representations are mapped into each other as

$$\underset{\sim}{\mathcal{D}}^{(2)}(g) = \underset{\sim}{C}\, \underset{\sim}{\mathcal{D}}^{(1)}(g)\, \underset{\sim}{C}^{-1} \quad , \quad \forall\, g \in G, \tag{2.27a}$$

$$\underset{\sim}{\mathcal{D}}^{(1)}(g) = \| \mathcal{D}^{(1)}_{\mu\mu'}(g) \| \,,\quad \underset{\sim}{\mathcal{D}}^{(2)}(g) = \| \mathcal{D}^{(2)}_{\rho\rho'}(g) \| \,,\quad \underset{\sim}{C} = \| C_{\mu\rho} \| \,, \tag{2.27b}$$

where the set of values which μ can take is the generalized spectrum of $\mathbb{H}^{(1)}$ in $L^2(R)$, and ρ in that of $\mathbb{H}^{(2)}$ in $L^2(R)$:

$$\underset{\sim}{H}^{(2)} = \underset{\sim}{C}\, \underset{\sim}{H}^{(1)}\, \underset{\sim}{C}^{-1},\quad \underset{\sim}{L}^{(2)} = \underset{\sim}{C}\, \underset{\sim}{L}^{(1)}\, \underset{\sim}{C}^{-1}, \qquad \forall\ \underset{\sim}{L} \in \rho\,(a). \tag{2.27c}$$

If $H^{(1)}, H^{(2)} \in a$ then C in (2.27c) is a (possibly exterior) automorphism of the algebra, mapping one subgroup basis into another subgroup basis. An example is provided by the two subgroup representations denoted by $\mathcal{Q}$ and P, as we shall see below.

If the $\mathbb{H}^{(\cdot)}$ are self-adjoint and have the same spectrum, $\underset{\sim}{C}$ will be a unitary transformation of $L^2(R)$, i.e. $\underset{\sim}{C}^{-1} = \underset{\sim}{C}^{\dagger}$. This is the for the transformation connecting the $\mathcal{Q}$ and P representations -the Fourier transform- which in fact maps $L^2(R)$ unitarily onto itself. There $C^{\lambda}(q,p) = (2\pi)^{-1/2} \exp(ipq/\lambda)$ is the transformation kernel. The $\mathcal{Q}$- and ℓ-representations were defined through eigenbases of the operators $\mathbb{Q}$ in (2.10a) and $\mathbb{H}^{\ell}$ in (2.17a), and have the same spectrum. The transformation kernel $\underset{\sim}{C}$ is then

$$C^{\lambda}_{q,\alpha} = (\chi^{\lambda}_{q}, \Lambda^{\lambda}_{\alpha}) = |\lambda|^{-1/2}\, \Lambda^{\lambda}_{\alpha}(q/\lambda), \tag{2.28}$$

i. e. the Airy transform (14, Sect. 8.5.3), which as expected is unitary in $L^2(R)$. This is a *point* transformation, where to the operator representing $\mathcal{Q}$ we may add any function of its canonically conjugate P, as $\mathcal{Q} \to \mathcal{Q} + f(P)$, while leaving P invariant. Thus we obtain, in particular $\underset{\sim}{Q} \overset{\underset{\sim}{C}}{\longrightarrow} \underset{\sim}{Q} + \frac{1}{2} \underset{\sim}{P}^2$. All such transformations are unitary (6, Sect. 4.2, 25).

Unitary transformations $\underset{\sim}{C}$ perform the equivalence between unitary transformations of groups: $\underset{\sim}{D}^{(2)}(g)$ is unitary iff $\underset{\sim}{D}^{(1)}(g)$ is unitary. Unitary transformations also preserve the spectra of self-adjoint operators, and hence two operators $\mathbb{H}^{(1)}$ and $\mathbb{H}^{(2)}$ with *different* spectra cannot be bridged in this way, at least not with a *unitary* operator.

We may well work with unitary transformations between different Hilbert spaces. For example, for $\mathbb{H}^{(1)} = \mathbb{Q}$ and $\mathbb{H}^{(2)} = \mathbb{H}^{h}$, the second Hilbert space is the space $\ell^2(Z^+)$ of sequences. The transformation matrix $\underset{\sim}{C}$ is then the 'rectangular' kernel

$$C^{\lambda}_{q,n} = (\chi^{\lambda}_{q}, \Psi^{\lambda}_{n}) = |\lambda|^{-1/2}\, \Psi^{\lambda}_{n}(q/\lambda), \tag{2.29}$$

with continuous rows, and columns numbered by $0,1,2,\ldots$. The transformation $\underset{\sim}{C}$ provides thus the passage from a function to its harmonic-oscillator series coefficients. It is unitary between $L^2(R)$ and $\ell^2(Z^+)$ and possesses a Parseval relation. Similarly, for $\mathbb{H}^{(1)} = \mathbb{Q}$ and $\mathbb{H}^{(2)} = \mathbb{D}$, our fifth example in (2.24), C is the bilateral Mellin transform kernel (14, Sect. 8.2.1), unitary between $L^2(R)$ and $L^2_{\sigma=+}(R) + L^2_{\sigma=-}(R)$. It should be emphasized that those transformations are unitary which do not change the spectrum of the operators involved. Thus, in the oscillator representation, $\underset{\sim}{Q}^{(h)} = \| Q^{(h)}_{nn'} \|$ is still an operator with R as its spectrum in $\ell^2(Z^+)$ and the same for $\underset{\sim}{Q}^{\lambda(D)}$ obtained from (2.24) in $L^2_{\sigma=+}(R) + L^2_{\sigma=-}(R)$. Through the unitary $\underset{\sim}{C}$ we are only changing the *number* operator property (the diagonality of the representing matrix), and hence only the basis employed to describe the representation.

2.15 Classical canonical transformations.

Classical Hamiltonian mechanics has a very general way of solving dynamical systems (26, Chapter 9). If $H(q,p)$ is the Hamiltonian, one poses the following problem: Can we implement a canonical transformation (i.e. one preserving Poisson brackets) which will map the canonically conjugate pair of observables of position q and momentum p onto the Hamiltonian $H(q,p)$ and a new observable $T(q,p)$ canonically conjugate to H? The latter is obtained through solving

$$T(q,p) = \int dq' \, [\, 2(H(q,p) - V(q')]^{-1/2}, \tag{2.30}$$

and has the meaning of time. When (2.30) has been performed, if one can invert to $q=q(H,T)$ and $p=p(H,T)$, (initial values included), then these latter variables describe the classical motion in phase space. As a close alternative, the action and angle variables may be introduced. Recall that the action of a periodic motion between q_+ and q_- (26, p. 293), with a factor of 2π, is

$$J\,(H) = \frac{1}{2\pi}\oint dq\quad p = \frac{1}{\pi}\int_{q_-}^{q_+} dq'[\,2(H(q,p)-V(q')]^{1/2}\ .$$

The angle variable is (26, p. 292) -with a minus sign-

$$\omega\,(H,T) = -\,(\partial H/\partial J)T. \tag{2.31b}$$

One can depart slightly from the usual notation: Since $q_\pm$ are the libration endpoints where $V(q_\pm) = H$ (while otherwise $H < V(q)$), the integral (2.31a) is positive. We defined thus

$$J = |\,\bar{q}\,|\,,\qquad \omega = \bar{p}\,\bar{q}\,/\,|\bar{q}|, \tag{2.31c}$$

where $\bar{q}$ and $\bar{p}$ is a canonically conjugate pair of variables related to q and p through a canonical transformation.

2.16 Quantum canonical transformations and discrete Hamiltonian spectra.

When Quantum mechanics tries to follow the methods of solution outlined above, the first problem it encounters is that the operators H and T or J and ω *cannot* be the generators -together with I- of a Heisenberg-Weyl algebra, when the spectrum of either of them is lower bound and/or discrete, and I a multiple of the identity operator. Discrete spectra are disallowed by the following contradiction: Let $\phi_m(y)$ be the eigenfunctions of a $\mathbb{Q}^?$ with eigenvalues $m \in Z$. Then, if there exists a conjugate companion $\mathbb{P}^?$ in a Heisenberg-Weyl algebra (1.1), the $(m\,,\,m')$ element of the commutation relation would be

$$(\phi_m, [\mathbb{Q}^?, \mathbb{P}^?]\,\phi_{m'}) = (\,\phi_m, [\,\mathbb{Q}^?\,\mathbb{P}^? - \mathbb{P}^?\,\mathbb{Q}^?]\,\phi_{m'}) =$$

$$= (\,\mathbb{Q}^{?\dagger}\,\phi_m, \mathbb{P}^?\,\phi_{m'}) - (\phi_m, \mathbb{P}^?\,\mathbb{Q}^?\,\phi_m) = (m-m')\,(\phi_m, \mathbb{P}^?\phi_{m'}). \tag{2.32a}$$

But on the other hand,

$$(\phi_m, [\,\mathbb{Q}^?\,,\,\mathbb{P}^?]\,\phi_{m'}) = i\,(\phi_m, \mathbb{I}^?\,\phi_{m'}) = i\delta_{m,m'}. \tag{2.32b}$$

For $m \neq m'$, $(\phi_m, \mathbb{P}^?\,\phi_{m'})$ is zero, while for $m = m'$ it is undefined. This is a very old and standard argument, dating back to Jordan in 1927 (27, Postulate D on p. 812, and statement of p. 819, 28 p. 2; 29-31). The same commutation relation also *cannot* be satisfied by any two bounded operators (32, Sect. 6.1.1).

2.17 Quantum canonical transformation to action and angle variables.

In the teeth of the above remarks, Moshinsky and Seligman (33-35) have looked at the quantum mechanical version of the canonical transformation $(q,p) \xrightarrow{C} (\bar{q},\bar{p})$. The first observation is that even in classical mechanics, this transformation is generally not bijective, i. e. the classical phase space motion may be subject to a group of transformations which leave the action-angle variables invariant, as the projection of the Riemann surface for a multivalued function over the complex plane, under exchange of sheets. This *ambiguity group* A is therefore an object which appears already in classical mechanics. Quantizing the system now means replacing q and p for self-adjoint operators $\mathbb{Q}$ and $\mathbb{P}$ in some appropiate representation, say, the Schrödinger representation on $L^2(R)$, while $\bar{q}$ and $\bar{p}$ are replaced by operators $\bar{\mathbb{Q}}$ and $\bar{\mathbb{P}}$ ... in what space? If we have constraining potentials giving rise to a closed classical orbit in phase space and were to propose simply $L^2(R)$, the spectrum of $\bar{\mathbb{Q}}$ would be lower-bound and discrete. In what space may we define $\bar{\mathbb{Q}}$ such that its spectrum be R, as it is for $\mathbb{Q}$? This is of interest since we desire that the quantum canonical transformation be unitary. As we have also the ambiguity group, we may build a space $L^2(R,A) = \sum_{g \in A} L^2_g(R)$ consisting of the ordinary $L^2(R)$ summed with itself once for every element g of the ambiguity group, so as to obtain discrete and perhaps infinite matrices with integral kernel elements. This is the space where $\bar{\mathbb{Q}}$ may act so as to have R for its spectrum, and here the canonical transformation to the original $L^2(R)$ may be bijective and unitary.

The space $L^2(R,A)$ is then classified in a way where linear combinations of the $L^2_g(R)$ spaces are taken so as to build the unitary irreducible representations of the ambiguity group A; in this way one defines an ambiguity *spin*. Some lower-bound discrete spectra allow for a dihedral ambiguity group of rotations by multiples of the period, and inversions. The corresponding ambiguity spin is a pair of numbers: a continuous one $\kappa \in [0,1)$ and a sign σ. The space L^2 is thus written as $\int_0^1 d\kappa \sum_{\sigma=\pm} L^2_{\kappa,\sigma}(R)$.

The operators $\bar{\mathbb{Q}}$ found by Moshinsky and Seligman (33) were such that they are a sum of a Hamiltonian operator with a lower bound and discrete spectrum, *times* the sign σ *plus* the representation index κ. In all, thus, the spectrum of $\bar{\mathbb{Q}}$ is R. Transform kernels are found which represent the above quantum canonical transformation.

2.18 Quantum mechanics on a continuous compact space.

There are other approaches to the problem of making the Hamiltonian of a contraining quantum potential to fit into a Heisenberg-Weyl algebra (3, Refs. 138-154). One of them follows a suggestion by Weyl (36) introducing a *mixed group* $W^\star$ (3, Sect. VI): It is a proper subgroup of W with the composition rule (2.2), and defined through

$$\mathcal{W}^{\star} = \{ g\,(x, y, z) \in \mathcal{W} \mid x = n_x/M,\ n_x \in Z;$$
$$y \equiv y \bmod L;\quad z \equiv z \bmod L/2M \}. \qquad (2.33)$$

It is a Lie group of transformations on the circle as in (2.7), where x is a discrete group coordinate for the subgroup elements H_Q while the subgroup manifold of H_P is a circle. For the latter, the infinitessimal generator $\mathbb{P} = -i\partial y$ has, discrete eigenvalues $p=n_p p_0$, $n_p \in Z$, $p_0 = 2\pi/L$. The former group H_Q does *not* have an infinitessimal generator. It has a *finite* generator instead: multiplication by e^{iy}. The eigenvalues of the H_I generator are similarly quantized to $\lambda=n_\lambda\lambda_0, n_\lambda \in Z$, $\lambda_0 = 4\pi M/L = 2\, p_0\, M$. In this model for Quantum mechanics on compact spaces one way work with a properly defined enveloping algebra, implement quantization procedures, classical limits and canonical transformations. Furthermore, as will be brought out in Chapter 3, in connection with the 2+1 Lorentz group, one can define a nonlocal inner product on the circle so that the operator $-id/dy$ have a lower-bound spectrum (37). On the other hand, one does *not* have a self-adjoint 'position' operator, and has in effect quantized on the Lorentz group level.

2.19 Left- and right-invariant Haar measure.

Having gained familiarity with the Heisenberg-Weyl algebra and group, we may state what a noncompact group is, and what its representations are like (20, Sect. 2.3).

Consider a Lie group G with a finite number D of continuous parameters $\{X_k\}_{k=1}^{D}$, $X \in R \subset R^D$, and functions $f(X)$ over G. A positive *Radon* measure is a positive linear form

$$\mu(f) = \int_G d\mu(g)\, f(g) = \int_G \omega(X)\, d^D X\, f(X) \geqslant 0 \qquad (2.34)$$

on the space $C_0^+(G)$ of continuous nonnegative functions f on G with support on a finite-radius sphere. The *left* (resp. *right*) invariant Haar measure is a positive Radon measure which is left (resp. right) invariant under the group action (2.4): $\mu^L(f) = \mu^L(f^L_{g'})$ (*resp.* $\mu^R(f) = \mu^R(f^R_{g'})$), for all $g' \in G$. A theorem may be proven (38, Sect. IV-15): Every Lie group has a *unique* left (and right) invariant Haar measure, up to multiplication by positive constants.

If we fix the multiplicative constant, we may define the *volume* of the group G, vol G, to be the limit of the sequence of values of (2.34) when f is taken as a sequence of characteristic functions over any nested growing sequence of group subsets. This is (2.34) for $f(g)=1$ when it exists. If the volume of G is a finite number, the group is said to be *compact*. If it is not finite, the group is said to be *noncompact*.

Let $\mu^L(f)$ be a given left Haar measure. We can produce a new left Haar measure through acting on the *right* of the argument with a fixed group element: $\mu^L_{g_0}(f) = \mu^L(f^R_{g_0})$. But since right and left actions commute, i. e. $(f^R_{g''})^L_{g'}(g) = f(g'^{-1}gg'') = (f^L_{g'})^R_{g''}(g)$, it follows that $\mu^L_{g_0}(f^L_{g'}) = \mu^L((f^R_{g_0})^L_{g'}) = \mu^L(f^R_{g_0}) = \mu^L_{g_0}(f)$, so $\mu^L_{g_0}$ is also a left-invariant measure. By the uniqueness theorem, we conclude that $\mu^L_{g_0}(f) = \Delta(g_0)\mu^L(f)$, i. e. $\mu^L_{g_0}$ may at most differ by a constant $\Delta(g_0)$ from μ^L.

2.20 Unimodular groups.

A *modular* function over G is a positive function $\Delta: G \longrightarrow R^+$ such that $\Delta(gg') = \Delta(g)\Delta(g')$. This implies $\Delta(e)=1$ and $\Delta(g^{-1})=1/\Delta(g)$. The constants $\Delta(g_0)$ seen above are modular functions, as can be verified acting with two elements from the right on the left-invariant measure. If $\Delta(g_0)=1$ for all $g_0 \in G$, then the left-invariant Haar measure is also invariant under right action, so that the left- and right-invariant measures are the same. Such groups are called *unimodular*.

The difference between right and left Haar measures never appears in compact group theory: Every compact group is unimodular. Proof: If G is compact, $f(X)=1$ is in $C^+_0(G)$ and we may normalize μ^L through asking for $\mu(1)=1$. The modular function of the group is then $\Delta(g) = \Delta(g)\mu(1) = \mu(1^R_g) = \mu(1) = 1$.

Abelian groups -compact or noncompact- are unimodular, since their left and right actions are the same. Noncompact groups may be non-unimodular, an example of this is the two-parameter solvable group of linear transformations $x \longrightarrow x' = a_1x+a_2$ seen in (39, p. 316). Noncompact groups which *are* unimodular are the following: (a) all abelian groups, (b) all semisimple groups, (c) all connected nilpotent groups -as the Heisenberg-Weyl group, (d) Lie groups for which the range of values of modular functions is compact. (e) direct products of unimodular groups.

2.21 The Haar measure and weight functions.

When the composition functions for the parameters of a group, are known, it is not difficult to build a weight function for a left-invariant Haar measure. Right-invariant Haar measures are very similar and will be given below. We require $\mu^L(f) = \mu^L(f^L_{g_0})$ in (2.34). This implies

$$\mu^L(f^L_{g_0}) = \int_G d\mu^L(g) f(g_0^{-1}g) = \int_{g_0 G} d\mu^L(g_0 g) f(g) = \mu^L(f). \qquad (2.35a)$$

Since $g_0 G = G$, we need $d\mu^L(g_0 g) = d\mu^L(g)$ for all $g_0 \in G$. If the coordinates of g are X and those of $g' = g_0 g$ are X' then the weight function and parameter volume element must satisfy

$$\omega^{L}(X)d^{D}X = \omega^{L}(X')d^{D}X \quad \omega^{L}(X')J\left(\frac{\partial X'(g_0 g(X))}{\partial X}\right) d^{D}X \tag{2.35b}$$

where $J(\partial\cdot/\partial\cdot) = J(g_0, X)$ is the transformation Jacobian. If the group identity is at $X=0$ so that $g(0)=e$, and $\omega^{L}(0)$ is a fixed number, (2.35b) gives us the appropiate weight function $\omega^{L}(X')$ at $g'=g_0$:

$$\omega^{L}(X') = \omega^{L}(0)\left[J\left(\frac{\partial X'(g_0 g(X))}{\partial X}\right)\Big|_{X=0}\right]^{-1} . \tag{2.36a}$$

This is the left-invariant weight function for the parameter volume element at $g'=g_0$, where the parameters are X' and which is invariant under left group translation.

A similar argument for right-invariant integration, measures and Jacobian leads to

$$\omega^{R}(X') = \omega^{R}(0)\left[J\left(\frac{\partial X'(g(X)g_0^{-1})}{\partial X}\right)\Big|_{X=0}\right]^{-1} \tag{2.36b}$$

as the appropiate right-invariant weight function at $g'=g_0^{-1}$ where the parameters are X'.

The measure (2.35) is also invariant under the inversion involution $g' \longrightarrow g'^{-1}$ of the group manifold, (20, p. 69, Prop. 3).

The Heisenberg-Weyl group W, being connected and nilponent, is unimodular. Hence, right and left invariant measures are the same. The Jacobian in (2.36) may be computed from the group composition functions (2.2) for $g'(X')=g_0(X_0)g(X)$ and $g''(X'') = g(X)g_0(X_0)^{-1}$,

$$x' = x_0 + x, y'=y_0+ y, z'=z_0 + z+(y_0 x - x_0 y)/2 \tag{2.37a}$$

$$x'' = x - x_0, y'' = y-y_0, z'' = z - z_0 - (yx_0 - xy_0)/2 . \tag{2.37b}$$

The Jacobians have 1's on the diagonal and are triangular, therefore the determinants are unity. Normalizing $\omega(0) = 1$, the invariant Haar measure for W is

$$d\mu(X) = dx \quad dy \quad dz. \tag{2.38}$$

The Heisenberg-Weyl group is very close to the abelian R^3 space. The 'twist' which makes the third parameter compose in a non-abelian way, is relatively minor.

2.22 Completeness of a set of representations.

We would like to insist on the distinction between a representation and its subgroup (or nonsubgroup) reduction. Different examples of the latter have been given in the first part of this Chapter, and refer to the row-column classification through the choice of basis for the homogeneous space of the group. Since we saw they are all unitarily equivalent, we can refer to any one fixed classification through a given basis in writing $\mathcal{D}^{\lambda}_{rr'}(g)$. The (in general, collective) indices r and r' are eigenvalues of a (maximal set of mutually commuting) operator(s) in the right- and left-action enveloping algebra of the group, and self-adjoint under the inner product defined by the Haar measure over G (i.e. in a Hilbert space $L^2(G)$). The eigenvalues may be discrete, continuous or mixed, and their range is assumed to resolve degeneracy completely. These representations compose as (2.23) in the discrete case, and as (2.12) in the continuous one. We shall write $\mathop{\mathsf{S}}_r \cdots$ *for* $\sum_r \cdots$ or $\int dr \ldots$. The *representation* index λ (which in general is a collective index) is usually given as eigenvalue(s) of a (maximal set of algebraically independent) right *and* left invariant operator(s) in the centre of the enveloping algebra, self-adjoint in $L^2(G)$. The range of the (generally collective) representation index $\lambda(\lambda=\{\lambda_1,\lambda_2,\ldots,\lambda_N\})$ is a subset $\hat{G}$ of R^N. In benign cases it is a measurable space, and a *Plancherel* measure and weight function $d\rho(\lambda)=\upsilon(\lambda)d\lambda$ exists.

For W, the centre of the enveloping algebras (2.9a) and (2.9b) is $\mathbb{I}^{W(R)} = -i\partial/\partial z = -\mathbb{I}^{W(L)}$, so this generator takes the place of the Casimir operator. The eigenvalue of $-i\partial/\partial z$ provides the representation label $\lambda \in R$.

The difficulty in treating the representations of noncompact groups vis-à-vis the same task for compact groups is the same as that of Fourier integral transforms over Fourier series. The space being noncompact in the former case will allow continuous spectra for certain operators whose eigenvectors are normalizable only in the Dirac sense. Casimir operators may also have *mixed* (continuous and discrete) spectra. In the W case, the difficulties will be handled through Fourier transform techniques. In general one has to invoke rigged Hilbert spaces (20, Chapter 14) and pay close attention to functional analysis arguments (40, 41). The problem for compact group are less, as there all unitary representations are finite dimensional and representations come in discrete series.

2.23 The orthogonality and completeness relations for the unitary irreducible representation matrix elements.

For unimodular groups (with some mild conditions), we can generally assert that the unitary irreducible representation matrices $\mathcal{D}^{\lambda}_{rr'}(g)$ are a generalized orthonormal set of functions over G, under the inner product of $L^2(G)$ given by the invariant Haar integral. We denote the latter through $\mathop{\mathsf{S}}_{g\in G} = \int_G d\mu(g)\cdots$, in order to include the finite and compact cases in our formulae. We may write

$$(\mathcal{D}^{\lambda}_{rr'}, \mathcal{D}^{\lambda'}_{r'r''})_G = \mathop{\mathbf{S}}_{g\in G} [\mathcal{D}^{\lambda}_{rr'}(g)]^* \mathcal{D}^{\lambda'}_{r''r'''}(g)$$

$$= \delta_{\hat{G}}(\lambda,\lambda')\, \delta_{r,r''}\, \delta_{r',r'''}\,, \qquad (2.39)$$

where $\delta_{r1,r2}$ means a collective Kronecker delta in discrete indices, (which may be generalized to a Dirac δ's over a continuous pair), $\delta_{\hat{G}}(\lambda,\lambda')$ should play the role of the Kronecker or Dirac δ's for $\lambda,\lambda' \in \hat{G}$, ($\hat{G}$ being endowed with a Plancherel measure and weight function $d\rho(\lambda) = \upsilon(\lambda)d(\lambda)$. Equation (2.39) is not hard to prove through integration by parts with the operators which determine the representation and row/column labels. Most interesting is the statement that the $\mathcal{D}^{\lambda}_{rr'}(g)$ are, moreover, *complete* in the $L^2(G)$ Hilbert space with inner product $(\cdot,\cdot)_{\hat{G}}$. This means that we may define an inner product in $\hat{G}$ through an integration $\mathbf{S}_{\lambda\in\hat{G}} \cdots = \int_{\hat{G}} d\rho(\lambda) \ldots$, which may contain a sum, if the Plancherel measure is a point measure in some domain. We have

$$(\mathrm{D}(g_1), \mathrm{D}(g_2))_{\hat{G}} = \mathop{\mathbf{S}}_{\lambda\in\hat{G}} \mathrm{Tr}\,[\mathrm{D}^{\lambda}(g_1)^{\dagger}\,\mathrm{D}^{\lambda}(g_2)]$$

$$= \delta_G\,(g_1,g_2)\,, \tag{2.40a}$$

where the trace of a product of matrices is

$$\mathrm{Tr}\,(\mathrm{M}\,\mathrm{N}) = \sum_{rr'} \mathrm{M}_{rr'}\mathrm{N}_{r'r}\,, \tag{2.40b}$$

For integral kernels an integral over r and r' is used. Equation (2.40a) defines an inner product between matrix or integral kernel functions over $\lambda \in \hat{G}$. The completion of this space of functions with finite norm defines a Hilbert space $L^2(\hat{G})$. The $\delta_G(g_1,g_2)$ is a Kronecker or Dirac delta for discrete or continuous groups: $\delta_G(g_1,g_2)=0$ for $g_1 \neq g_2$, and under sum or integration, it satisfies

$$\mathop{\mathbf{S}}_{g\in G} f(g)\delta_G(g,g') = f(g') \tag{2.41a}$$

for any continuous function $f(g)$. Similarly, $\delta_{\hat{G}}(\lambda,\lambda')$ is such that

$$\mathop{\mathbf{S}}_{\lambda\in\hat{G}} \hat{\mathrm{F}}(\lambda)\;\delta_{\hat{G}}(\lambda,\lambda') = \hat{\mathrm{F}}(\lambda') \tag{2.41b}$$

for continuous matrices or integral kernel functions $\hat{\mathrm{F}}$ of $\lambda \in \hat{G}$. The defining properties (2.41) imply a relation between weight functions and δ's, as

$$\omega(g(X))\,\delta_G(g(X),g(X')) = \delta^{\mathcal{D}}(X-X'), \tag{2.42a}$$

$$\upsilon(\lambda)\ \delta_{\hat{G}}(\lambda,\lambda') = \delta(\lambda-\lambda'). \tag{2.42b}$$

In the case of compact groups the Plancherel measure is a point measure and $S_{\lambda\in\hat{G}} \cdots = \sum_{\lambda} \dim(\lambda)/vol\, G \cdots$ where $\dim(\lambda)$ is the dimension of the λ unitary irreducible representation, and vol G is the volume of the group. The corresponding $\delta_{\hat{G}}$ is thus a Kronecker $\delta_{\hat{G}}(\lambda,\lambda') = \delta_{\lambda\lambda'}$ vol $G/\dim(\lambda)$.

For compact groups, the Peter-Weyl theorem (20, Sect. 7.2, 42) supports the proof of (2.39)-(2.40). For finite groups this can be found in (39, Eqs. (3.143), completeness appears only as (3.178)). The general case of unimodular noncompact groups appears and is referenced to in Barut and Raczka's book (20, Chapter 14).

2.24 The Heisenberg-Weyl case.

We return to our example $\mathcal{W}$ and choose for simplicity the $\mathcal{Q}$-eigenbasis where $\mathcal{D}^{\lambda}_{qq'}(g(x,y,z))$ is given by (2.11). The Haar measure is given by (2.38) with unit weight function and hence δ_G in (2.42a) is an ordinary Dirac δ in x, y and z. We do not yet know the unitary irreducible representation space $\hat{\mathcal{W}}$, but since $\lambda \in \mathcal{R}$, we want to determine if $\mathcal{R}$ is or acts as the full representation space. To this end we have avail to (2.39). If $\mathcal{R}$ were not $\hat{\mathcal{W}}$, we would not get a Dirac δ in this variable. We perform

$$(\mathcal{D}^{\lambda}_{qq'},\mathcal{D}^{\lambda'}_{q''q'''})_{\mathcal{W}} = \int_{-\infty}^{\infty} dx \int_{-\infty}^{\infty} dy \int_{-\infty}^{\infty} dz\ \{\delta(q-q'+\lambda y)\exp i\,[\lambda(z+xy/2)+xq]\}^{*} \times$$

$$\times\{\delta(q''-q'''+\lambda' y)\ \exp i\,[\lambda'(z+xy/2)+xq'']\} =$$

$$=(|\lambda||\lambda'|)^{-1}\delta((q-q')/\lambda-(q''-q''')/\lambda)\int_{-\infty}^{\infty} dz\ \ \exp i\,[(\lambda-\lambda')z]\ \times$$

$$\times\int_{-\infty}^{\infty} dx\ \ \exp[(-i\,x)((\lambda'-\lambda)(q''-q''')/2\lambda+(q-q''))] = \tag{2.43}$$

$$= 4\pi^2|\lambda|^{-1}\delta(\lambda-\lambda')\delta(q-q'')\delta(q'-q''') = \delta_{\hat{\mathcal{W}}}(\lambda,\lambda')\delta(q-q'')\delta(q'-q''').$$

So we do, indeed, obtain a Dirac δ. From (2.42b), the Plancherel measure for $\mathcal{W}$ is

$$d\rho(\lambda) = (|\lambda|/4\pi^2)\, d\lambda\,, \qquad \lambda \in \mathcal{R}. \tag{2.44}$$

Completeness is verified as

$$(\mathbb{D}(g_1), \mathbb{D}(g_2))_W = (4\pi^2)^{-1} \int_R |\lambda| d\lambda \int_R dq \int_R dq'\, \mathcal{D}^\lambda_{qq'}(g_1)^* \mathcal{D}^\lambda_{q'q}(g_2)$$

$$= \delta(x_1-x_2)\delta(y_1-y_2)\delta(z_1-z_2) = \delta_W(g_1,g_2). \tag{2.45}$$

The orthogonality and completeness relations (2.43) and (2.45) are valid *whatever* subgroup or nonsubgroup row/column classification we choose. Hence, orthogonality and completeness relations similar to these follow not only for (2.11), but for (2.15), (2.18), (2.22) and (2.25).

2.25 Harmonic analysis on a group.

Having the complete and orthonormal generalized basis $\mathcal{D}^\lambda_{rr'}(g)$ over G allows us to perform *harmonic analysis* over the manifold G, expressing any function within a wide class $f(g)$ over G, through a series or integral over these functions, as

$$f(g) = \mathop{\mathbf{S}}_{\lambda \in \hat{G}} \sum_{r'r'} \hat{f}_{rr'}(\lambda)\, \mathcal{D}^\lambda_{rr'}(g)^*, \tag{2.46a}$$

The 'linear combination coefficients' $\hat{f}_{rr'}(\lambda)$ can be obtained through performing the G-inner product of (2.46a) with $\mathcal{D}^{\lambda'}_{r''r'''}(g)$, exchanging $\mathop{\mathbf{S}}_g$ with $\mathop{\mathbf{S}}_\lambda \sum_{rr'}$ and using (2.39) so as to obtain

$$\hat{f}_{rr'}(\lambda) = \mathop{\mathbf{S}}_{g\in G} f(g)\, \mathcal{D}^\lambda_{rr'}(g). \tag{2.46b}$$

We can think of $\hat{\mathbb{F}}(\lambda)$ as a matrix- or integral-kernel-valued function on $\hat{G}$ and write (2.46) as

$$f(g) = \mathop{\mathbf{S}}_{\lambda \in G} \mathrm{Tr}\,[\mathbb{D}(g)\hat{\mathbb{F}}] = (\mathbb{D}(g^{-1}), \hat{\mathbb{F}})_{\hat{G}}\,, \tag{2.47a}$$

$$\hat{\mathbb{F}}(\lambda) = \mathop{\mathbf{S}}_{g\in G} f(g)\, \mathbb{D}^\lambda(g) = (\mathbb{D}^{\lambda^*}, \mathbb{F})_G. \tag{2.47b}$$

Finally, the Parseval relation

$$(F,H)_G = \mathbf{S}_{g\in G}\, f(g)^* h(g) =$$

$$= \mathbf{S}_{\lambda\in\hat{G}} \sum_{rr'} \hat{f}_{rr'}(\lambda)^* \hat{h}_{rr'}(\lambda) = \mathbf{S}_{\lambda\in\hat{G}} \mathrm{TR}\,[\hat{F}^{\dagger}(\lambda)\hat{h}(\lambda))] = (\hat{F},\hat{H})_{\hat{G}}, \qquad (2.48)$$

holds, telling us that the harmonic transform (2.47) is *unitary* between $L^2(G)$ and $L^2(\hat{G})$.

The subject of orthogonal and complete sets of functions over G and $\hat{G}$ spaces extends to all coset spaces $H\backslash G$ or G/H together with right- or left-invariant measures on these, and a corresponding reduction in the row- and column-indices (43). It also extends to other, more general equivalence sets on the group called bilateral classes (44). Rather than delve on the general theory, we shall give some results for the 2+1 Lorentz group in the next Chapter.

CHAPTER 3: A FURTHER EXAMPLE.

In this Chapter we shall present in some detail the case of the semisimple noncompact Lie group of lowest dimension. In the former two Chapters we dealt with the Heisenberg-Weyl group, which was 'as abelian as possible': Only one of the three commutators in the algebra was nonzero. Now we work on the 'least abelian' of three-parameter groups: The 2+1 Lorentz group SO(2,1) and its covering group $\overline{SL(2,\mathbb{R})}$.

The 2+1 Lorentz group has many resemblances as well as definite differences with the three-dimensional rotation group SO(3), which is probably most familiar for physicists who have worked with quantum mechanical systems such as atoms and nuclei. We shall find all irreducible unitary representations of the algebra and covering group. We then realize both the algebra and the group on a coset space related to the Iwasawa decomposition: the circle, and functions and differential operators thereupon. As a final development, we find the in general nonlocal measure defining Hilbert spaces of functions on the circle, for the various representation series.

3.1 The SO(2,1) group and algebra.

Consider a three-dimensional space $\mathbb{R}^3$ with metric (+,-,-), where the distance is $d^2 = x_0^2 - x_1^2 - x_2^2$. This is invariant under the '2+1' Lorentz transformations. These transformations are represented by (pseudo-) rotations around each of the three axes,

$$\exp(i\,\psi\,J_1^o) = \begin{pmatrix} 1 & 0 & 0 \\ 0 & ch\,\psi & sh\,\psi \\ 0 & sh\,\psi & ch\,\psi \end{pmatrix} \Leftrightarrow J_1^o = \begin{pmatrix} 0 & 0 & 0 \\ 0 & 0 & -i \\ 0 & -i & 0 \end{pmatrix}, \quad (3.1a)$$

$$\exp(i\,\chi\,J_2^o) = \begin{pmatrix} ch\,\chi & 0 & sh\,\chi \\ 0 & 1 & 0 \\ sh\,\chi & 0 & ch\,\chi \end{pmatrix} \Leftrightarrow J_2^o = \begin{pmatrix} 0 & 0 & -i \\ 0 & 0 & 0 \\ -i & 0 & 0 \end{pmatrix}, \quad (3.1b)$$

$$\exp(i\,\phi\,J_0^o) = \begin{pmatrix} \cos\phi & -\sin\phi & 0 \\ \sin\phi & \cos\phi & 0 \\ 0 & 0 & 1 \end{pmatrix} \Leftrightarrow J_0^o = \begin{pmatrix} 0 & i & 0 \\ -i & 0 & 0 \\ 0 & 0 & 0 \end{pmatrix}. \quad (3.1c)$$

We have expressed the rotation around the kth axis as $\exp(i\,\alpha\,J_k^o)$ in order to produce a 3×3 representation of the associated Lie algebra. The distance is also invariant under the inversions $x_0 \leftrightarrow -x_0$ and $x_1 \leftrightarrow -x_1$; these compound (in semidirect product) with (3.1). We shall not consider them in what follows, as we are interested primarily in reaching every element with a Lie generator, and they lie in group components not connected to the identity.

From (3.1) we define the 3×3 faithful group representation of the 2+1 Lorentz group as

$$G^{o}(\alpha,\beta,\gamma) = exp\ (i\alpha J_0^{o})\ exp\ (i\beta J_2^{o})\ exp\ (i\gamma J_0^{o})\ , \qquad (3.2a)$$

$$\alpha \equiv \alpha \bmod 2\pi,\ \gamma \equiv \gamma \bmod 2\pi\ ,\quad \beta \in R, \qquad (3.2b)$$

thereby parametrizing it through the *Euler* angles α,β,γ adapted for the nonpositive metric. The matrices have unit determinant while the generators J_k^{o} are all traceless.

The abstract Lorentz group composition law is obtained from the composition law of the matrix representation (3.2) in the same way as done for the Heisenberg-Weyl group in (2.1)-(2.2). The matrices (3.2) are '2+1' pseudo-orthogonal, i. e. orthogonal with respect to the Lorentz metric L as

$$G^{o}\,L\,G^{oT} = L\ , \qquad L = \begin{pmatrix} -1 & 0 & 0 \\ 0 & -1 & 0 \\ 0 & 0 & 1 \end{pmatrix}. \qquad (3.3a)$$

We call this group, Special (i.e. of unit determinant) Orthogonal group in 2+1 real dimensions: SO(2,1). The fact that no isochorous space-time inversions are included is usually denoted through a zero subscript, but we shall omit it understanding that we deal with the connected Lie group only.

Correspondingly, the algebra representation matrices are pseudo-skew-symmetric:

$$J_k^{o}\ L = -\ L\ J_k^{o\dagger}\ , \qquad k=1,\ 2,\ 0. \qquad (3.3b)$$

and satisfy the SO(2,1) algebra give by

$$[J_1,J_2] = -iJ_0,\ [J_2,J_0] = i\,J_1,\ [J_0,J_1] = i\,J_2. \qquad (3.4)$$

Notice the minus sign in the first commutator: The ordinary three-dimensional rotation group algebra so(3) has a plus sign in its stead. one cannot bring so(3) to (3.4), no matter how we redefine generator signs, unless, of course, we introduce i's ($J_1 \to iJ_1$, $J_2 \to iJ_2$); but as we work with real groups, its representation structure would be readically changed.

3.2 The SU(1,1) algebra and group.

We are familiar with Pauli matrices, so we may look for another representation of the Lie algebra so(2,1) in (3.4) through 2×2 matrices. Indeed, we can associate the representation

$$J_1 \to \mathrm{J}_1^u = \frac{1}{2}\begin{pmatrix} 0 & -1 \\ 1 & 0 \end{pmatrix} \Leftrightarrow exp(i\psi \mathrm{J}_1^u) = \begin{pmatrix} ch\psi/2 & -ish\psi/2 \\ ish\psi/2 & ch\psi/2 \end{pmatrix}, \tag{3.5a}$$

$$J_2 \to \mathrm{J}_2^u = \frac{-i}{2}\begin{pmatrix} 0 & 1 \\ 1 & 0 \end{pmatrix} \Leftrightarrow exp(i\chi \mathrm{J}_2^u) = \begin{pmatrix} ch\chi/2 & sh\chi/2 \\ sh\chi/2 & ch\chi/2 \end{pmatrix}, \tag{3.5b}$$

$$\mathrm{J}_0 \to \mathrm{J}_0^u = \frac{1}{2}\begin{pmatrix} -1 & 0 \\ 0 & 1 \end{pmatrix} \Leftrightarrow exp(i\phi \mathrm{J}_0^u) = \begin{pmatrix} e^{-i\phi/2} & 0 \\ 0 & e^{i\phi/2} \end{pmatrix}. \tag{3.5c}$$

The matrices to the right define a matrix group

$$\mathrm{G}^u\ (\bar{\alpha},\bar{\beta},\bar{\gamma}) = exp(i\bar{\alpha}\mathrm{J}_0^u)\ exp(i\bar{\beta}\mathrm{J}_2^u\ )\ exp(i\bar{\gamma}\mathrm{J}_0^u)\ , \tag{3.6a}$$

$$\bar{\alpha} \equiv \bar{\alpha} \bmod 4\pi,\quad \bar{\gamma} \equiv \bar{\gamma} \bmod 2\pi,\quad \bar{\beta} \in R. \tag{3.6b}$$

In particular notice that

$$\mathrm{G}^u(\bar{\alpha},\bar{\beta},\bar{\gamma}) = -\mathrm{G}^u(\bar{\alpha},+2\pi,\bar{\beta},\bar{\gamma}) = -\ \mathrm{G}^u(\bar{\alpha},\bar{\beta},\bar{\gamma}+2\pi) = \mathrm{G}(\bar{\alpha}+4\pi,\bar{\beta},\bar{\gamma}). \tag{3.6c}$$

The 2×2 matrices (3.6) have unit determinant and the generator matrices in (3.5) are traceless. The abstract group with the composition law obtained from (3.6) constitutes the *special pseudo-unitary group* in 1+1 dimensions SU(1,1):

$$\mathrm{G}^u\ \underset{\sim}{\sigma}_3\ \mathrm{G}^{u\dagger} = \underset{\sim}{\sigma}_3 \qquad\qquad \underset{\sim}{\sigma}_3 = \begin{pmatrix} 1 & 0 \\ 0 & -1 \end{pmatrix}, \tag{3.7a}$$

$$\mathrm{J}_k^u\ \underset{\sim}{\sigma}_3 = \underset{\sim}{\sigma}_3\ \mathrm{J}_k^{u\,\dagger}\ , \qquad k = 1,2,0. \tag{3.7b}$$

The Lie algebra of SU(1,1), su(1,1), is dentical to so (2,1) in (3.4); the groups SU(1,1) and SO(2,1) are not. As in spin angular momentum theory, SU(1,1) *covers* SO(2,1) twice, since between $\mathrm{G}^u(\bar{\alpha},\bar{\beta},\bar{\gamma})$ and $\mathrm{G}^o(\alpha,\beta,\gamma)$ we can establish a 2:1 mapping given by $\mathrm{G}^u(\alpha,\beta,\gamma) \to \mathrm{G}^o(\alpha,\beta,\gamma)$ and $\mathrm{G}^u(\alpha+2\pi,\beta,\gamma) \to \mathrm{G}^o(\alpha,\beta,\gamma)$.

An alternative parametrization of the SU(1,1) group which is sometimes preferable to the Euler angles in (3.6) is found through demanding (3.7a) for any 2×2 complex matrix. This leads to

$$\mathrm{G}^u(\eta,\Theta) = \begin{pmatrix} \eta & \Theta^* \\ \Theta & \eta^* \end{pmatrix}, \tag{3.8a}$$

$$|\eta|^2 - |\Theta|^2 = 1, \qquad \eta,\Theta \in C. \tag{3.8b}$$

In this parametrization, the SU(1,1) manifold is a two-dimensional complex hyperboloid. The ralation between the Euler angle parameters in (3.6) and the parameters in (3.8) favoured by Bargmann (23, Sect. 3b) is easily found (reinterpreted by Sally: 45, p. 3) as

$$\eta = e^{-i(\alpha+\gamma)/2}\,\mathrm{ch}\,\beta/2\,, \quad \Theta = e^{i(\alpha-\gamma)/2}\,\mathrm{sh}\,\beta/2. \tag{3.8c}$$

3.3 The group SL(2,R).

Another parametrization of SU(1,1) is obtained through a similarity transformation of (3.8) as

$$G^{S}(a,b,c) = \begin{pmatrix} a & b \\ c & d \end{pmatrix} = \begin{pmatrix} 1 & -1 \\ -i & -i \end{pmatrix} G^{u}(\eta,\Theta) \begin{pmatrix} 1 & -1 \\ -i & -i \end{pmatrix}^{-1} =$$

$$= \begin{pmatrix} \mathrm{Re}\,\eta - \mathrm{Re}\,\Theta & -\mathrm{Im}\,\eta + \mathrm{Im}\,\Theta \\ \mathrm{Im}\,\eta + \mathrm{Im}\,\Theta & \mathrm{Re}\,\eta + \mathrm{Re}\,\Theta \end{pmatrix}, \tag{3.9a}$$

$$ad-bc=1, \qquad a,b,c,d \in R. \tag{3.9b}$$

The corresponding 2×2 algebra representation is

$$J_1^S = \frac{-i}{2}\begin{pmatrix} 0 & 1 \\ 1 & 0 \end{pmatrix} \Leftrightarrow \exp\,(i\Psi J_1^S) = \begin{pmatrix} \mathrm{ch}\Psi/2 & \mathrm{sh}\Psi/2 \\ \mathrm{sh}\Psi/2 & \mathrm{ch}\Psi/2 \end{pmatrix}, \tag{3.10a}$$

$$J_2^S = \frac{i}{2}\begin{pmatrix} 1 & 0 \\ 0 & -1 \end{pmatrix} \Leftrightarrow \exp\,(i\chi J_2^S) = \begin{pmatrix} e^{-\chi/2} & 0 \\ 0 & e^{\chi/2} \end{pmatrix}, \tag{3.10b}$$

$$J_0^S = \frac{i}{2}\begin{pmatrix} 0 & -1 \\ 1 & 0 \end{pmatrix} \Leftrightarrow \exp\,(i\phi J_0^S) = \begin{pmatrix} \cos\phi/2 & \sin\phi/2 \\ -\sin\phi/2 & \cos\phi/2 \end{pmatrix}. \tag{3.10c}$$

In this form, two further group isomorphisms are displayed: Since $G^S(a,b,c)$ is the most general unimodular (unit determinant) 2×2 real matrix, the group represented here is clearly SL(2,R).

The 2×2 unimodular real matrices (3.10) have the further property:

$$G^S\;S\;G^{S^T} = S\,, \quad S = \begin{pmatrix} 0 & 1 \\ -1 & 0 \end{pmatrix}, \tag{3.11}$$

where S is the *symplectic* metric matrix. The relation (3.11) defines the two-dimensional real symplectic group Sp(2,R). This group and its 2N -dimensional versions Sp(2N,R) are important as dynamical groups -rather, algebras- for the N -dimensional harmonic oscillator.

3.4 The covering group $\overline{SL(2,R)}$.

We saw $SO(2,1) \overset{1:2}{\approx} SU(1,1) \overset{1:1}{\approx} SL(2,R) \overset{1:1}{\approx} Sp(2,R)$. What are their connectivity properties? A convenient handle is provided by the complex hyperboloid in the η-Θ plane of $SU(1,1)$. Although we need four dimensions, we can look at the surface described by $(\mathrm{Re}\eta)^2+(\mathrm{Im}\eta)^2-(\mathrm{Re}\Theta)^2=1+(\mathrm{Im}\Theta)^2$. For fixed $\mathrm{Im}\,\Theta=0$, the remaining three parameters are constrained to a one-sheeted equilateral hyperboloid with a circular waist in the $(\mathrm{Re}\,\eta, \mathrm{Im}\,\eta)$ plane. The group unit $g^u(1,0)$ is on that waist. As we let $\mathrm{Im}\,\Theta$ range over R, the outside of the hyperboloid fills. The group manifold of $g^u(\eta,\Theta)$ is thus the full Θ complex plane, times the η complex punctured by a round hole of radius $1+|\Theta|^2$. We may surround that hole any number of times describing a path which may *not* be continuously deformed to a vanishing loop. The argument of η increases, but whether or not an increase by 2π is considered to bring us back to the starting point depends on the number of distinct Riemann sheets we provide for $g^u(\eta,\Theta)$ in η.

The simply connected universal covering group of $SU(1,1) \approx$ $\approx SL(2,R) \approx Sp(2,R)$, denoted by $\overline{SL(2,R)}$, is that where the argument of η may take any real value without repeating any group element.

A convenient parametrization for $\overline{SL(2,R)}$ is provided by (3.5)-(3.8) with new parameters as

$$\bar{G}^u(\gamma,\omega) = (|\gamma|^2-1)^{-1/2}\begin{pmatrix} e^{i\omega} & \gamma^* e^{-i\omega} \\ \gamma e^{i\omega} & e^{-i\omega}\end{pmatrix} \tag{3.12a}$$

$$\gamma \in C, \qquad |\gamma| < 1, \qquad \omega \in R, \tag{3.12b}$$

where

$$\gamma = \Theta/\eta, \qquad \omega = \arg\eta . \tag{3.12c}$$

The rotation subgroup $exp\,(i\phi J_0^u)$ in particular, unwinds from the twice-covered circle (3.6b) to the real line.

3.5 Raising, lowering and Casimir algebra elements.

In studying the self-adjoint representations of the Lie algebra $sl(2,R)$ we search for matrices or integral kernels to represent the three algebra generators J_k, $k=0,1,2$ in (3.4). It proves convenient to define the complex linear combinations

$$J_+ = J_1 + i\,J_2, \quad J_- = J_1 - iJ_2, \tag{3.13}$$

which together with J_0 also define the algebra though their commutation relations

$$[J_0, J_\pm] = \pm J_\pm, \quad [J_+, J_-] = -2J_0. \tag{3.14}$$

If the J_k are represented by self-adjoint matrices/integral kernels $\mathrm{J}_k = \mathrm{J}_k^\dagger$, then $\mathrm{J}_+^\dagger = \mathrm{J}_-$ and $\mathrm{J}_-^\dagger = \mathrm{J}_+$. They are the raising and lowering operators.

An operator in the enveloping algebra of sl(2,R) and in the centre of the right- and left- acting algebra, is the second-order Casimir operator

$$C = J_1^2 + J_2^2 - J_0^2 = J_\mp J_\pm - J_0^2 \mp J_0 . \tag{3.15}$$

3.6 Eigenbases for a representation.

We follows a classical approach which parallels the standard treatment of angular momentum (c.f. 46, p. 24-26), noting that if we construct eigenfunctions of $\mathbb{J}_0$ and $\mathbb{C}$

$$\mathbb{J}_0 \ \phi_\mu^k = \mu \, \phi_\mu^k , \tag{3.16a}$$

$$\mathbb{C} \ \phi_\mu^k = q \, \phi_\mu^k , \quad q = k(1-k), \tag{3.16b}$$

then the $\mathbb{J}_\pm$ will act as raising and lowering operators:

$$\mathbb{J}_+ \phi_\mu^k = c_{k,\mu}^+ \ \phi_{\mu+1}^k , \tag{3.17a}$$

$$\mathbb{J}_- \phi_\mu^k = c_{k,\mu}^- \ \phi_{\mu-1}^k . \tag{3.17b}$$

The value of k will characterize the representation, while the range of μ will determine its rows.

We make two remarks on μ and k. First, (3.17) tells us that if some ϕ_μ^k is a basis function for some representation, then -unless the proportionality constant $c_{k,\mu}^\pm$ in (3.17) be zero at some point- all other $\phi_{\mu+n}^k$, n integer, will be involved. We can thus write the $\mathbb{J}_0$ -eigenvalue as $\mu = m + \varepsilon$, where m is integer and $\varepsilon \in (-1/2, 1/2]$. If we work with the global representations of SO(2,1), $\mathbb{J}_0$ will allow only for integer μ and hence only $\varepsilon = 0$ is allowed. For SU(1,1) $\approx$ SL(2,R) $\varepsilon = 0$ and $\varepsilon = 1/2$ are allowed, as analogues of vector and spinor representations of SU(2). If it is $\overline{\mathrm{SL}(2,R)}$ we are working with, all ε in (-1/2, 1/2] are appropiate. Second, we have followed Bargmann (23) in writing the Casimir operator eigenvalue $q \in R$ as $k(1-k)$. Its convenience will be seen below. The number $k(1-k)$ is invariant under $k \leftrightarrow 1-k$ i. e. an inversion through the $k=1/2$ point in the complex plane. Also, q is real for k real ($k \geq 1/2$ in order not to double-count), but also for $k=1/2+i\rho, \rho \in R$. We thus have

$$q \leq 1/4 \quad \Leftrightarrow \quad k \in R. \tag{3.18a}$$

$$q = \frac{1}{4} + \rho^2 \geqslant \frac{1}{4} \quad \Leftrightarrow \quad k = \frac{1}{2} + i\rho, \ \rho \in R. \qquad (3.18b)$$

3.7 Self-adjoint representations.

The third assumption in searching for self-adjoint representations is that $\{\phi^k_{m+\varepsilon}\}_{m\varepsilon Z}$ or a proper irreducible subset thereof, constitute an orthonormal and complete basis under an inner product $(\cdot,\cdot)^{k,\varepsilon}$ which defines some Hilbert space $\mathcal{H}^{k,\varepsilon}$ still to be determined. The generator representation $\mathbb{J}_j$, $j=0,1,2$ will be self-adjoint in that space; it will allow us first to fix the proportionality constants $c^{\pm}_{k\mu}$ in (3.17) as

$$|c^{\pm}_{k,\mu}|^2 (\phi^k_{\mu\pm 1}, \phi^k_{\mu\pm 1}) = (\mathbb{J}_{\pm}\phi^k_{\mu}, \mathbb{J}_{\pm}\phi^k_{\mu}) =$$

$$= (\phi^k_{\mu}, \mathbb{J}_{\mp}\mathbb{J}_{\pm}\phi^k_{\mu}) = (\phi^k_{\mu}, [\mathbb{C} + \mathbb{J}^2_0 \pm \mathbb{J}_0]\phi^k_{\mu}) = \qquad (3.19)$$

$$= (q + \mu^2 \pm \mu)\, (\phi^k_{\mu}, \phi^k_{\mu}).$$

As we demand that $(\phi^k_{\mu}, \phi^k_{\mu})$ be unity for all μ in the spectrum of $\mathbb{J}_0$, $|c^{\pm}_{k\mu}|^2$ is obtained as

$$|c^{\pm}_{k,\mu}|^2 = q + \mu^2 \pm \mu = (\mu \pm k)(\mu \mp k \pm 1), \qquad (3.20)$$

which must be a positive quantity. This only allows the absolute value of $c^{\pm}_{k\mu}$ to be determined, but if we denote by $\gamma^{\pm}_{k\mu}$ the phase of $c^{\pm}_{k\mu}$, then (3.17) become

$$\mathbb{J}_{\pm}\phi^k_{\underset{\sim}{\mu}} = c^{\pm}_{k,\mu}\,\phi^k_{\underset{\sim}{\mu}\pm 1}, \qquad (3.21a)$$

$$c^{\pm}_{k\mu} = \gamma^{\pm}_{k\mu}\,[(\mu \pm k)(\mu \mp k \pm 1)]^{1/2}. \qquad (3.21b)$$

Eqs. (3.20) and (3.21) contain the information we need to find the self-adjoint representations of sl(2,R). Non-self-adjoint ones will be commented upon later.

3.8 The continuous representation series.

If both ϕ^k_{μ} and $\phi^k_{\mu\pm 1}$ are to be nonzero, (3.20) must be a positive quantity, i. e. $q > -\mu^2 \mp \mu$. The maximum of $-\mu^2 \mp \mu$ occurs at $\mu_{max} = \mp 1/2$ and has a value of 1/4; so recalling (3.18b) for $q > 1/4$, the representation $k = 1/2 + i\rho, \rho \neq 0$ contains all values of $\mu = m + \varepsilon$, $m \in Z$, for any fixed $\varepsilon \in (-1/2, 1/2]$.

The conditions $q > -\mu^2 \mp \mu = 1/4 - (\mu - \mu_{max})^2$ may be satisfied also for a subset of $\varepsilon \in (1/2,1/2]$ such that the maximum of the right-hand side does not fall in the μ-content of the representation: If $\mu = m + \varepsilon$, the points falling nearest to the maxima $\mu_{max} = \mp 1/2$ do so at a distance $\mu - \mu_{max} = 1/2 - |\varepsilon|$ from it, so that only $q > |\varepsilon|(1-|\varepsilon|) \geqslant 0$ is required. For any fixed q in $(0, 1/4]$, thus, we can find a range of ε in $(-1/2,1/2]$ which allows for unbounded values of μ. Writing $q = k(1-k)$, adding $-1/4$ to both sides of the inequality to complete squares, the condition on ε for fixed real k is given by $|k-1/2| < 1/2-|\varepsilon|$. In particular, for $k = 0$ all ε but $\varepsilon = 1/2$ allow for the unbounded range of μ.

Bargmann (23, see also 45) considers only SU(1,1) representations: $C_q^{\varepsilon=0}$ the *principal (integer) series* for $1/4 \leqslant q$, and $C_q^{\varepsilon=1/2}$ the *principal (half-integer) series* for $1/4 < q$. The *exceptional* (or *supplementary*) *series* called $C_q^{\varepsilon=0}$ occurs for $0 < q < 1/4$. These together are called the *continuous series*. We summarize them pictorically in Figs. 1a and 1b.

3.9 The 'Discrete' representation series.

We consider next the representations $q \leqslant 1/4$ for which the positivity of (3.20) is violated for some values of μ, which must be absent from the range of eigenvalues of $\mathbb{J}_0$. From (3.18a), $(\mu \pm k)(\mu \mp k \pm 1)$ has two zeros: $\mu_1^{\pm} = \mp k$ and $\mu_2^{\pm} = \pm (k-1)$. Hence $|c_{k\mu}^{+}|^2$, as given by (3.20)-(3.21), has a formal negative value for $\mu \in (k-1, -k)$ and similarly $|c_{k\mu}^{-}|^2$ for $\mu \in (k, -k+1)$. No point in the spectrum of $\mathbb{J}_0$ may be in those intervals. See Figs. 2a and 2b.

Consider first the case $|k-1/2| > 1/2$ i. e. $k \in (-\infty, 0) \cup (1, \infty)$, where $q < 0$. In those intervals, the distance between $\mu_1^{\pm}$ and $\mu_2^{\pm}$ is larger than 1, so no ε can be chosen such that the sequence $\mu = \varepsilon + m$, $m \in z$ avoids falling in the negative-$|c|^2$ region *unless* it falls on the boundary zeros. The row index μ will then have an upper or lower bound. For, consider what happens when $\mathbb{J}_-$ acts on the 'lowest state' $\phi^{k>1}_{\mu_2^-}$, $\mu_2^- = k$: It gives zero since $c^-_{k,\mu_2^-} = 0$, hence ϕ^k_{k-1} does not belong to the irreducible eigenfunction set of the algebra, which consists then of $\{\phi^{k>1}_{k+m}\}_{m=0}^{\infty}$. The same statement holds for $\phi^{k<0}_{\mu_2^-}$, $\mu_2^- = -k+1$, and only reflects the inversion symmetry under $k \leftrightarrow 1-k$. Correspondingly, for $\mathbb{J}_+$, the representation must contain a 'highest state' $\phi^{k>1}_{\mu_2^+}$, $\mu_2^+ = k-1$ which the raising operator will turn to zero. The irreducible eigenfunction set will then be $\{\phi^{k>1}_{k-m-1}\}_{m=0}^{\infty}$, and the analogue statement holds for $\phi^{k<0}_{\mu_1^+}$, $\mu_1^+ = -k$.

These lower- and upper-bound representations were called 'Discrete' series D_k^+ and D_k^- by Bargmann (23). Since he worked with SU(1,1), he allowed only for integer or half-integer values in the spectrum $\{\mu\}$ of $\mathbb{J}_0$, so k was only allowed to be an integer or half-integer. Here, it may be any real number.

When $|k-1/2| < 1/2$ i. e. $k \in (0,1)$, q covers the interval (0,1/4) twice, and the point $q = 1/4$ once (corresponding to $k = 1/2$). In this interval, the distance between $\mu_1^{\pm}$ and $\mu_2^{\pm}$ is less than 1, so that the sequence $\mu = \varepsilon + m$, $m \in Z$ may 'jump' the forbidden range of μ. This constitutes the exceptional interval of the continuous series seen in the last Section. Again, lower- and upper-bound representations occur when the lowest μ falls on μ_1^- or μ_2^- and

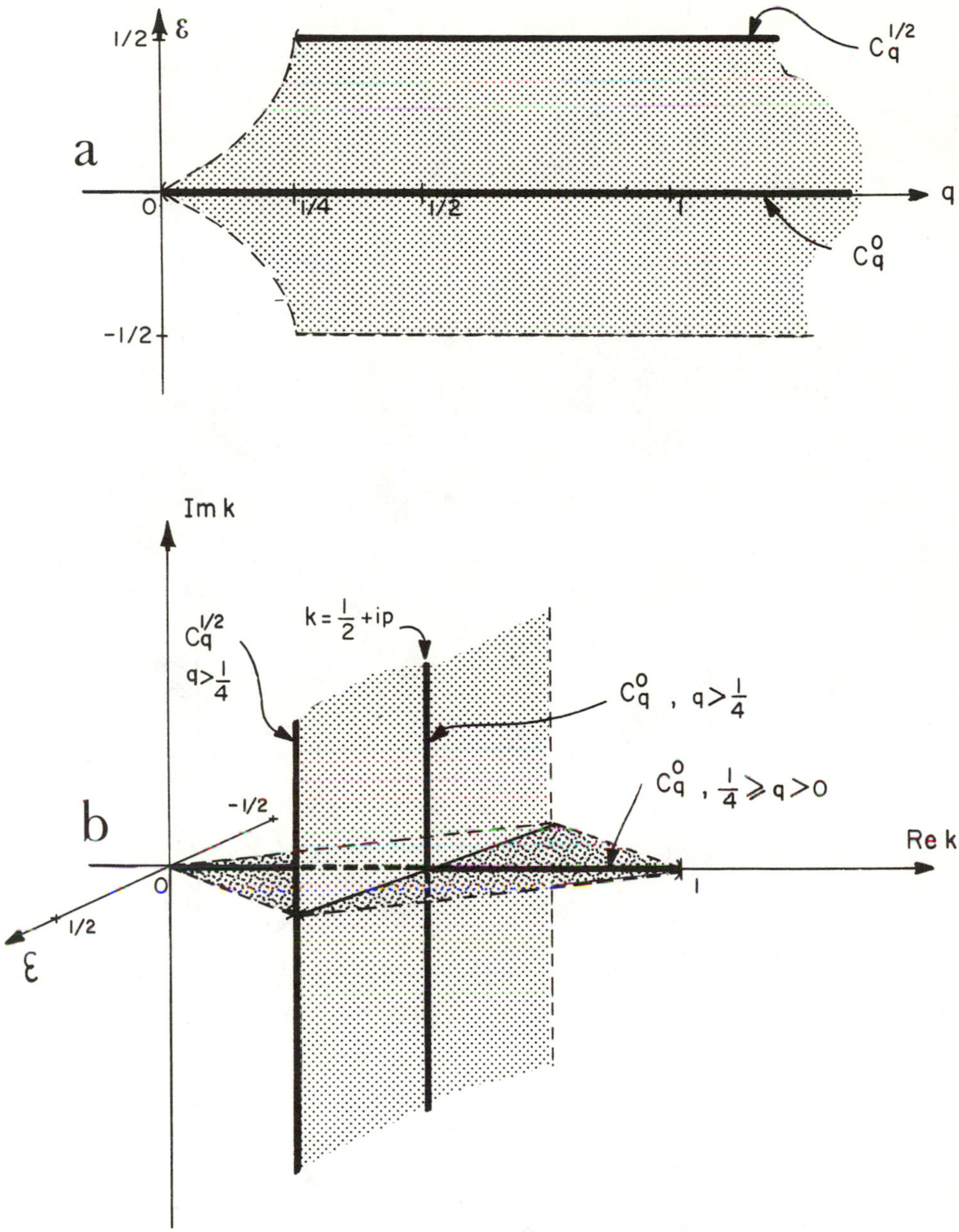

FIGURES 1 The continuous C_q^ε irreducible representation series of $sl(2,\mathcal{R})$. (a) as function of real q. (b) as function of complex k for $q=k(1-k)$. Shaded regions correspond to allowed self-adjoint representations. Dashed lines indicate that the region does not include the boundary. The parallel boundaries at $\varepsilon=-1/2$ are to be identified. Heavy lines indicate Bargmann's continuous series C_q^0 and $C_q^{1/2}$.

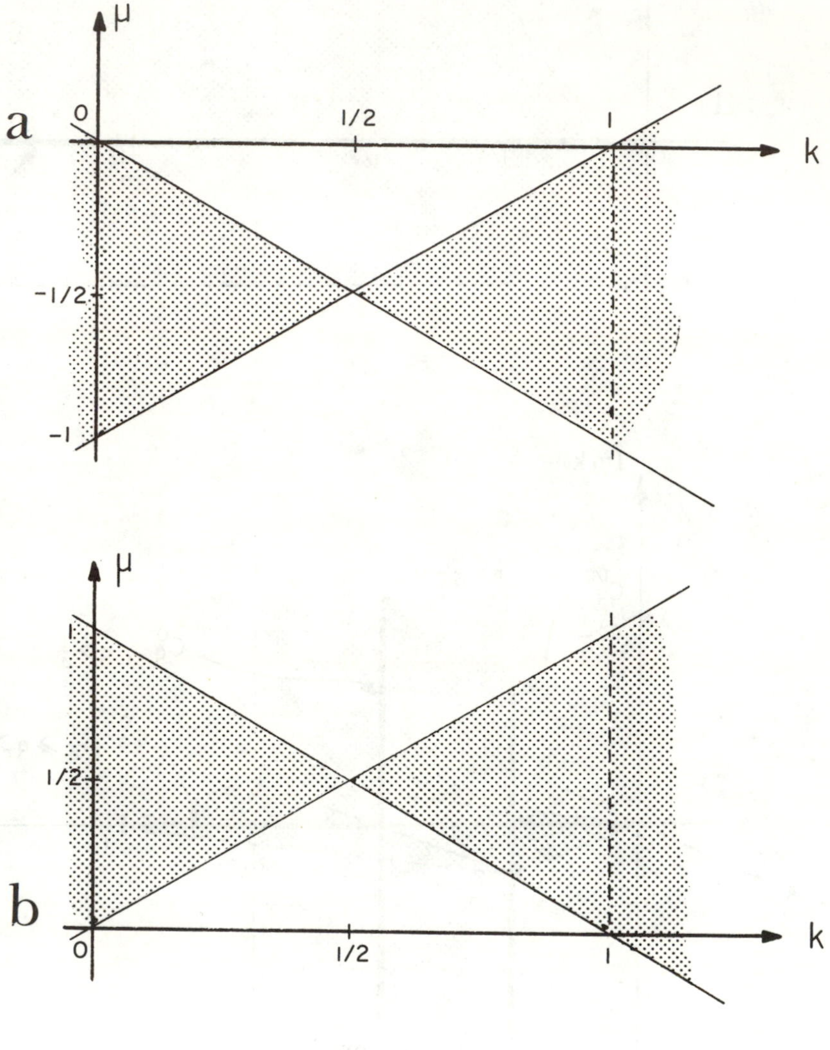

FIGURES 2 (a) the zeros of $|c^+_{k\mu}|^2$. (b) the zeros of $|c^-_{k\mu}|^2$. Shaded regions correspond to negative values of these quantities which must be excluded from the range of μ.

the highest on μ_1^+ or μ_2^+. We can make use of the inversion symmetry $k \leftrightarrow 1-k$ in order to ascribe the lowest μ of the lower-bound representation to $\mu_1^- = k$, for $k \in (0,1)$ and, correspondingly, the highest μ of the upper-bound representation to $\mu_1^+ = -k$ for $k \in (0,1)$.

3.10 Some isolated points.

The points $k = 1$ and $k = 0$, both mapping on $q = 0$ deserve special attention. The first, $k = 1$, is on par with the neighbouring points along k, it corresponds to $\mathcal{D}_1^\pm$ containing $\mu = \pm(1+m), m=0,1,2,\ldots$. The point $k = 0$, however, is unique: It lies in the origin of Figs. 2a and 2b, on zeros of both c_{00}^+ and c_{00}^-. In consequence, $\mathbb{J}_\pm \phi_0^0 = 0$, and so the eigenfunction ϕ_0^0 is the basis for the one-dimensional trivial representation of the algebra by zero. As a representation for the group, ϕ_0^0 constitutes the trivial one-dimensional unitary representation.

In summary, the lower- and upper-bound self-adjoint representations of sl(2,$\mathcal{R}$) are $\mathcal{D}_k^\pm$, $k > 0$ containing the eigenfunctions ϕ_μ^k of $\mathbb{J}_0$ and $\mathbb{C}$ for $\mu = \pm(k+m)$, $m=0,1,2,\ldots$. If $k \in (0,1]$ we can write $\mu = \varepsilon \pm m$ as for the continuous series, but keeping the congruence interval of ε to be $(-1/2,1/2]$. The discrete series provide the boundaries of the open regions of the exceptional interval in Figs. 1a and 1b. In particular, the C_q^0 series convers the interval $q > 0$: The point $q = 0$ belongs to the discrete series $\mathcal{D}_1^\pm$ and to the trivial $\mathcal{D}_0^\pm$. Similarly, the $C_q^{1/2}$ series covers the interval $q > 1/4$, while the point $q = 1/4$ belongs to $\mathcal{D}_{1/2}^\pm$.

For $0 < q < 1/4$ both the exceptional continuous and discrete series coexist. Their basis functions have the same eigenvalue under $\mathbb{C}$ but different spectra under $\mathbb{J}_0$. This underlines the need of having two different Hilbert spaces to accomodate them.

In order to express these facts in the manner of Figs. 1a and 1b, we may let ε range outside the interval $(-1/2,1/2]$, so that the μ-content of $\mathcal{D}_k^\pm$ be $\mu = \varepsilon \pm m$, $m = 0,1,2,\ldots$ (i. e. $\varepsilon = \pm k$). We may then, superpose them as in Fig. 3. In order to display the representations clearly in the k-plane, we refer the reader to Figs. 4, where we mark those representations of so(2,1) which contain the eigenvalues $\mu = \varepsilon + m$, i. e. plotting vector representations of SO(2,1) ($\varepsilon = 0$), spinor representations ($\varepsilon = 1/2$), four-fold covering representations ($\varepsilon = 1/4$), etc.

Having classified all self-adjoint representations of the algebra sl(2,$\mathcal{R}$) in terms of eigenbases of the Casimir operator $\mathbb{C}$ labelled by its eigenvalues $q = k(1-k)$, and having specified their representation content with respect to $\mathbb{J}_0$, we can give the generator matrix elements

$$J_j^{k,\varepsilon} = \| (J_j)_{m\,m'}^{k,\varepsilon} \|, \quad (J_j)_{m,m'}^{k\varepsilon} = (\phi_{m+\varepsilon}^k, \mathbb{J}_j\, \phi_{m'+\varepsilon}^k), \; j=1,2,0, \text{ or } \pm 0, \tag{3.22}$$

as

$$(J_0)_{m\,m'}^{k,\varepsilon} = \delta_{m\,m'}(\varepsilon + m) = \delta_{m\,m'}\,\mu(m, \varepsilon), \tag{3.23a}$$

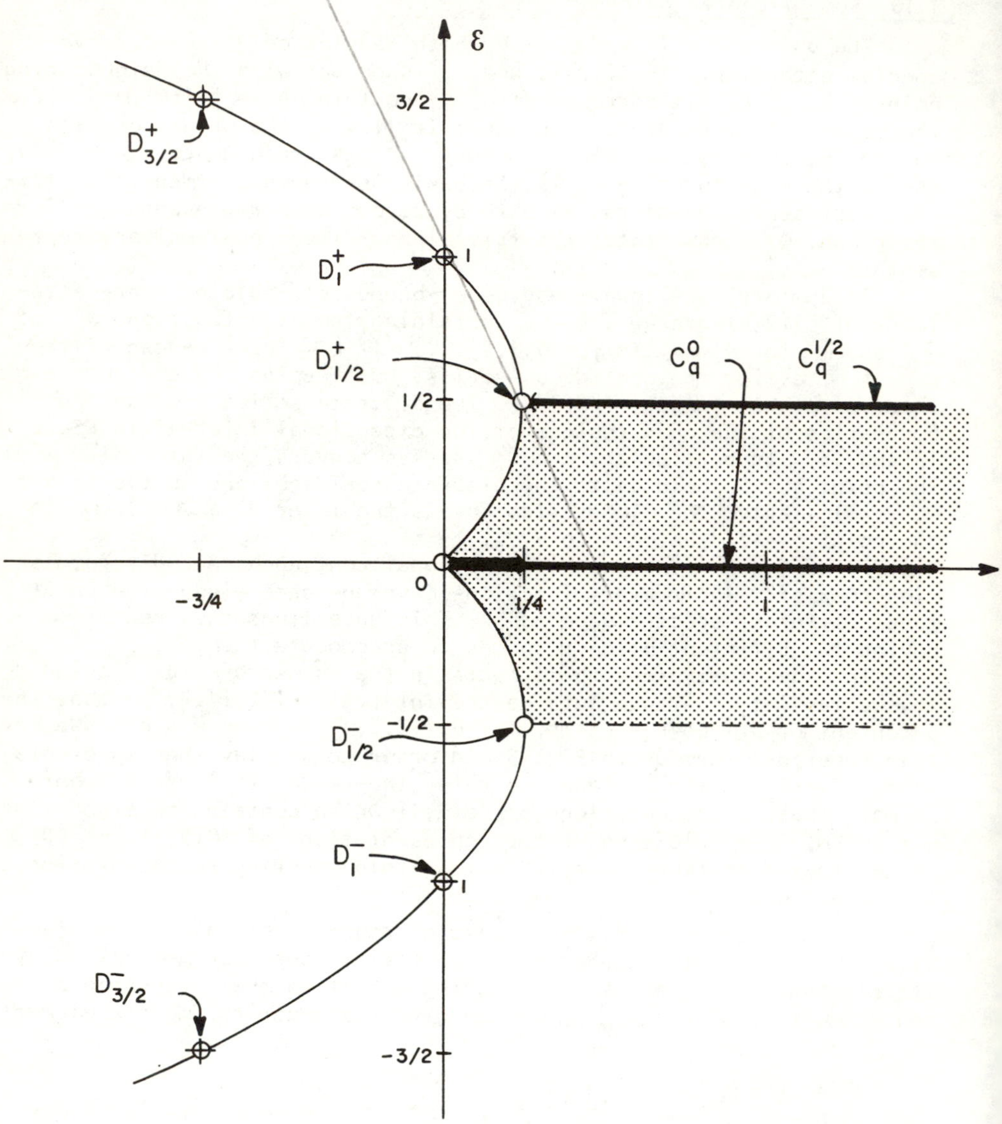

FIGURE 3. Discrete $\mathcal{D}^{\pm}_k$ and continuous $\mathcal{C}^{\varepsilon}_q$ irreducible representation series.

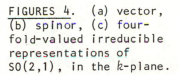

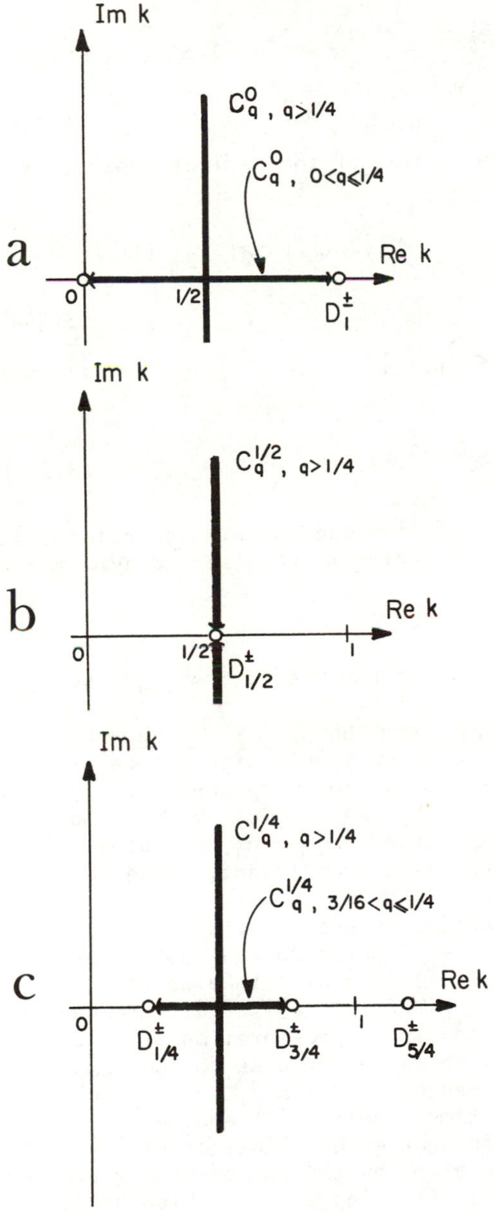

FIGURES 4. (a) vector, (b) spinor, (c) four-fold-valued irreducible representations of SO(2,1), in the k-plane.

$$(J_+)^{k,\varepsilon}_{m\,m'} = \delta_{m+1,m'}[(m'+\varepsilon+k)(m'+\varepsilon-k+1)]^{1/2} = \delta_{m+1,m'}c^{+}_{k,\mu(m'\varepsilon)},$$

(3.23b)

$$(J_-)^{k,\varepsilon}_{m\,m'} = \delta_{m-1,m'}[(m'+\varepsilon-k)(m'+\varepsilon+k-1)]^{1/2} = \delta_{m-1,m'}c^{-}_{k,\mu(m',\varepsilon)}\ .$$

(3.23c)

We may give a résumé of the ranges of the various indices as

$$C^{\varepsilon}_{q}:\ m,m' \in Z,\ k \in \{\tfrac{1}{2}+i\rho, \rho \in R\} \cup \{(0,1), |k-\tfrac{1}{2}| < \tfrac{1}{2} - |\varepsilon|\}, \varepsilon \in (-\tfrac{1}{2},\tfrac{1}{2}],$$

(3.24a)

$$D^{+}_{k}:\ m,m' \in \{0,1,2,\ldots\},\ k \in (0,\infty),\ \varepsilon = k\ , \tag{3.24b}$$

$$D^{-}_{k}:\ m,m' \in \{0,-1,-2,\ldots\},\ k \in (0,\infty),\ \varepsilon = k. \tag{3.24c}$$

These are infinite matrices which follow the commutation relations (3.14) and for which the Casimir operator is represented by a multiple q of the unit matrix.

3.11 "Other" representations.

Some words about non-self-adjoint representations. The self-adjointness of (3.22)-(3.24) stemmed out of asking for $q+\mu^2\pm\mu$ in (3.20) to be positive, which in turn came from the $\mathbb{J}_+{}^{\dagger} = \mathbb{J}_-$ requirement in the second equality of (3.19). If we work only with the raising and lowering action of $\mathbb{J}_\pm$ on some set of vector components as given in (3.21), accepting imaginary $c^{\pm}_{k\mu}$'s, then we may let k be complex. Choosing any given complex μ as a starting point, we obtain all other $\mu+n$, $n \in Z$ unless we meet a zero coefficient, in which case indecomposability will occur.

Suppose we start with a negative integer k, with ϕ^k_0. We may apply $\mathbb{J}_\pm$ to obtain $\phi^k_{\pm 1}, \phi^k_{\pm 2}, \ldots$. Figures 2a and 2b tell us that applying $\mathbb{J}_\pm$ a sufficient number of times will take us to $\mu^{\pm}_{1} = \mp k = \pm|k|$, but then as $c^{\pm}_{k,\pm|k|} = 0$, $\mathbb{J}_\pm \phi^k_{\pm|k|} = 0$. We have thus a $(2|k|+1)$-dimensional non-self-adjoint representation of sl(2,R). The elements of the algebra will be represented as follows: J_0 will be diagonal, real and hence self-adjoint, while $J_+{}^{\dagger} = -J_-$ so J_1 and J_2 will be skew-adjoint and i times their so(3) analogues. The same procedure applies for any integer or half-integer $k = \mp|k|$, as there the upper and lower bounds given by the zeroes of $c^{\pm}_{k\mu}$ are separated by integers. Moreover, the offending factors which produce the zero barriers may be removed by a simple norm redefinition of the basis elements ϕ^k_μ, as they are no longer subject to any orthonormality conditions. We may thus build representations with 'one-way' barriers, through which one raise, but not lower μ, or vice-versa. This implies a structure of the type (1.17) at each of two barriers, leading to indecomposable representations with the structures

$$\begin{pmatrix} X & X & X \\ 0 & X & X \\ 0 & 0 & X \end{pmatrix}, \begin{pmatrix} X & 0 & 0 \\ X & X & 0 \\ X & X & X \end{pmatrix}, \begin{pmatrix} X & X & 0 \\ 0 & X & 0 \\ 0 & X & X \end{pmatrix} \text{and} \begin{pmatrix} X & 0 & 0 \\ X & X & X \\ 0 & 0 & X \end{pmatrix}.$$

Indecomposable representations of so(2,1) have been described by Chacón, Levi and Moshinsky (47), and those of arbitrary semisimple groups by Gruber and Klimyk (48).

3.12 Subgroups and coset spaces.

We have not yet given any *realization* of the so(2,1) algebra generators (3.4)-(3.14) as differential operators on a manifold, nor have we provided the functions ϕ^k_μ or specified the Hilbert spaces for which these will provide complete and orthonormal bases. A general way of providing these is in following the procedure of Chapter 2, namely we let the group act on functions of coset spaces of the group and thus find the infinitesimal generators as differential operators on the coset manifolds. In order to construct these, we should know all nonequivalent subgroups of $\overline{SL(2,R)}$ and preferably divide by the largest proper subgroup so that the coset space have as low dimension as possible, without loosing information. For su(1,1) ≈ sl(2,R) this is relatively simple and the results in the 2×2 representation (3.9)-(3.10), and up to equivalence $g \cdot g^{-1}$ are as follows:

a) The *elliptic* subalgebra (subgroup) E

$$J_0 \to J_0^S = \frac{i}{2}\begin{pmatrix} 0 & -1 \\ 1 & 0 \end{pmatrix}, \quad \exp(i\phi J_0^S) = \begin{pmatrix} \cos\phi/2 & \sin\phi/2 \\ -\sin\phi/2 & \cos\phi/2 \end{pmatrix} \in E \quad (3.25a)$$

where the parameter ϕ ranges over $[0, 4\pi)$ for $SU(1,1) \approx SL(2,R)$, and may be identified through (3.9) and (3.12) with the parameter ω for $\overline{SL(2,R)}$, ranging over the full real line.

b) The *hyperbolic* subalgebra (subgroup) H

$$J_2 \to J_2^S = \frac{i}{2}\begin{pmatrix} 1 & 0 \\ 0 & -1 \end{pmatrix}, \quad \exp(i\chi J_2^S) = \begin{pmatrix} e^{-\chi/2} & 0 \\ 0 & e^{\chi/2} \end{pmatrix} \in H, \ \chi \in R. \quad (3.25b)$$

c) The *parabolic* subalgebra (subgroup) P

$$J_0 + J_1 \to J_0^S + J_1^S = \begin{pmatrix} 0 & -i \\ 0 & 0 \end{pmatrix}, \ \exp(i\xi[J_0^S + J_1^S]) = \begin{pmatrix} 1 & \xi \\ 0 & 1 \end{pmatrix} \in P, \ \xi \in R. \quad (3.25c)$$

d) There is a single two-parameter *solvable* subalgebra (subgroups) S obtained from H and P generated by J_2 and $J_0 + J_1$, with group elements

$$s(\chi,\xi) = \begin{pmatrix} 1 & \xi \\ 0 & 1 \end{pmatrix}\begin{pmatrix} e^{-\chi/2} & 0 \\ 0 & e^{\chi/2} \end{pmatrix} = \begin{pmatrix} e^{-\chi/2} & \xi\, e^{\chi/2} \\ 0 & e^{\chi/2} \end{pmatrix} \in S. \qquad (3.26)$$

We can now select among a number of homogeneous spaces $G_1\backslash G$ where G is SO(2,1) or SL(2,R), and G_1 may be any of (3.25)-(3.26). The $S\backslash$ SL(2,R) coset space is a one-dimensional manifold, the covering of the circle S_1. It turns out that this manifold by itself is too 'small' to contain all representations, although we shall come back to it below with a new interpretation.

Another most natural choice for coset space is $E \backslash$ SL(2,R), as the ensuing description closely resembles the description of the rotation group SO(3) through functions on the sphere S_2 = SO(2) \ SO(3), and the Euler angle decomposition of SO(2,1) in (3.2) is convenient for its description. The problem with this coset space is that although we do indeed get the expression for the generators $\mathbb{J}_k$, $k=0,1,2$ as in (2.8)), as differential operators of the first degree in β and γ, the Casimir operator will be a second-order differential operator resembling the angular part of the Laplacian (46, Eqs. (III-6)-(III-8), except for its eigenvalues. These functions are rather tedious to work with in integrations. A similar outcome awaits H\SL(2,R) (49, the functions $\mathfrak{B}_{0n}$ $(ch\beta)e^{in\alpha}$ in Chapter VI, Eqs. 3.2(6) and 4.5(9')).

3.13 The Iwasawa decomposition.

The simplest approach by far is to use the coset space P \SL(2,R) with a particularly fortunate choice of group parameters. This is given by the *Iwasawa* decomposition of $g \in$ SL(2,R)

$$G = \begin{pmatrix} a & b \\ c & d \end{pmatrix} = \begin{pmatrix} 1 & \xi \\ 0 & 1 \end{pmatrix}\begin{pmatrix} \exp(-\chi/2) & 0 \\ 0 & \exp(\chi/2) \end{pmatrix} \begin{pmatrix} \cos\phi/2 & \sin\phi/2 \\ -\sin\phi/2 & \cos\phi/2 \end{pmatrix}$$

$$= \begin{pmatrix} \cos\phi/2\; e^{-\chi/2} - \xi \sin\phi/2\; e^{\chi/2} & \sin\phi/2\; e^{-\chi/2} + \xi \cos\phi/2\; e^{\chi/2} \\ -\sin\phi/2\; e^{\chi/2} & \cos\phi/2\; e^{\chi/2} \end{pmatrix}, \qquad (3.27a)$$

$$\tan\frac{\phi}{2} = -\frac{c}{d}, \quad e^{\chi} = c^2+d^2, \quad \xi = \frac{b}{d} + \frac{c/d}{c^2+d^2}. \qquad (3.27b)$$

The Iwasawa decomposition of a noncompact group (20, Sect. 1.6.C) or algebra (50) expresses an arbitrary group element $g \in G$ as $g=nak$ where $k \in K$, the maximal compact subgroup - in this case SO(2) generated by J_0 - times $a \in A$, a maximal abelian subgroup, times $n \in N$, a nilpotent subgroup. The product NA is the solvable group in (3.26).

Functions on the coset space $M=P$ \SL(2,R) are two-variable functions $f(\chi,\phi)$ whose transformation properties under right action of the group [c. f. (2.4a)] by some element $g_0 \in$ SL(2,R) is

$$f(\chi,\phi) \xrightarrow{g_0\ (R)} f(\chi'(\chi,\phi,g_0),\phi'(\chi,\phi,g_0)) \, , \tag{3.28a}$$

and can be obtained through multiplying (3.27a) on the right by such a matrix

$$\begin{pmatrix} a & b \\ c & d \end{pmatrix} \xrightarrow{g_0} \begin{pmatrix} a' & b' \\ c' & d' \end{pmatrix} = \begin{pmatrix} a & b \\ c & d \end{pmatrix}\begin{pmatrix} a_0 & b_0 \\ c_0 & d_0 \end{pmatrix} = \begin{pmatrix} aa_0+bc_0 & ab_0+bd_0 \\ ca_0+dc_0 & cb_0+dd_0 \end{pmatrix}. \tag{3.28b}$$

Making use of (3.27b) we find that the coset parameters transform as

$$\tan\frac{\phi}{2} \to \tan\frac{\phi'}{2} = -\frac{ca_0+dc_0}{cb_0+dd_0} = \frac{a_0\tan\phi/2 - c_0}{d_0 - b_0\tan\phi/2} \, , \tag{3.29a}$$

$$e^{\chi} \to e^{\chi'} = (ca_0+dc_0)^2 + (cb_0+dd_0)^2$$

$$= e^{\chi}\cos^2\left(\frac{\phi}{2}\right)\left[\left(a_0\tan\frac{\phi}{2} - c_0\right)^2 + \left(d_0-b_0\tan\frac{\phi}{2}\right)^2\right]. \tag{3.29b}$$

3.14 sl(2,R) algebras of differential operators.

Global transformations will be futher analyzed below; here we are interested in infinitessimal transformations which in 3×3 matrix form (3.10) appear as

$$\begin{pmatrix} a_0 & b_0 \\ c_0 & d_0 \end{pmatrix} = 1 + i\left[\delta_1 J_1^S + \delta_2 J_2^S + \delta_0 J_0^S\right] + o(\delta^2) = \begin{pmatrix} 1-\frac{1}{2}\delta_2 & \frac{1}{2}(\delta_1+\delta_0) \\ \frac{1}{2}(\delta_1-\delta_0) & 1+\frac{1}{2}\delta_2 \end{pmatrix}. \tag{3.30a}$$

On a dense subspace of the space of functions (3.28a) on the coset space $M = P \setminus SL(2,R)$, they are given by

$$f(\chi',\phi') = [1+i(\delta_1\mathbb{J}_1^M + \delta_2\mathbb{J}_2^M + \delta_0\mathbb{J}_0^M) + o(\delta^2)]\, f(\chi,\phi), \tag{3.30b}$$

where $\mathbb{J}_k^M$, $k=0,1,2$ will be first-order differential operators in χ and ϕ. Replacement of (3.30a) in (3.29), Taylor expansion and collection of δ_k's allows us to find them:

$$\mathbb{J}_1^M = i(\cos\phi\,\partial_\phi + \sin\phi\,\partial_\chi), \tag{3.31a}$$

$$\mathbb{J}_2^M = i(\sin\phi\,\partial_\phi - \cos\phi\,\partial_\chi), \tag{3.31b}$$

$$\mathbb{J}_0^M = -i\partial_\phi \quad . \tag{3.31c}$$

They satisfy the commutation relations (3.4). From here we can find the raising and lowering operators (3.13) as

$$\mathbb{J}_+^M = \mathbb{J}_1^M + i\mathbb{J}_2^M = i\, e^{i\phi}\,(\partial_\phi - i\partial_\chi), \tag{3.32a}$$

$$\mathbb{J}_-^M = \mathbb{J}_1^M - i\mathbb{J}_2^M = i\, e^{-i\phi}(\partial_\phi + i\partial_\chi). \tag{3.32b}$$

Most important, we find the Casimir operator (3.14) in this realization to be

$$\mathbb{C}^M = \mathbb{J}_1^{M^2} + \mathbb{J}_2^{M^2} - \mathbb{J}_0^{M^2} = -\,\partial_\chi(1+\partial_\chi). \tag{3.33}$$

One can see the special feature of the Iwasawa decomposition (3.27a) and the parabolic coset space $M=P\backslash SL(2,R)$: That $\mathbb{J}_0^M$ is a differential operator in ϕ and $\mathbb{C}^M$ in χ_M. Functions on M belonging to a given eigenvalue $k(1-k)$ under $\mathbb{C}^M$ can be characterized through

$$f_k\,(\chi,\phi) = e^{-k\chi}\; f_k(\phi). \tag{3.34}$$

On these spaces of functions, ∂_χ may be replaced by $-k$, so as to obtain from (3.31) and (3.32),

$$\mathbb{J}_\pm^{(k)} = i\, e^{\pm i\phi}\;(\partial_\phi \pm ik), \tag{3.35a}$$

$$\mathbb{J}_0^{(k)} = -i\partial_\phi \; . \tag{3.35b}$$

This is the $sl(2,R)$ algebra realization employed by Bargmann (23). Equation (3.34) will be used below to examine the global action of $SL(2,R)$ on this space of functions.

3.15 Basis functions on the circle.

In Bargmann's realization, the normalized eigenfunctions ϕ_μ^k in Eq. (3.16) are proportional to the exponential functions $e^{i\mu\phi}$. The eigenvalue μ we saw, must range in integer steps $\mu=\varepsilon\pm m, m=0,1,2,\ldots$ passing through ε(for C_q^ε) or starting from a value $\mu=\varepsilon$, for $D_k^\pm$, as specified in Eqs. (3.24), so that $e^{i\mu\phi} = e^{i\varepsilon\phi}e^{im\phi}$. We let ϕ range on the unit circle S_1(i.e. $\phi\equiv\phi \bmod 2\pi$) and write the(possibly multivalued) function $f(\phi)$ on a space associated with k and ε as

$$f(\phi) = e^{i\varepsilon\phi} \sum_n{}^{(k)} \; f_n\, e^{in\phi} \quad , \tag{3.36a}$$

where we are indicating the sum over the proper μ-set as

$$\sum_{n}^{(k)} = \sum_{n=-\infty}^{\infty} [\text{for } C^{\varepsilon}_{k(1-k)}] , = \sum_{n=0}^{\pm\ \infty} [\text{for } \mathcal{D}^{\pm}_{k} \text{ , with } \varepsilon = \pm k]. \quad (3.36b)$$

The possible multivaluation will not be a problem when calculating inner products - with a measure still to be found- , as any sesquilinear inner product involving $(e^{i\mu_1\phi})^*$ and $e^{i\mu_2\phi}$ with $\mu_1=\varepsilon+m_1$ and $\mu_2=\varepsilon+m_2$ will cancel out the $e^{i\varepsilon\phi}$ factors, or otherwise render them innocuous.

We have not yet written down the *normalization* constants for the eigenfunctions. If we assume that $L^2(S_1)$ is the proper space, these constants will be $(2\pi)^{-1/2}$ for all m. But we may have other spaces. In general their normalization will be prescribed by the condition (3.19)-(3.21).

We may fix the normalization of $\phi^k_{\mu_0}$ for $\mu_0=\varepsilon$ and from this obtain all other $\phi^k_{\varepsilon+m}$, with ε and m ranging as in (3.36), through ascending or descending by means of $\mathbb{J}^{(k)}_{\pm}$:

$$(\mathbb{J}^{(k)}_{\pm})^m \phi^k_{\varepsilon}(\phi) = c^{\pm}_{k\varepsilon} (\mathbb{J}^{(k)}_{\pm})^{m-1} \phi^k_{\varepsilon\pm 1}(\phi) =$$

$$= \left[\prod_{n=0}^{m-1} c^{\pm}_{k,\varepsilon+n}\right] \phi^k_{\varepsilon\pm m}(\phi). \quad (3.37a)$$

We have the realization (3.35a) for $\mathbb{J}^{(k)}_{\pm}$ and $\phi^k_{\varepsilon}(\phi)=K^k_{\varepsilon}\, e^{i\varepsilon\phi}$, hence

$$\phi^k_{\varepsilon\pm m}(\phi) = K^k_{\varepsilon}\left[\prod_{n=0}^{m-1} c^{\pm}_{k,\varepsilon\pm n}\right]^{-1} [ie^{\pm i\phi}(\partial_\phi \pm ik)]^m e^{i\varepsilon\phi}$$

$$= K^k_{\varepsilon}\left[\prod_{n=0}^{m-1} c^{\pm}_{k,\varepsilon\pm n}\right]^{-1} (-\varepsilon\mp k)[ie^{\pm i\phi}(\partial_\phi \pm ik)]^{m-1} e^{i(\varepsilon\pm 1)\phi}$$

$$= K^k_{\varepsilon}\left[\prod_{n=0}^{m-1} \frac{-\varepsilon\mp k\mp n}{c^{\pm}_{k,\varepsilon\pm n}}\right] e^{i(\varepsilon\pm m)\phi} \quad (3.37b)$$

$$= K^k_{\varepsilon}\, \eta^k_{\varepsilon\pm m}\left[\prod_{n=0}^{m-1}\left|\frac{\varepsilon\pm n\pm k}{\varepsilon\pm n\pm(1-k)}\right|\right]^{1/2} e^{i(\varepsilon\pm m)\phi} ,$$

where $\eta^k_{\varepsilon\pm m}$ is the product of the inverse of the phases $\gamma^{\pm}_{k\mu}$ of the $c^{\pm}_{k\mu}$ in (3.21) for $\mu=\varepsilon,\varepsilon\pm 1,\ldots,\varepsilon\pm(n-1)$.

We shall not need a definite phase convention in these notes. It should only be noted that phases *can* be arranged so that the $\gamma^{\pm}_{k\mu}=1$, that is, that the raising and lowering operators have purely positive matrix matrix elements. This is the phase definition followed in (23, Sect. 6d) and known as Bargmann's convention.

We use the Pochhammer product symbol

$$\prod_{n=0}^{m-1}(a+n) = a(a+1)\cdots(a+m-1) = (a)_m = \Gamma(a+m)/\Gamma(a), \tag{3.37c}$$

$$\prod_{n=0}^{m-1}(b-n) = b(b-m+1)_m = \Gamma(b+1)/\Gamma(b+1-m), \tag{3.37d}$$

so as to write (3.37b) in the form

$$\phi^{k}_{\varepsilon\pm m}(\phi) = K^{k}_{\varepsilon}\eta^{k}_{\varepsilon\pm m}\ \sigma^{k,\varepsilon}_{\pm m}\ e^{i(\varepsilon\pm m)\phi} \tag{3.38a}$$

$$\sigma^{k\varepsilon}_{\pm m} = \left|\frac{(k\pm\varepsilon)_m}{(1-k\pm\varepsilon)m}\right|^{1/2} = \left|\frac{\Gamma(\varepsilon+1-k)}{\Gamma(\varepsilon+k)}\frac{\Gamma(\varepsilon+k\pm m)}{\Gamma(\varepsilon+1-k\pm m)}\right|^{1/2} = \left|\frac{\Gamma(1-k\pm\varepsilon)}{\Gamma(k\pm\varepsilon)}\frac{\Gamma\ (k\pm\varepsilon+m)}{\Gamma(1-k\pm\varepsilon+m)}\right|^{1/2}. \tag{3.38b}$$

We now examine the characteristics of the $\sigma^{k\varepsilon}_{\pm m}$ square root factor for the various representation series. We refer to (3.24)-(3.36b) for notation and index ranges. For the principal series C^{ε}_{q} in the nonexceptional interval, $k=1/2+i\rho$, $\rho\in R$ and hence the ratio of functions of the form $\gamma(1-k)/\gamma(k)=\gamma(1/2-i\rho)/\gamma(1/2+i\rho)$ has absolute value unity.

Thus

$$C^{\varepsilon}_{q},\ q > \frac{1}{4},\quad \sigma^{k,\varepsilon}_{\pm m} = 1 \quad \text{(continuous nonexceptional).} \tag{3.39a}$$

For the exceptional interval $k\in(0,1)$, on the other hand, for $k \lesseqgtr 1/2$ we have $(k+\alpha)/(1-k+\alpha) \lesseqgtr 1$. The ratio of Γ-functions becomes a product of such factors for $\alpha=\pm\varepsilon,\pm\varepsilon+1,\ldots,\pm\varepsilon+m-1$, so that

$$C^{\varepsilon}_{q=k(1-k)},\quad k\in\begin{cases}(0,1/2)\\(1/2,1)\end{cases},\quad \sigma^{k\varepsilon}_{\pm m} \lesseqgtr 1 \quad \text{(continuous exceptional).} \tag{3.39b}$$

Finally, for the discrete series $\mathcal{D}^{\pm}_{k}$, where $\varepsilon=\pm k$, we may write explicitly

$$\mathcal{D}^{\pm}_{k},\quad k\in\begin{cases}(0,1/2)\\(1/2,\infty),\end{cases}\quad \sigma^{k,\pm k}_{\pm m} = \left[\frac{\Gamma(2k+m)}{\Gamma(2k)\Gamma(m+1)}\right]^{1/2} \lesseqgtr 1,\text{(discrete)}. \tag{3.39c}$$

The inversion symmetry $k \leftrightarrow 1-k$ is not explicitly present in this formula, but a unitary intertwining operator may be set up to bridge these two representations (45, Sects. 2.3 and 2.4).

3.16 In search of Hilbert spaces.

We have now sets of functions $\phi^{k}_{\mu}(\phi)$ where μ is in the spectrum of $\mathbb{J}_0$ in a given, definite representation, (C^{ε}_{q} or $D^{\pm}_{k}$) $\phi\in S_1$ (the circle) given by (3.8). We search for inner products where these are orthonormal and complete. Completeness of these denumerable bases, when proven, will define the Hilbert spaces as the closure of linear combinations of these functions.

The most general sesquilinear inner product on S_1 may be written as

$$(f,g)^{(k,\varepsilon)} = \int_{S_1} d\phi \int_{S_1} d\phi'\; f(\phi)^* \,\Omega^{(k,\varepsilon)}(\phi,\phi')\; g(\phi'). \qquad (3.40)$$

Under this inner product the algebra generators (3.35) must be hermitian: $(\mathbb{J}^{(k)}_{j} f,g)^{(k,\varepsilon)} = (f,\mathbb{J}^{(k)}_{j} g)^{(k,\varepsilon)}$. We have seen before the requirements imposed by $j=1,2$, i. e. by $J^{(k)\dagger}_{+} = J^{(k)}_{-}$. We now impose the hermiticity of $\mathbb{J}^{(k)}_0$, integrating by parts:

$$\begin{aligned}(\mathbb{J}^{(k)}_0 f,g)^{(k,\varepsilon)} &= \int_{S_1} d\phi \int_{S_1} d\phi'\,(-i\partial_\phi f(\phi))^* \,\Omega^{(k,\varepsilon)}(\phi,\phi')\; g(\phi')\\ &= \int_{S_1} d\phi \int_{S_1} d\phi'\; f(\phi)^*[-i\partial_\phi \Omega^{(k,\varepsilon)}(\phi,\phi')]g(\phi')\\ =(f,\mathbb{J}^{(k)}_0 g)^{(k,\varepsilon)} &= \int_{S_1} d\phi \int_{S_1} d\phi'\; f(\phi)^* \,\Omega^{(k,\varepsilon)}(\phi,\phi')[-i\partial_{\phi'} g(\phi')]\\ &= \int_{S_1} d\phi \int_{S_1} d\phi'\; f(\phi)^*[i\partial_{\phi'} \Omega^{(k,\varepsilon)}(\phi,\phi')]g(\phi').\end{aligned} \qquad (3.41)$$

This leads to

$$\int_{S_1} d\phi \int_{S_1} d\phi'\; f(\phi)^*[(\partial_\phi-\partial_{\phi'})\Omega^{(k,\varepsilon)}(\phi,\phi')]g(\phi') = 0 \qquad (3.42)$$

which is to hold for all $f(\phi)$ and $g(\phi')$ in a space $\mathcal{H}^{(k,\varepsilon)}$. There is a rather subtle point to the ensuing reasoning, and that is since $\mathcal{H}^{(k,\varepsilon)}$ may be different from $L^2(S_1)$, what (3.42) actually implies is that $z(\phi,\phi')=(\partial_\phi-\partial_{\phi'})\Omega^{(k,\varepsilon)}(\phi,\phi')$, as a function of ϕ, is orthogonal in the sense of $L^2(S_1)$ to $f(\phi)$, and similarly orthogonal to $g(\phi')$ in the same sense. If we were to put $z(\phi,\phi')$ as an inner product measure in place of $\Omega(\phi,\phi')$ in (3.40), the result would be zero for all f,g in $\mathcal{H}^{(k,\varepsilon)}$, and hence $z(\phi,\phi')$ is equivalent to the zero function itself. On the basis of this equivalence, we may replace $z(\phi,\phi')$ by zero and thus conclude

$$(\partial_\phi-\partial_{\phi'})\Omega^{(k,\varepsilon)}(\phi,\phi') = 0 \Rightarrow \Omega^{(k,\varepsilon)}(\phi,\phi') = \Omega^{k,\varepsilon}(\phi-\phi'), \qquad (3.43a)$$

i. e. Ω is a function only of the angle difference $\phi-\phi'$. As the basis functions $e^{i\mu\phi}$ with μ as specified by the representation C_q^ε or $\mathcal{D}_k^\pm$, is assumed complete in $\mathcal{H}^{(k,\varepsilon)}$, we may expand $\Omega^{(k,\varepsilon)}(\psi)$ as

$$\Omega^{(k,\varepsilon)}(\psi) = \sum_m^{(k)} \omega_m^{k,\varepsilon}\, e^{i(\varepsilon\pm m)\psi} \quad , \tag{3.43b}$$

with $\Sigma^{(k)}$ as in (3.36), and $\omega_m^{k,\varepsilon}$ as yet unknown and put it in (3.40) with the orthonormality condition imposed on the basis functions, using the Fourier series orthogonality results for the exponential functions. We thus develop

$$\begin{aligned}\delta_{m,m'} &= (\phi^k_{\varepsilon\pm m}, \phi^k_{\varepsilon\pm m'})^{(k,\varepsilon)} = \int_{S_1} d\phi \int_{S_1} d\phi' [K^k_\varepsilon\, \eta_{\varepsilon\pm m}\sigma^{k\varepsilon}_{\pm m}\, e^{i(\varepsilon\pm m)\phi}]^* \times \\ &\times [\sum_{m''}^{(k)} \omega^{k\varepsilon}_{m''}\, e^{i(\varepsilon+m'')(\phi-\phi')}][K^k_\varepsilon\, \eta_{\varepsilon\pm m'}\sigma^{k\varepsilon}_{\pm m'} e^{i(\varepsilon\pm m')\phi'}] \\ &= |K^k_\varepsilon|^2 [\eta_{\varepsilon\pm m}\, \sigma^{k\varepsilon}_{\pm m}]^* [\eta_{\varepsilon\pm m'}\, \sigma^{k\varepsilon}_{\pm m'}] \sum_{m''}^{(k)} \Omega^{k\varepsilon}_{m''} \times \\ &\times [\int_{S_1} d\phi\, e^{i(n''\mp n)\phi}][\int_{S_1} d\phi'\, e^{i(\pm n'-n'')\phi'}] = \\ &= \delta_{mm'}\, (2\pi)^2 |K^k_\varepsilon|^2 |\sigma^{k\varepsilon}_{\pm m}|^2\, \omega^{k\varepsilon}_{\pm m} .\end{aligned} \tag{3.44a}$$

It follows that the proper coefficients for the weight function $\Omega^{k,\varepsilon}(\phi-\phi')$ are

$$\begin{aligned}\omega^{k\varepsilon}_{\pm m} &= [2\pi |K^k_\varepsilon\, \sigma^{k\varepsilon}_{\pm m}|]^{-2} = \frac{1}{(2\pi|K^k_\varepsilon|)^2} \left| \frac{(1-k\pm\varepsilon)_m}{(k\pm\varepsilon)_m} \right| \\ &= \frac{1}{(2\pi|K^k_\varepsilon|)^2} \left| \frac{\Gamma(\varepsilon+1-k)}{\Gamma(\varepsilon+k)} \, \frac{\Gamma(\varepsilon+k\pm m)}{\Gamma(\varepsilon+1-k\pm m)} \right| \\ &= \frac{1}{(2\pi|K^k_\varepsilon|)^2} \left| \frac{\Gamma(k\pm\varepsilon)}{\Gamma(1-k\pm\varepsilon)} \, \frac{\Gamma(1-k\pm\varepsilon+m)}{\Gamma(k\pm\varepsilon+m)} \right| .\end{aligned} \tag{3.44b}$$

3.17 Local and nonlocal weight functions.

Let us now examine the various representation series in order to give the description of the $\mathcal{H}^{(k,\varepsilon)}$ spaces.

For the continuous nonexceptional series C_q^ε we have $m\in\mathbb{Z}$ and (3.39a), since $\sigma^{k\varepsilon}_{\pm m}$ is only a phase. Then

$$\Omega^{1/2+i\rho,\varepsilon}(\psi) = (2\pi|K^k_\varepsilon|)^{-2} \sum_{m\in\mathbb{Z}} e^{i(m+\varepsilon)\psi} = e^{i\varepsilon\psi}[2\pi|K^k_\varepsilon|^2]^{-1}\delta(\psi) . \tag{3.45a}$$

We have hence

$$\mathcal{H}^{k=1/2+i\rho,\varepsilon} = L^2(S_1) \tag{3.45b}$$

and the inner product is the ordinary L^2-product on the circle.

For the continuous exceptional series,

$$\Omega^{k,\varepsilon}(\psi) = \frac{1}{(2\pi|K^k_\varepsilon|)^2}\left[\sum_{m=0}^{\infty}\omega_m^{k\varepsilon}\, e^{i(\varepsilon+m)\psi} + \sum_{m=1}^{\infty}\omega_{-m}^{k\varepsilon}\, e^{i(\varepsilon-m)\psi}\right]$$

$$= \frac{e^{i\varepsilon\psi}}{(2\pi|K^k_\varepsilon|)^2}\left[\sum_{m=0}^{\infty}\frac{(1-k+\varepsilon)_m(1)_m}{(k+\varepsilon)_m}\frac{e^{im\psi}}{m!} + \sum_{m=0}^{\infty}\frac{(1-k-\varepsilon)_m(1)_m}{(k-\varepsilon)_m}\frac{e^{-im\psi}}{m!} - 1\right]$$

$$= \frac{e^{i\varepsilon\psi}}{(2\pi|K^k_\varepsilon|)^2}\left[{}_2F_1(1,1-k+\varepsilon;k+\varepsilon;e^{i\psi}) + {}_2F_1(1,1-k-\varepsilon;k-\varepsilon;e^{-i\psi}) - 1\right]. \tag{3.46a}$$

When $\varepsilon=0$ it was proven by Bargmann (23, Eqs. (8.7), (8.9) and (8.11)) that this expression reduces to

$$\Omega^{k,0}(\psi) = \frac{1}{(2\pi|K^k_0|)^2}\sum_{m=-\infty}^{\infty}\frac{(1-k)_{|m|}}{(k)_{|m|}}e^{im\psi} = \frac{2^{-1-k}}{\pi\sqrt{\pi}|K^k_0|^2}\frac{\Gamma(k)}{\Gamma(k-1/2)}(1-\cos\psi)^{k-1}. \tag{3.46b}$$

Bargmann (23, Sect. 8c) also proves that $\mathcal{H}^{k,0}$ is a Hilbert space when $k\in(1/2,1)$, while Sally (45, Sect. 2.4) proves this for the remaining cases.

For the discrete series $\mathcal{D}^{\pm}_k$, finally, we can use (3.39c)-(3.44) in order to find

$$\begin{aligned}\Omega^{k,\pm k}(\psi) &= \sum_{m=0}^{\infty}\omega_{\pm m}^{k,\pm k}\, e^{\pm i(k+m)\psi}\\ &= \frac{e^{\pm ik\psi}}{(2\pi|K^k_{\pm k}|)^2}\;{}_2F_1(1,1;2k;e^{i\psi}).\end{aligned} \tag{3.47}$$

This series converges absolutely for $k \geqslant 1$, conditionally for $1/2 < k < 1$ (excluding $\psi=0, 2\pi, \ldots$ although its integral with bounded functions is bounded), and diverges for $k=1/2$. For $0 < k < 1/2$ this series diverges but may still be used to define an inner product between appropiately convergent functions.

The nonlocal inner product (3.47) has been examined in (51) where it is proven that it defines a Hilbert space. There (37, Appendix) it is also proven that this representation is equivalent to the description of $\mathcal{D}^{\pm}_k$ by Bargmann and Sally (23, 45) as an integral over the complex unit disk, and by Gel'fand (51, Chapter VII) for single-valued representations of $SO(2,1)$ on the upper complex half-plane. In the Hilbert space $\mathcal{H}^{k,\pm k}$, the operator $\mathbb{J}_0^{(k)} = -i\partial_\phi$ has thus a spectrum given by $\pm k, \pm(k+1), \pm(k+2), \ldots$, in contrast with the usual $L^2(S_1)$, where it is Z.

3.18 Unitary group action and multipliers.

The measures found above also make the exponentiated action of the algebra generators unitary. Let us make this explicit. The $\overline{SL(2,R)}$ action on functions belonging to $\mathcal{H}^{k,\varepsilon}$ is found from (3.29) and (3.34). It is convenient to parametrize the acting group through Euler angles (3.6). The action of $\exp(i\alpha\mathbb{J}_0)$ and $\exp(i\beta\mathbb{J}_2)$ on the coset parameters of $M=P\backslash SL(2,R)$ are found giving in (3.29) the g_0 parameters through their values (3.10), namely

$$e^{i\alpha\mathbb{J}_0^M} \quad : \tan\frac{\phi}{2} \longrightarrow \tan\frac{\phi_0(\phi,\alpha)}{2} = \tan\frac{\phi+\alpha}{2} \tag{3.48a}$$

$$e^{i\alpha\mathbb{J}_0^M} \quad : e^{\chi} \longrightarrow e^{\chi_0(\chi,\phi,\alpha)} = e^{\chi} \tag{3.48b}$$

$$e^{i\beta\mathbb{J}_2^M} \quad : \tan\frac{\phi}{2} \longrightarrow \tan\frac{\phi_2(\phi,\beta)}{2} = e^{-\beta}\tan\frac{\phi}{2} \tag{3.49a}$$

$$e^{i\beta\mathbb{J}_2^M} \quad : e^{\chi} \longrightarrow e^{\chi_2(\chi,\phi,\beta)} = e^{\chi}[\operatorname{ch}\beta+\cos\frac{\phi}{2}\operatorname{sh}\beta] = e^{\chi}\frac{\sin\phi}{\sin\phi_2(\phi,\beta)}. \tag{3.49b}$$

From here, it follows through (3.34) that the $\overline{SL(2,R)}$ group action on functions in the (k,ε)-irreducible representation spaces $\mathcal{H}^{k,\varepsilon}$ is

$$e^{-k\chi} f_k(\phi) \xrightarrow{e^{i\gamma\mathbb{J}_\bullet^M}} e^{-k\chi'(\chi,\phi,\gamma)} f_k(\phi'(\phi,\gamma)), \tag{3.50a}$$

and hence, on functions on S_1,

$$\begin{aligned} f_k(\phi) \xrightarrow{e^{i\gamma\mathbb{J}_\bullet^{(k)}}} & e^{-k\chi'(\chi,\phi,\gamma)} e^{k\chi} f_k(\phi'(\phi,\gamma)) \\ &= [\mu_\bullet(\phi,\gamma)]^k f_k(\phi'(\phi,\gamma)). \end{aligned} \tag{3.50b}$$

The group action has thus become a *multiplier* action. As in Chapter 2, the function's argument is transformed, and the function itself is multiplied by a factor

$$\mu_\bullet(\phi,\gamma) = e^{-\chi'(\chi,\phi,\gamma)} e^{\chi}. \tag{3.50c}$$

For the two generating Euler subgroups of (3.48), we have

$$e^{i\alpha \mathbb{J}_0^{(k)}} \quad : \quad \mu_0(\phi,\alpha) = 1 \tag{3.51a}$$

$$e^{i\beta \mathbb{J}_2^{(k)}} \quad : \quad \mu_2(\phi,\beta) = \text{ch}\beta + \cos\frac{\phi}{2}\,\text{sh}\beta = \frac{\sin\phi}{\sin\phi_2(\phi,\beta)} = \left[\frac{\partial\phi_2(\phi,\beta)}{\partial\phi}\right]^{-1}.$$

Putting together (3.48a), (3.50b) and (3.51a) we see that the action of $exp(i\alpha\mathbb{J}_0)$ is a pure translation in S_1. This is a manifestly unitary transformation

$$(e^{i\alpha\mathbb{J}_0^{(k)}} f\,,\; e^{i\alpha\mathbb{J}_0^{(k)}} g)^{(k,\varepsilon)} = (f,\, g)^{(k,\varepsilon)}, \tag{3.52a}$$

in any of the $\mathcal{H}^{k\varepsilon}$ spaces seen above, as shown by a simple change of variables in the integral.

As to the action of $\exp(i\beta\mathbb{J}_2)$, we see that it *deforms* the circle parameter in the sense that $\partial\phi_2(\phi,\beta)/\partial\phi \neq 1$. Interestingly, as the last equality in (3.51b) shows, this is exactly given by the multiplier function. Now from (3.49a), (3.50b) and (3.51b), we may show that

$$(e^{i\beta\mathbb{J}_2^{(k)}} f,\; e^{i\beta\mathbb{J}_2^{(k)}} g)^{(k,\varepsilon)} = (f,\, g)^{(k,\varepsilon)}\,, \tag{3.52b}$$

for all the spaces $\mathcal{H}^{k,\varepsilon}$. This is clear for the nonexceptional continuous series, where in the $L^2(S_1)$ inner product, the change in $d\phi$ is $[\mu_2(\phi,\beta)]^{-1}$ which is exactly offset by the product of the multipliers: $([\mu_2(\phi,\beta)]^{1/2+i\rho})^*(\mu_2(\phi,\beta))^{1/2+i\rho}$. The element which is pleasing in this approach is that unitarity (3.52b) holds for all series when the appropiate inner product and weight function (3.46) and (3.47) are used.

Once appropiate Hilbert spaces have been found -and for SL(2,$\mathcal{R}$) we have seen there are three families of them- the irreducible representation matrix elements may be readily calculated.

3.19 Closing remarks.

Since we must stop at some point, we would like to insist on the fact that SL(2,$\mathcal{R}$), for all its 'simplicitly', is a rather richly structured object of which we have only given an overall view. Books devoted to SL(2,$\mathcal{R}$) exist (53). We should bear in mind that this is just one example of a noncompact group. There exist others in higher dimensions, nevertheless, the systematics in their treatment are not very different from what we have seen for their smallest representative member.

We shall resist the temptation to close thise lectures notes with a barrage of references, and simply point out that group theory, in spite of its simple and very compact four defining axioms, appears to contain a noncompact body of results★

REFERENCES

(1) D. R. Reuben "Evething you always wanted to know about Sex, but were afraid to ask" (Mc Kay, 1969).

(2) M. Moshinsky, J. Patera, R. T. Sharp and P. Winternitz, Everything you always wanted to know about SU(3)$\supset$0(3), Ann. Phys. 95, 139-169 (1975).

(3) K. B. Wolf, The Heisenberg-Weyl ring in quantum mechanics, in "Group Theory and its applications" III Ed. E. M. Loebl (Academic, 1975).

(4) J. E. Moyal, Quantum mechanics as a statistical theory, Proc. Cambridge Phil. Soc. 45, 99-124 (1949).

(5) C. L. Mehta, Phase-space formulation of the dynamics of canonical variables. J. Math. Phys. 5, 677-686 (1964).

(6) K. B. Wolf and A. García, The Weyl group -a survey. Rev. Mex. Fís. 21, 191-218 (1972).

(7) P. A. M. Dirac, The Fundamental equations of quantum mechanics. Proc. Roy. Soc. 109, 642-653 (1925).

(8) P. A. M. Dirac, "The principles of quantum mechanics" 4th. ed. (Oxford Univ. Press, 1958).

(9) J. von Neumann, Die Eindeutigkeit der Schrödingerschen Operatoren, Math. Ann. 104, 570-578 (1931).

(10) V. Bargmann, On a Hilbert space on analytic functions and an associated integral transform. Part I. Commun. Pure Appl. Math. 14, 187-214 (1961).

(11) A. O. Barut and L. Girardello, New "Coherent" states associated with non-compact groups. Commun. Math. Phys. 21, 41-55 (1971).

(12) K. B. Wolf, Canonical Transforms, I. Complex linear transforms. J. Math. Phys. 15, 1295-1301 (1974).

(13) K. B. Wolf, Canonical Transforms. II. Complex radial transforms. J. Math. Phys. 15, 2102-2111 (1974).

(14) K. B. Wolf, "Integral Transforms in Science and Engineering" (Plenum Press, 1979).

(15) J. D. Louck, private communication.

(16) M. Abramowitz and I. Stegun, editors, "Handbook of Mathematical Functions" National Bureau of Standards, Applied Mathematics Series No. 55 (1964).

(17) I. D. Ado, The representations of Lie algebras by matrices. (in Russian) Usp. Mat. Nauk, N. S. 2, 159-173 (1947).

(18) H. Weyl, Quantenmechanik und Gruppentheorie. Z. Phys. 46, 1-46 (1928).

(19) S. Helgason, "Differential Geometry and Symmetric Spaces" (Academic Press, 1962).

(20) A. O. Barut and R. Rączka, "Theory of Group Representations and Applications". (Polish Scientific Publishers, 1977).

(21) R. Hermann, "Lie Groups for Physicists" (Benjamin, 1966).

(22) S. Lie and F. Engel, "Theorie der Transformationsgruppen" (Teubner, Leipzig; Vol I, 1888; Vol. II, 1890; Vol. III, 1893).

(23) V. Bargmann, Irreducible unitary representations of the Lorentz group. Ann. Math. 48, 568-640 (1947).

(24) W. Miller Jr., "Lie theory and special functions" Mathematics

in Science and Engineering, Vol. 43 (Academic, 1968).
(25) K. B. Wolf, Point transformations in quantum mechanics. Rev. Mex. Fís. 22, 45-74 (1973).
(26) H. Goldstein, "Classical Mechanics" (Addison-Wesley, 1950).
(27) P. Jordan, Über eine Neue Begründung der Quantenmechanik, I. Z. Phys. 40, 809-838 (1927).
(28) P. Jordan, Über eine Neue Begründung der Quantenmechanik II. Z. Phys. 38, 513-517 (1926).
(29) F. London. Über die Jacobischen Transformationen der Quantenmechanik. Z. Phys. 37, 915-928 (1926).
(30) P. Jordan, Über Kanonische Transformationen in der Quantenmechanik. I. Z. Phys. 37, 383-386 (1926).
(31) P. Jordan, Über Kanonische Transformationen in der Quantenmechanik. II. Z. Phys. 38,513-517 (1926).
(32) Xia Dao-Xing, "Measure and integration theory on infinite-dimensional spaces" (Academic Press, 1972).
(33) M. Moshinsky and T. H. Seligman, Canonical transformations to action and angle variables, and their representations. J. Phys. A 12, L135-L139 (1979).
(34) M. Moshinsky and T. H. Seligman, Canonical transformations to action and angle variables, and their representations in quantum mechanics. Ann. Phys. 114, 243-272 (1978).
(35) M. Moshinsky and T. H. Seligman, Canonical transformations to action and angle variables and their representations in quantum mechanics. II. The Coulomb Problem. Ann. Phys. 120, 402-422 (1979).
(36) H. Weyl, "The theory of groups and quantum mechanics' (2nd ed. Dover, 1930).
(37) C. P. Boyer and K. B. Wolf, Canonical transforms. III. Configuration and phase descriptions of quantum systems possessing an sl(2,R) dynamical algebra. J. Math. Phys. 16, 1493-1502 (1975).
(38) E. Hewitt and K. A. Ross, "Abstract Harmonic Analysis I" (Springer Verlag, 1963).
(39) M. Hamermesh, "Group theory and its applications to physical problems" (Addison-Wesley, 1962).
(40) G. Warner, "Harmonic Analysis on semi-simple Lie groups" (Springer -Verlag, 1972).
(41) D. P. Želopenko, "Harmonic Analysis on semi-simple complex Lie groups" (Nauka, 1974).
(42) F. Peter and H. Weyl, Die Vollständigkeit der primitiven Darstellungen einer geschlossenen kontinuirlischen Gruppe. Math. Ann. 97, 737-755 (1927).
(43) R. L. Anderson and K. B. Wolf, Complete sets of functions on homogeneous spaces with compact stabilizers. J. Math. Phys. 11, 3176-3183 (1970).
(44) K. B. Wolf and T. H. Seligman, Harmonic analysis on bilateral classes. Comunicaciones Técnicas IIMAS No. 207 (1979), to appear in SIAM J. Math. Anal.
(45) P. J. Sally Jr, Analytic continuation of the irreducible unitary representations of the universal covering group of SL(2,R). Mem. Amer. Math. Soc. 69 (Providence R. I., 1967).
(46) M. E. Rose, "Elementary theory of angular momentum" (Wiles, 1957).

(47) E. Chacón, D. Levi and M. Moshinsky, Lie algebras in the Schrödinger picture and radial matrix elements. J. Math. Phys. 17, 1919-1929 (1976).

(48) B. Gruber and A. U. Klimyk, Properties of linear representations with a highest weigh for the semisimple Lie algebras. J. Math. Phys. 16, 1816-1832 (1975).

(49) N. Ja. Vilenkin, "Special functions and the theory of group representations. Trans. Amer. Math. Soc. Vol. 22 (1968).

(50) J. F. Cornwell, Direct determination of the Iwasawa decomposition for noncompact semisimple Lie algebras. J. Math. Phys. 16, 1992-1999 (1975).

(51) I. M. Gel'fand, M. I. Graev and N. Ya. Vilenkin, "Generalized Functions" Vol. 5, (Academic, 1966).

(52) S. Lang, "$sl_2(R)$" (Addison-Wesley, 1975).

ACKNOWLEDGEMENT.

I would like to thank Mrs. Elizabeth Derbez for her invaluable help in typing this manuscript.

Lectures on MAGNETIC INTERACTIONS OF STABLE PARTICLES AND MAGNETIC RESONANCES

A. O. Barut
Department of Physics, The University of Colorado, Boulder, CO 80309

Contents

INTRODUCTION

Atomic, nuclear and particle physicists all have to deal with spin effects at one time or another. The main attitudes of physicists towards spin and magnetic moment interactions seem to be the following. Either, they are considered to be small, unimportant and causing only "inessential complications" (except for the Pauli principle), hence could hardly change the phenomena in a qualitativly new way. Or, they are very complicated (complex Chlebsch-Gordon coefficients and all that), hence assume and hope that they are not important and concentrate on more basic problems. While in the final

ISSN:0094-243X/81/710073-36$1.50

analysis it is true that the magnitude of magnetic moment interactions is small in atomic and molecular physics, the situation turns out to be quite different at nuclear and particle energies and distances.

One of the objectives of these lectures is to show that there is a new regime of distances and energies beyond atomic phenomena where magnetic interactions dominate, provide strong forces and give rise to new states of matter. Even the reason of the smallness of the spin effects in atomic physics has been incorrectly justified before (due to the cancellations of two incorrect operations), which one can do now rigorously.

The second main objective of these lectures is to actually identify these new magnetic states of matter formed out of the stable particles, electron, proton and neutrino, with the nuclear and hadronic states. The magnetic forces had been lumped up to now together with the so called strong and weak interactions. For we subtract Coulomb forces between, for example, two protons, and everything else we call summarily strong forces. We wish now to subtract also magnetic forces in a nonperturbative fashion to see what is left, if any. The surprising strength and richness of magnetic forces at short distances will then lead us to the hypothesis that there is nothing left and that electromagnetism alone may be sufficient to account for strong and weak interactions. These ideas are further elaborated in Appendix I. The main part of these lectures deals with dynamical models of spin-orbit and spin-spin forces between particles. In the second part, we discuss a number of striking applications of the dynamical models in hadron and nuclear physics.

I. DYNAMICAL MODELS

1) Charge-Dipole Interactions in Classical relativistic Mechanics

We start from the relativistic Hamiltonian of a charged particle of charge e, rest mass m in an external electromagnetic field with potentials $(A_o, \vec{A})$:

$$H = eA_o + m\left[1 + \frac{1}{m^2}(\vec{p} - e\vec{A})^2\right]^{\frac{1}{2}} \quad . \tag{1}$$

The classical motion of the charge in the magnetic field is called the Störmer's problem,[1] and has applications in the trapping of charged cosmic ray particles in earth's magnetic field.

In this case, $A_o = 0$, and

$$A = \nabla x\left(\frac{\vec{u}}{r}\right) = \mu\frac{\hat{\mu}\ x\ \vec{r}}{r^3} \quad . \tag{2}$$

In polar coordinates, with $\hat{\mu}$ in the direction $\theta = 0$,

$$A_r = A_\theta = 0 \ , \quad A_\varphi = \mu \sin^2\theta/r \ ,$$

hence

$$H = m\left[1 + \frac{1}{m^2}\left[P_r^2 + \frac{1}{r^2}P_\theta^2 + \frac{1}{r^2\sin^2\theta}\left(P_\phi - e\mu\frac{\sin^2\theta}{r}\right)^2\right]\right]^{\frac{1}{2}} \ . \tag{3}$$

Clearly, ϕ is a cyclic coordinate, hence P_ϕ is a constant of the motion. Thus the problem has essentially two degrees of freedom (r, θ), i. e. the meridian plan.

Instead of the Hamiltonian H, we can consider the function $F \equiv \frac{1}{2M}(H^2 - m^2)$ as the "Hamiltonian." Indeed

$$\frac{\partial F}{\partial q} = \frac{\partial F}{\partial H}\frac{\partial H}{\partial q} = \frac{H}{M}\frac{\partial H}{\partial q} \quad ,$$

and we have the same Hamilton's equations, with F as with H, if H = M. Note that H is a constant of the motion, hence M is the relativistic mass. Then

$$F = \frac{1}{2M}\left[P_r^{\;2} + \frac{1}{r^2}P_\theta^{\;2} + \frac{1}{r^2\sin^2\theta}\left(P_\phi - e\mu\frac{\sin^2\theta}{r}\right)^2\right] \quad . \qquad (4)$$

The quantity F is also a constant of the motion and is equal to $\vec{P}^2/2M$, where P is the relativistic momentum of the particle: $H^2 = m^2 + \vec{p}^2$. We have finally a two-dimensional "effective non-relativistic" Hamiltonian

$$P_r^{\;2} + \frac{1}{r^2}P_\theta^{\;2} + \frac{1}{r^2\sin^2\theta}\left(K_1 - e\mu\frac{\sin^2\theta}{r}\right)^2 = K_2 = \vec{p}^2 \qquad (5)$$

with two integration constants K_1 and K_2. The third term is the "effective potential" $V(r, \theta)$ in the meridian plane,

$$V(r, \theta) = \left(\frac{K_1}{r\sin\theta} - \frac{e\mu\sin\theta}{r^2}\right)^2 \quad , \qquad (6)$$

and is plotted in Fig. 1.

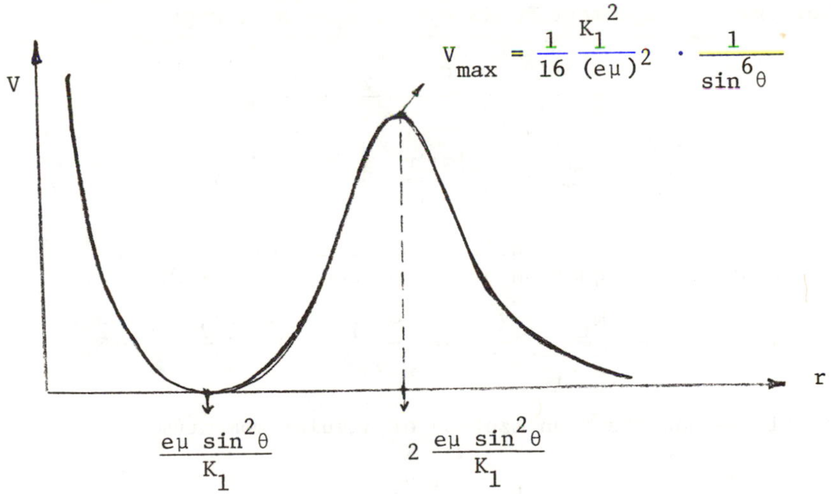

Fig. 1

Fig. 1 shows the essentials of a phenomenon that we shall use throughout this work, namely the existence of a potential well, hence the trapped orbits. In quantum mechanics these will be replaced by the resonance energy levels.

For future references we note that the size of the trapped orbits are of the order of $r \approx \frac{e\mu}{P_\phi}$ and the maximum (non-relativistic) energy (for $P_r = P_\theta = 0$) is of the order of $\frac{1}{2m} P_\phi{}^2/(e\mu)^2$. If we measure μ in units of $\frac{e}{2m}$ with value g, we have $r \approx \frac{\alpha}{2m}\frac{g}{P_\phi}$ and $E \sim \frac{2m}{\alpha^2}(P_\phi/g)^2$.

Equation (5) is not separable, nor the associated Hamilton-Jacobi equations. We shall see that the corresponding quantum mechanical Hamiltonian is separable. The reason for this is that in the latter case $\hat{\mu}$ in eq. (2) is not a fixed dipole moment vector, but is a dynamical variable and precesses. It seems possible that for a properly rotating magnetic dipole moment $\hat{\mu}$ also the classical problem would be separable, but this has not been done yet. The chemical problem of Störmer is actually an interesting example of a non-integrable dynamical system with homoclinic points.[2]

Another difference with the quantum theory is that the potential well in quantum case is much deeper and becomes negative at the minimum (see Sect. 3).

A further open problem, to my knowledge, is the classical theory of dipole-dipole interactions.

2) Semi-Classical Theory of Magnetic Interactions

In this section we treat the motion of a charged particle in the field of a charged dipole-moment according to Bohr-Sommerfeld quantization.

The force on the charged particle e_1 due to the magnetic field $\vec{B}$ of the dipole moment $\vec{\mu}_2$ is $\vec{F} = e_1 \vec{v}_1 \times \vec{B}$, where

$$\vec{B} = \frac{3\hat{r}(\vec{\mu}_2 \cdot \vec{r}) - \vec{\mu}_2}{r^3} + \frac{8\pi}{3}\vec{\mu}_2\,\delta(\vec{r}) \quad . \qquad (1)$$

Here

$$\vec{\mu}_2 = g_2 \frac{|e|\hbar}{2m_1}\vec{S} \, . \quad (c = 1) \qquad (2)$$

For simplicity we shall consider circular orbits, in which case the relativistic equation of motion becomes ($r \neq 0$)

$$\frac{m_1}{\sqrt{1 - v_1^2}}\frac{v_1^2}{r_1} = \frac{e_1 v_1 \mu_2}{r^3} - \frac{e_1 e_2}{r^2} \quad . \qquad (3)$$

To this we add the quantization of angular momentum

$$L = \frac{m_1 v_1 r_1}{\sqrt{1 - v_1^2}} = n\hbar \quad . \qquad (4)$$

If we eliminate r_1 between (3) and (4), we obtain $r_1 = n\hbar \frac{\sqrt{1 - v_1^2}}{m_1 v_1}$ and

$$v_1 + \frac{e_1 e_2}{n\hbar} = \frac{e_1 \mu_2 m_1}{n^2 \hbar^2} \frac{v_1^2}{\sqrt{1 - v_1^2}} \qquad (5)$$

It is instructive to solve this equation graphically (Fig. 2). The right and left hand sides of eq. (5) are plotted for different signs of e_1, e_2 and μ_2, with ε = sign $(e_1 e_2)$ and η = sign $(e_1 \mu_2)$.

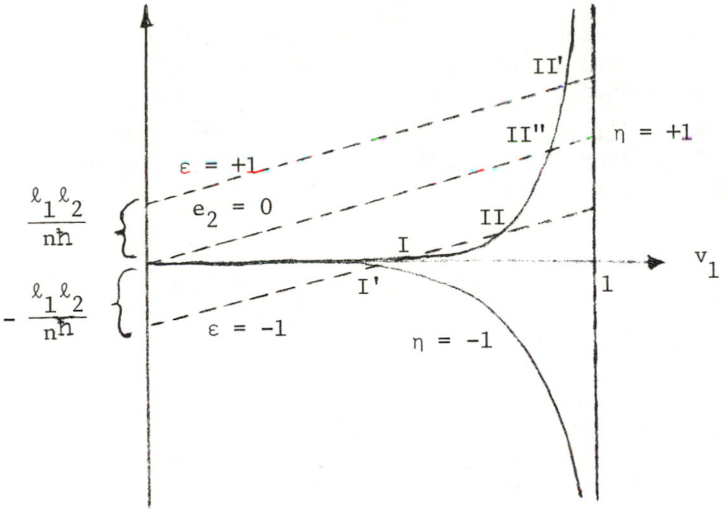

Fig. 2

Solutions I, I´ correspond to the ordinary Coulomb levels (v_1 small), the difference between I and I´ due to the influence of the sign of the dipole moment of the nucleus (thus a hyperfine effect). Solutions II, II´ and III" are the new magnetic resonance states for opposite, equal or zero signs of the charges $e_1 e_2$; the difference is the Coulomb-perturbation. Whereas the Coulomb bound levels of course exist between opposite charges, the charge-dipole bound levels exist in both cases (II and II´), and even if $e_2 = 0$, (II"), provided the sign of the nuclear dipole moment is the same as that of the charge e_1.

If we include the magnetic moment $\vec{\mu}_1$ of the moving particle, there are two other terms in the force equation (3),

$$\nabla X\ (\vec{B}^{(2)} \times \vec{\mu}_1) + \nabla X\ ((-\vec{v}_1 \times \vec{E}^{(2)}) \times \vec{\mu}_1) \ . \qquad (6)$$

The second term is again the spin-orbit term. For circular orbits, the first term, the dipole-dipole interactions, $\nabla(\vec{\mu}_1 \cdot B^{(2)}) = \nabla\left(\frac{3\vec{\mu}_1 \cdot \hat{r}\,\vec{\mu}_2 \cdot \hat{r} - \vec{\mu}_1 \cdot \vec{\mu}_2}{r^3}\right)$, becomes

$$-\vec{\mu}_1 \cdot \vec{\mu}_2 \; \nabla\left(\frac{1}{r^3}\right) = 3\vec{\mu}_1 \cdot \vec{\mu}_2 \, \frac{\vec{r}}{r^5} \quad . \tag{7}$$

This gives an additional term

$$3\vec{\mu}_1 \cdot \vec{\mu}_2 \, \frac{m^2 v_1^{\,2}}{n^3 \hbar^3 (1 - v_1^{\,2})} \tag{8}$$

on the right hand side of eq. (5). Its effect on Fig. 2 is small for Coulomb bound states I and I′, but becomes large at high energies near v_1 1, depending on the sign δ of $(\mu_1 \mu_2)$. As an example, in the case $\eta = +1$, $\delta = -1$, Eq. 5 and 8 can have the form shown in Fig. 3. We now have solutions of the third type (III, III′) in which dipole-dipole interactions dominate.

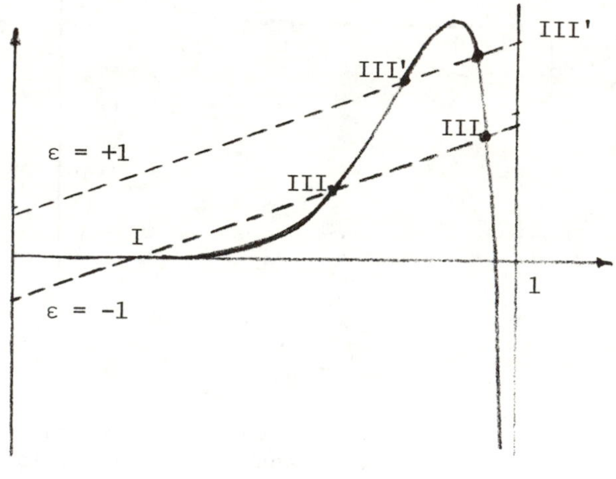

Fig. 3

We shall see the counterpart of all these solutions in the quantum-mechanical case.

In Table I we list some of the properties of these three types of solutions, if each existed alone. These are obtained from eq. (5) and the equation in the line above it. There will be of course modifications when all act together.

Table I: Properties of three types of solutions ($c = \hbar = 1$)

	I Charge-charge	II Charge-dipole	III Dipole-dipole
Velocity v	α/n	$1 - \alpha^2/8n^4$	$1 - \frac{3\alpha}{2n^3}$
Size r	$\frac{n^2}{n\alpha}\sqrt{1 - \frac{\alpha^2}{n^2}}$	$\frac{\alpha}{m}\,\frac{1}{2n}$	$\frac{\sqrt{3\alpha}}{m\sqrt{n}}$
Kinetic energy	$m/\sqrt{1 - \alpha^2/n^2}$	$\frac{2mn^2}{\alpha}$	$\frac{mn^{3/2}}{\sqrt{3\alpha}}$
Potential energy	$-\frac{\alpha^2 m}{n^2\sqrt{1 - \alpha^2/n^2}}$	$-2\,\frac{mn^2}{\alpha}$	$-\frac{1}{4}\,\frac{mn^{3/2}}{\sqrt{3\alpha}}$
Total energy	$m\sqrt{1 - \alpha^2/n^2}$	0	$\frac{3}{4}\,\frac{mn^{3/2}}{\sqrt{3\alpha}}$
Uncertainty Principle	$\Delta p \sim p \sim \sqrt{2mE}$ $\sqrt{2mE}\; r \sim n$	$\Delta p \sim p \sim E$,	$Er \sim n$

In order to estimate the energy of the resonances, we take $\mu = \frac{e}{2m}$, then velocities for the charge-dipole states are $v\;\; 1 - \frac{\alpha^3}{\delta n^4}$, and for dipole-dipole states $v \approx 1 - \frac{3\alpha}{2n^3}$. These given for the kinetic energies $\frac{m}{\sqrt{1-v^2}} = \frac{2mn^2}{\alpha^{3/2}}$ and $\frac{mn^{3/2}}{\sqrt{3}\alpha^{\frac{1}{2}}}$, respectively.

The characteristic magnetic energies can also be obtained through an application of the uncertainty principle.

It is well-known that for the nonrelativistic Hamiltonian $H = \frac{p^2}{2m} - \frac{\alpha}{r} = E$, assuming the uncertainties $\Delta p \Delta r \sim 1$ with $\Delta p \sim p$, $\Delta r \sim r$, we obtain the minimum of the energy, $E \approx \frac{1}{2mr^2} - \frac{\alpha}{r}$, using $dE/dr = 0$ at a distance $r = \frac{1}{m\alpha}$ (Bohr radius), with the value $E = -\frac{1}{2}m\alpha^2$.

In the magnetic case, we have

$$E = \sqrt{p^2+m^2} + \tfrac{1}{2}\,\frac{e_1 v \cdot \mu_2}{r^2} - \mu_1\mu_2\,\frac{1}{r^3}$$

or

$$E \sim p - \left(\tfrac{1}{2}\,\frac{e_1\mu_2 p}{mr^2} + \frac{\mu_1\mu_2}{r^3}\right)$$

Setting again $p \sim \frac{1}{r}$ and $dE/dr = 0$, we find,

$$r \sim \frac{3}{2}\,\frac{\sqrt{\alpha}}{m} \quad \text{and} \quad E \sim \frac{4}{9}\,\frac{m}{\sqrt{\alpha}}\cdot (g_1 = g_2 = 2).$$

3) Magnetic Interactions in Nonrelativistic Quantum Theory.

We consider again two charged particles e_1, e_2 with magnetic moments $\vec{\mu}_1$ and $\vec{\mu}_2$. The Hamiltonian is given by

$$H = \frac{1}{2m_1}\left(\vec{P}_1 - e_1\,\frac{\vec{\mu}_2 X(\vec{r}_1-\vec{r}_2)}{|r_1-r_2|^3}\right)^2 + \frac{1}{2m_2}\left(\vec{p}_2 - e_2\,\frac{\vec{\mu}_1 X(\vec{r}_2-\vec{r}_1)}{|r_1-r_2|^3}\right)^2 + \frac{e_1 e_2}{|r_1-r_2|}$$

$$- \mu_1\mu_2 S_{12}(\vec{r}_1-\vec{r}_2), \tag{1}$$

where S_{12} is the so called tensor and dipole-dipole interaction potential:

$$S_{12}(\vec{r}) = \frac{3\sigma_1\cdot\hat{r}3\sigma_2\cdot\hat{r}-\sigma_1\cdot\sigma_2}{r^3} + \frac{\delta\pi}{3}\,\vec{\sigma}_1\cdot\vec{\sigma}_2\delta(\vec{r}). \tag{2}$$

This 2-body problem with magnetic forces is separable, like the 2-body Coulcomb problem. With

$$\vec{r} = \vec{r}_1-\vec{r}_2,\; R = \frac{m_1\vec{r}_1+m_2\vec{r}_2}{M},\; p = \frac{m_2\vec{p}_1-m_1\vec{p}_2}{M},\; P = \vec{p}_1+\vec{p}_2 \quad ,$$

$$\frac{1}{\mu} = \frac{1}{m_1} + \frac{1}{m_2} \tag{3}$$

we obtain after some calculation

$$H = \frac{1}{2M}\,P^2 + \frac{1}{2\mu}\,p^2 - \vec{p}\cdot\frac{\vec{a}\times\vec{r}}{r^3} - \vec{P}\,\frac{\vec{b}\times\vec{r}}{r^3} + \frac{A}{r^4} + \frac{e_1 e_2}{r} - \mu_1\mu_1 S_{12}(\vec{r}) \quad , \tag{4}$$

where

$$\vec{a} = \frac{e_1}{m_1}\,\vec{\mu}_2 + \frac{e_2}{m_2}\,\vec{\mu}_1 \,,$$

$$\vec{b} = \frac{e_1}{m_1+m_2}\,\vec{\mu}_2 - \frac{e_2}{m_1+m_2}\,\vec{\mu}_1 \,,$$

$$A = \frac{e_1^2}{m_1}\mu_2^2 + \frac{e_2^2}{m_2}\mu_1^2 \quad .$$

Note that in addition to the center of mass kinetic energy we have another magnetic term $-\vec{P}\cdot\frac{\vec{b}\times\vec{r}}{r^3}$ for a moving atom, which vanishes both for identical particles and identical particle-antiparticle systems.

In the center of mass frame ($\vec{P} = 0$) we have

$$H_{rel} = \frac{1}{2\mu}p^2 - \vec{p}\cdot\frac{\vec{a}\times\vec{r}}{r^3} + \frac{A}{r^4} + \frac{e_1e_2}{r} - \mu_1\mu_2 S_{12}(\vec{r}) \quad . \qquad (5)$$

Using $-\vec{p}\cdot(\vec{a}\times\vec{r}) = \vec{p}\cdot(\vec{r}\times\vec{a}) = (\vec{p}\times\vec{r})\cdot\vec{a} = -\vec{a}\cdot(\vec{r}\times\vec{p}) = -\vec{a}\cdot\vec{L}$, we see that the second term correspons to a charge-dipole potential, and the last term to a dipole-dipole potential.

In the <u>special case</u>: $m_2 >> m_1$, $\vec{\mu}_1 = 0$ hence $\vec{a} = \frac{e_1}{m_1}\mu_2$, $\vec{b} = 0$, $A = \frac{e_1^2}{m_1}\mu_2^2$, we have the simpler Hamiltonian

$$H_{rel} = \frac{1}{2\mu}p^2 + \frac{e_1e_2}{r} - \frac{e_1\mu_2}{m_1}\frac{\vec{\sigma}\cdot\vec{L}}{r^3} + \frac{e_1^2\mu_2^2}{m_1}\frac{1}{r^4} \quad . \qquad (6)$$

One usually solves the Coulomb part of this Hamiltonian, $\left(\frac{p^2}{2\mu} + \frac{e_1e_2}{r}\right)$, exactly and treats the remaining two terms as perturbations. We shall see that this procedure is not only deficient and incomplete even if the perturbation theory would be valid, but also masks an important phenomenon, namely the existence of new quasi-bound states that we have seen in the two previous sections. We now try to solve exactly the eigenvalue problem $H\psi = E\psi$ with H given in eq. (6). For simplicity we take the spin of the particle 2 to be ½.

Choosing for ψ the eigenstates of $\vec{\sigma}\cdot L$ with eigenvalues

$$c_\ell = \begin{cases} -(\ell+1) = -(j+3/2), & \text{for } (j = \ell - \tfrac{1}{2}) \\ \ell = j - \tfrac{1}{2}, & \text{for } (j = \ell + \tfrac{1}{2}) \end{cases}$$

we have the radial equation

$$\left(\frac{1}{2\mu}p^2 + \frac{e_1e_2}{r} - \frac{e_1\mu_2}{m_1}\frac{C_\ell}{r^3} + \frac{e_1^2\mu_1^2}{m_1}\frac{1}{r^4}\right)\psi = E\psi \qquad (7)$$

The effective radial potential is then given by

$$V = \frac{\ell(\ell+1)}{2\mu r^2} + \frac{e_1e_2}{r} - \frac{e_1\mu_2 C_\ell}{m_1 r^3} + \frac{e_1^2\mu_2^2}{m_1}\frac{1}{r^4}$$

$$= \frac{\ell(\ell+1)}{2\mu r^2} + \varepsilon \frac{\alpha}{r} - g\varepsilon \frac{\alpha}{2m_1 m_2} \frac{C_\ell}{r^3} + g^2 \frac{\alpha^2}{4m_1 m_2^2} \frac{1}{r^4} \tag{8}$$

where $\varepsilon = \text{sign}(e_1 e_2)$, g is the g-factor of particle 2 (positive or negative). The typical form of V is shown in Fig. 4 when the sign of the third term is negative.

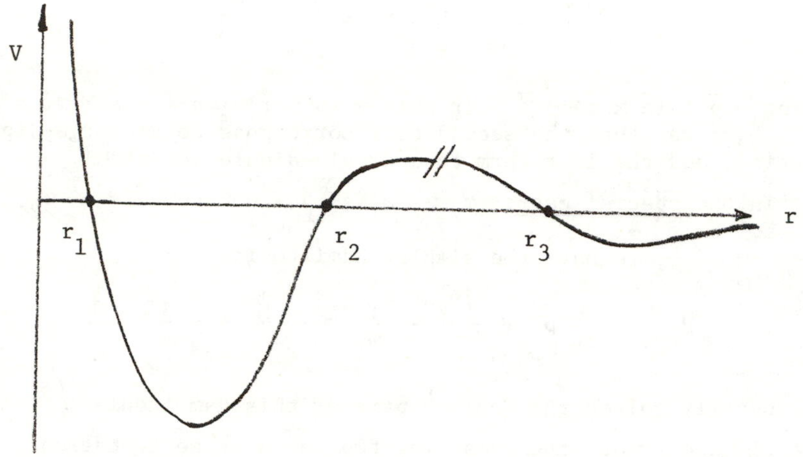

Fig. 4

In the regions of r_1, r_2 and r_3 the terms in the potential with $\frac{1}{r^4}$, $1/r^3$ and $1/r^2$ respectively dominated, hence we estimate these ranges to be

$$r_1 \sim \frac{g}{2} \frac{\alpha}{m_2} \cdot \frac{1}{|C_\ell|} ,$$

$$r_2 \sim \frac{g}{2} \frac{\alpha}{M} \frac{|C_\ell|}{\ell(\ell+1)} , \quad M = m_1 + m_2 , \quad \ell \neq 0 \tag{9}$$

$$r_3 \sim \frac{\ell(\ell+1)}{2} \frac{1}{\mu\alpha} , \quad \mu = (m_1 m_2)/M , \quad \ell \neq 0 .$$

The last one is clearly the order of magnitude of the Bohr radius, the others agree exactly with the semi-classical ranges given Table I.

The attractive spin-orbit case, hence the possibility of magnetic resonances in the deep potential well occurs in the following cases

($\ell \neq 0$):

$j = \ell - \frac{1}{2}$ (C_ℓ neg.)

g\ε	-1	+1
+	yes	no
-	no	yes

$j = \ell + \frac{1}{2}$ (C_ℓ pos.)

g\ε	-1	+1
+	no	yes
-	yes	no

Thus for electron-proton-like systems ($\varepsilon = -1$, g pos.) we can have resonances (called superpositronium) in the states $j = \ell - \frac{1}{2}$ only, i. e. $P_{\frac{1}{2}}$, $D_{3/2}$, etc. For electron-anti proton-like systems ($\varepsilon = +1$, g pos.) only in the states $j = \ell + \frac{1}{2}$, i. e. $P_{3/2}$, $D_{5/2}$, etc.

Special case II: $m_1 = m_2$; particle-particle, or particle-antiparticle case (e. g. pp or $p\bar{p}$ or e^-e^+, e^-e^-, . . .).

$$H_{rel} = \frac{1}{m} p^2 - \varepsilon g \frac{\alpha}{2m^2} \frac{(\vec{\sigma}_2 + \vec{\sigma}_1) \cdot \vec{L}}{r^3} + \varepsilon \frac{\alpha}{r} + \frac{\alpha^2}{4m^2} g^2 \frac{1}{r^4} - \varepsilon \frac{\alpha g^2}{4m^2} S_{12}$$

$$= \frac{1}{m} p^2 + \varepsilon \frac{\alpha}{r} - \varepsilon g \frac{\alpha}{2m^2} \frac{1}{r^3} \{(\vec{\sigma}_2 + \vec{\sigma}_1) \cdot \vec{L} + g \frac{1}{2} S_{12}\} + \frac{g^2\alpha^2}{4m^2} \frac{1}{r^4}$$

$$H_{rel} = \frac{1}{m} p^2 + \varepsilon \frac{\alpha}{r} - \varepsilon g \frac{\alpha}{2m^2} \frac{1}{r^3} [2\vec{S} \cdot \vec{L} + \frac{g}{2} \hat{S}_{12}] + \frac{g^2\alpha^2}{4m^2} \frac{1}{r^4} ,$$

where for $r \neq 0$ we have dropped the $\vec{\sigma}_1 \cdot \vec{\sigma}\, \delta^3(r)$ - term.

The states of spin $\frac{1}{2}$-particles are shown in Table II. In the above ℓ-basis the matrix $\hat{S}_{12}$ is given by[3]

$$\hat{S}_{12} = \frac{-2}{2J+1} \begin{vmatrix} J - 1 & -3\sqrt{J(J+1)} \\ -3\sqrt{J(J+1)} & J + 2 \end{vmatrix}$$

From Table II we can then ascertain the possibility of resonances in the following cases:

For singlet S = 0 states, the magnetic force can only come from the $\delta(\vec{r})\vec{\sigma}_1 \cdot \vec{\sigma}_2$ term (superimposed on the $1/r^4$- term!) which is attractive for $\varepsilon = -1$, $g > 0$ (e^-e^+, $p\bar{p}$) [as in the standard treatment

Table II

	S = 0	S = 1		
J	ℓ	$\ell - 1$	ℓ	$\ell + 1$
	$^1S_0, {}^1P_1, {}^1D_2, \ldots$	$^3P_0, {}^3D_1, {}^3F_2;$	$^3P_1, {}^3D_2, {}^3F_1;$	$^3S_1, {}^3P_2, {}^3D_3, \ldots$
$2\vec{S} \cdot \vec{L}$	0	$-2(J + 2)$	-2	$2(J - 1)$
$\hat{S}_{12}$ (without the $\delta^3(r)$ - part)	0	Mixes $(\ell - 1)$ and $(\ell + 1)$ states $^3S_1$ + $^3D_1$	2	Mixes $(\ell - 1)$ and $(\ell + 1)$ states $^3P_2$ + $^3F_2$, etc.
$\left(\frac{8\pi}{3}\right) \vec{\sigma}_1 \cdot \vec{\sigma}_2$	$-3\left(\frac{8\pi}{3}\right)$		$+1\left(\frac{8\pi}{3}\right)$	
L^2	$J(J + 1)$	$(J + 1)(J + 2)$	$J(J + 1)$	$J(J - 1)$

of hyperfine splitting of $^1S_0$, $^3S_1$ levels in H-atom]; in this case there is no negative potential well.

For triplet S = 1 states, (a) $\varepsilon = -1$, $g > 0$ ($e^- e^+$, $p\bar{p}$), the potential well can exist for $^3P_0$, $^3D_1$, $^3F_2$, . . . ; $^3P_1$, $^3D_2$, $^3F_3$, . . . (b) $E = + 1$, $g > 0$ (ee, pp), the potential well can exist for $^3S_1$, $^3P_2$, $^3D_3$, . . . and for some mixtures of $^3S_1 + {}^3D_1$, etc. .

In addition the Pauli principle restricts the pp- states to $^3P_2$ only among those lowest states listed. The np-case is the same, as for as the sign of the magnetic moment is concerned, as the $p\bar{p}$ case. This selects $^3P_0$, $^3F_2$, $^3P_1$ for isospin I = 1- states (np + pn); and $^3D_1$ . . ., $^3D_2$ . . . for I = 0 states.

In general the potential well becomes more pronounced with increasing angular momentum.

4). Spin-Orbit Coupling in the Klein-Gordon Equation

The motion of a spin-0 charge in a dipole field $\vec{\mu}$ fixed at origin can be reduced exactly to the non-relativistic equation (6) of Sect. 3:

$$\sqrt{(\vec{p} - e\vec{A})^2 + m^2}\,\phi = E_R\,\phi\ ,$$

$$[(\vec{p} - e\vec{A})^2 + m^2]\,\phi = E_R\left(\sqrt{(\vec{p} - e\vec{A})^2 + m^2}\right)\phi = E_R^{\,2}\,\phi$$

$$(\vec{p} - e\vec{A})^2\,\phi = (E_R^{\,2} - m^2)\,\phi \qquad .$$

If we compare this with the non-relativistic equation we have

$$2m\,E = E_R^{\,2} - m^2$$

Hence if we have solution E of the equation (6), Sect. 3, then there exists also a resonance at the relativistic energy

$$E_{Rel} = (2m\,E + m^2)^{\frac{1}{2}}$$

for the Klein-Gordon equation. Thus, with the same dipole potential $\vec{A}$, an energy $E = \frac{2mn^2}{\alpha}$ (see Table I) would give $E_{Rel} = mn\left(\frac{4}{\alpha} + 1\right)^{\frac{1}{2}}$ $\cong \frac{2mn}{\sqrt{\alpha}}$.

5. Discussion of the Perturbative versus Non-perturbative solutions

The spin-orbit and spin-spin terms in the Schrödinger (or Dirac eq.) have always been treated as perturbations, even up to the present day charmonium quark models. One solves an "unperturbed" problem, e. g. $H_o = \frac{p^2}{2m} + V(r)$, with some scalar or Coulomb-type potential V(r), and then takes the matrix elements of the spin terms between these unperturbed states. Even in atomic physics, where magnetic effects are small, this procedure is in fact, incorrect. First of all the $\frac{1}{r^4}$ - potential cannot be used in first order perturbation analysis, it is too singular at r = 0, and its matrix elements do not exist. One simply drops this term with the argument that its coefficient, proportional to e^2, is small. To quote one reference ". . . Thus its neglect (namely the e^2A^2 - term) can in no way be held responsible for the conflicts which appear further along."[4] The conflicts do indeed appear. For if one drops the $1/r^4$ - term, the next term is the singular and attractive term: $-1/r^3$. And with such a potential, the Hamiltonian is no longer essentially self-adjoint. For this reason, the magnetic forces have always been an enigma in in the rigorous theory of dynamical systems. Not only can these difficulties be avoided if the problem is treated non-perturbatively, for the Hamiltonian (6), Sect. 3 can be solved exactly, but also we find new resonance states not accessible by any order of perturbation theory.

Here is a lesson in the unmindful use of perturbation theory. The "perturbation" terms are not uniformly small, in fact they become dominant terms at short distances. If the wave function is concentrated at about the Bohr radius, there the magnetic terms are small, and the results of atomic physics can be justified without neglecting the $1/r^4$ - term. But there is another region, another regime where the magnetic terms dominate and where the Coulomb interaction is but a small perturbation. We saw these two types of solutions in classical and semi-classical treatments of our problem in Sections 1 and 2.

Historically the myth has been established that magnetic forces are always small and this has been carried over from atomic physics to nuclear and particle physics; "We may recall that in its spin dependence <u>this</u> [namely the tensor force S_{12}, eq. (2), Sect. 3] is a force similar to the one between two magnetic dipoles but its strength in the deuteron is too large to be accounted by the interaction between the magnetic moments of the proton and the neutron."[5] Actually the order of magnitude of the size and strength of magnetic interactions is exactly that of nuclei and of nuclear forces.

6. Exactly soluble Models. Complex Eigenvalue Problem

For the following class of potentials

$$V(r) = \frac{v_2}{r^2} + \frac{v_3}{r^3} + \frac{v_4}{r^4} \tag{1}$$

the resonances and bound-states can be given in analytical form.[6] This is precisely the form of our magnetic potentials.

We thus consider the eigenvalue problem

$$\left[\frac{d^2}{dr^2} + \lambda^2 - V(\lambda, r)\right] \phi(r) = 0, \quad r \geq 0 \quad , \tag{2}$$

in the space of functions

$$L = \left\{ \phi(r);\ \phi(0) = 0,\ \frac{\phi'(r)}{\phi(r)} \xrightarrow[r \to \infty]{} i\lambda ,\ \text{for some } \lambda \in C \right\}. \tag{3}$$

This space includes the L_2 - space of bound-states wave function if λ is pure imaginary, $\lambda = i\kappa$, $\kappa > 0$. Otherwise we have an outgoing wave boundary condition; the wave functions are no longer normalizable. Such states correspond to the poles of the S-matrix defined by the asymptotic expansion

$$\phi(r) \xrightarrow[r \to \infty]{} N \left[e^{-i\lambda r} - S(\lambda)\, e^{i\lambda r} \right] .$$

It is well-known that the poles of the S-matrix along the positive imaginary axis correspond to bound states, along the negative imaginary axis to the so called anti-bound states, and a pair of conjugate complex poles in the lower half plane to resonance states (or Gamow-state).

The importance of exactly soluble bound-state eigenvalue problems in quantum theory is evident. It is remarkable that we can also get a discrete set of resonances as eigenvalues.

Theorem:[6] For potentials given in (1) with $v_4 = 1$ (Eq. (2) can always be scaled that $v_4 = 1$) such that

$$v_3 = -2(M+1) , \quad M = 0, 1, 2, 3, \ldots \tag{4}$$

the eigenvalue problem (1) has (M + 1) discrete complex eigenvalues given by the vanishing of the tridiagonal determinant of order (M + 1):

$$\begin{vmatrix} D & 2 & 0 & 0 & 0 & 0 & \cdot \\ -2i\lambda M & D-2M & 4 & 0 & 0 & 0 & \cdot \\ 0 & -2i\lambda(M-1) & D+2(1-2M) & 6 & 0 & 0 & \cdot \\ 0 & 0 & -2i\lambda(M-2) & D+3(2-2M) & 8 & 0 & \cdot \\ \cdot & \cdot & \cdot & \cdot & \cdot & \cdot & \cdot \\ \cdot & \cdot & \cdot & \cdot & \cdot & \cdot & \cdot \\ & & & \cdot & -2i\lambda & D+M(-M-1) & \cdot \end{vmatrix} = 0 , \tag{5}$$

where

$$D = M^2 + M + 2i\lambda - v_2 \quad . \tag{6}$$

Equation (5) appears now as the quantization condition for resonances. The proof of the theorem goes along similar lines as the real eigenvalue problems. One factorizes the asymptotic behavior of $\phi(r)$ at $r \to \infty$ (eq. (3)) and at $r \to 0$, where the $1/r^4$ - term dominates: set $\phi(r) = F(r)\, e^{-1/r}\, e^{i\lambda r}\, r^{\nu}$. Then one makes a power series ansatz for F(r) and obtains recursion relations for the coefficients. Conditions (4) and (5) then solves these recursion relations. For details we refer to Ref. 6.

As an example, consider M = 1. The $D = 2 + 2i\lambda - v_2$, and the 2×2 - determinant (5) gives in this case a pair of complex-conjugate poles at

$$\lambda = \frac{1}{2} [-i(v_2 - 2) \pm \sqrt{2(v_2 - 2)}] \quad . \qquad (7)$$

For $v_2 < 0$, the two poles move along the imaginary axis, collide at origin at $v_2 = 0$ and then become two complex poles in the lower half plane for $v_2 > 0$.

The resonances are narrow if the poles are close to the real axis.

7. Numerical Solutions for Resonances

When the coefficient v_3 in the potential (1) does not have exactly the values given in (4) - this is the case for example if we include spin-spin terms as well - then numerical methods can be used to locate the resonances.

This has indeed been done. Very narrow resonances have been obtained in potentials of the form

$$V(r) = \frac{A}{r} + \frac{B}{r^2} + \frac{C}{r^3} + \frac{D}{r^4} \quad .$$

The phase shift jumps sharply by about π at the resonance energy as it should corresponding to the so called narrow anomalous resonances. (Fig. 5)

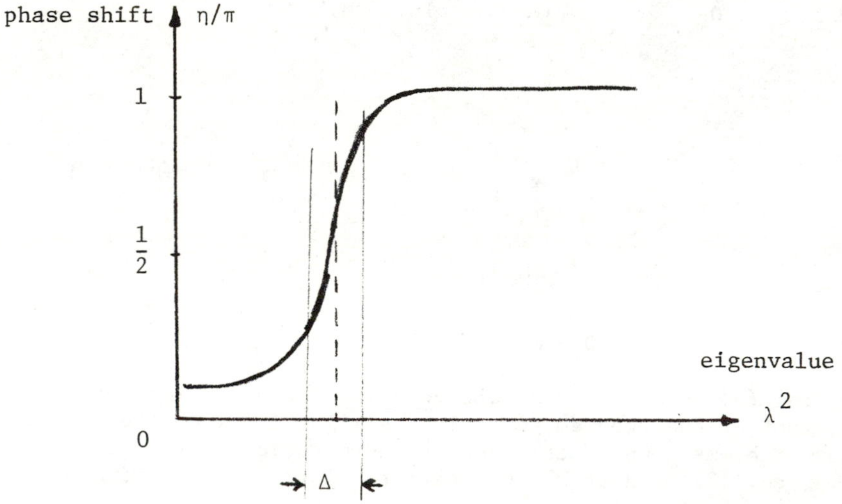

Fig. 5

One obtains from these calculations the width Δ of the resonance

(i. e. its lifetime), and also the Regge trajectories by evaluating the resonance energies as a function of B and C which depend on the angular momentum.[7]

8. The Dirac Particle with anomalous Magnetic Moment in Dipole-Field

In this section we consider a relativistic Dirac particle with an anomalous magnetic moment a, charge e in the field of a fixed quantum dipole moment μ_2, charge e_2. The potential $\vec{A}$ is the same as in eq. (2), Sect. 1. The equation is (in units $c = \hbar = 1$)

$$\left[\gamma^\mu (p_\mu - e_1 A_\mu) - m\right] \psi = -a \frac{e_1}{4m} \gamma^\mu \gamma^\nu F_{\mu\nu} \psi \quad , \tag{1}$$

with

$$A_\mu = \left(\frac{e_2}{r} , \mu_2 \frac{\vec{\sigma} \times \vec{r}}{r^3} \right) \quad .$$

The Hamiltonian form of this equation is[8]

$$[\vec{\alpha} \cdot (\vec{p} - e_1\vec{A}) - (E - \varepsilon \frac{\alpha}{r}) + \beta m] \psi = -a \varepsilon \frac{\alpha}{2m} \frac{1}{r^2} i \beta \alpha_r \psi, \tag{2}$$

where $\varepsilon = \text{sign } (e_1 e_2)$ and $\alpha_r = \vec{\alpha} \cdot \vec{r}/r$. When $\vec{A} = 0$, the new Hamiltonian H, eq. (2), commutes with J^2, J_z and $K = \beta(1 + \vec{\sigma} \cdot \vec{L})$ as in the ordinary Dirac equation with Coulomb potential alone. Eq. (2) separates into angular and radial parts with

$$\psi^\kappa_{jm} = \begin{pmatrix} \phi \\ \chi \end{pmatrix} = \begin{pmatrix} g(r)\, \Omega(j, j + \frac{1}{2}) \\ if(r)\, \Omega(j, j - \frac{1}{2}) \end{pmatrix} , \tag{3}$$

where $\Omega_{j\ell}$ are the spherical spinors, and $-\kappa$ is the eignevalue of K

$$\kappa = \pm (j + \frac{1}{2}) \quad .$$

The radial equations are

$$\begin{aligned} \frac{df}{dr} &= \frac{\kappa - 1}{r} f + (m + V_s - E) g + V_e g + V_m f \\ \frac{dg}{dr} &= - \frac{\kappa + 1}{r} g + (m + V_s + E) f - V_e f - V_m g \; . \end{aligned} \tag{4}$$

Here we have inserted, for generality, a possible scalar potential V_s (to be included in the mass term in Eq. (1)), an electric potential V_e which in our case equals $\varepsilon \frac{\alpha}{r}$, and a magnetic potential V_m equals in our case to $V_m = a\,\varepsilon \frac{\alpha}{2mr^2}$.

The coupled radial equations can be transformed[8, 9] to two uncoupled second order Sturm-Liouville eigenvalue equations:

$$\left[\frac{d^2}{dr^2} + k^2 - V_i(r) \right] \psi_i = 0, \quad i = 1, 2 \ . \quad , \tag{5}$$

where

$$k^2 = E^2 - m^2$$

and the effective energy and angular momentum dependent potentials are given by

$$V_1^{eff}(\kappa, E, r) = \frac{\kappa(\kappa+1)}{r^2} + 2E\,V_e - V_e^2 + V_s^2 + 2m\,V_s +$$

$$V_m^2 + 2\frac{\kappa V_m}{r} - V_m' + \frac{1}{2}\frac{V_e'' - V_s'' + 2(V_s' - V_e')(\frac{\kappa}{r} + V_m)}{(m + E + V_s - V_e)} +$$

$$\frac{3}{4}\frac{(V_s' - V_e')^2}{(m + E + V_s - V_e)^2} \quad , \tag{6}$$

and

$$V_2^{eff}(\kappa, E, V_e, V_m) = V_1^{eff}(-\kappa, -E, -V_e, -V_m) \ .$$

The effective potential (6) has considerable structure at short distances; it is energy and angular-momentum dependent. For each E and κ one has to plot the potential and look for a resonance and then vary these parameters. For very large E it becomes approximately energy independent and its form is shown in Fig. 6. We see in this case that the range and the magnitude of the deep potential-well is of the order of

$$r \sim \frac{\alpha^2}{m} \quad \text{and} \quad E \sim \frac{m}{\alpha^2} = 9.6 \text{ GeV} \ .$$

These scales might be identified with the regime of the so called ψ - resonances in the e^+e^- - system. (See part II).

Next we must include the dipole field $\vec{A}$ as in eq. (2), <u>and</u> the spin-spin terms due to anomalous magnetic moment. In this more complicated case the radial and angular parts can still be separated, but we obtain 4 coupled first order equations, instead of the two of eqs.

(4). The Hamiltonian is now

$$H = H^{BK} - e\,\mu_2\,\vec{\alpha}\cdot\frac{\vec{\sigma}\times\vec{r}}{r^3} - a\,\frac{e_1\mu_2}{2m_1}\,\beta\left[\frac{\hat{S}_{12}}{r^3} + \frac{8\pi}{3}\,\vec{\sigma}_1\cdot\vec{\sigma}_2\,\delta^3(r)\right.$$

$$\left. + \frac{e_2}{r^3}\,i\,\vec{\alpha}\cdot\vec{r}\right] , \qquad (7)$$

where H^{BK} referes to the Hamiltonian of (2).[8]

The 4-coupled radial equations[10] have not yet been investigated in detail.

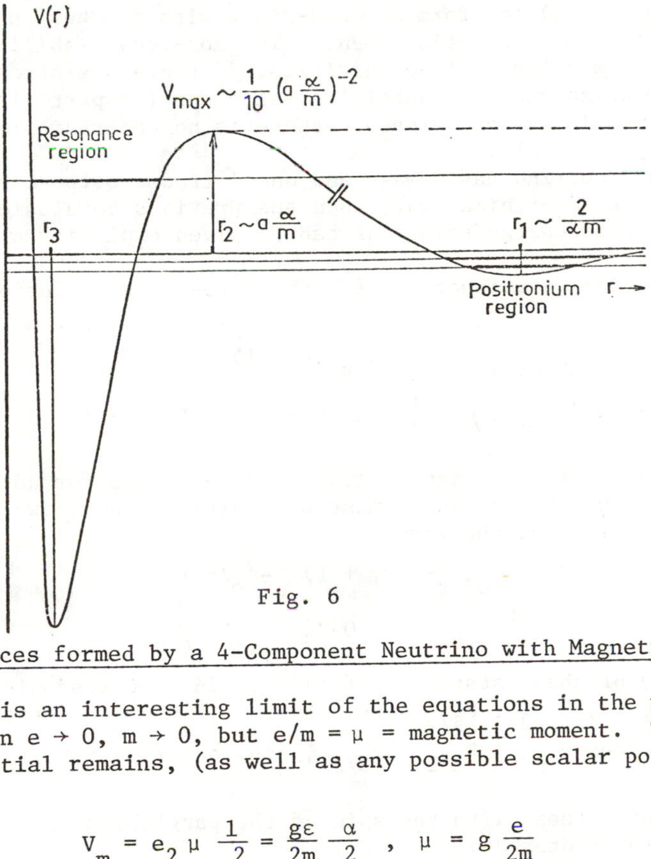

Fig. 6

9. <u>Resonances formed by a 4-Component Neutrino with Magnetic Moment</u>

There is an interesting limit of the equations in the previous section when $e \to 0$, $m \to 0$, but $e/m = \mu$ = magnetic moment. The magnetic potential remains, (as well as any possible scalar potential V_s).

$$V_m = e_2\,\mu\,\frac{1}{r^2} = \frac{g\varepsilon}{2m}\,\frac{\alpha}{r^2} \quad , \quad \mu = g\,\frac{e}{2m}$$

and we find for $V_s = 0$ (and neglecting the $\vec{\alpha}\cdot\vec{E}$ term in eq. (7), Sect. 4),

$$V_1^{eff} = \frac{\kappa(\kappa+1)}{r^2} + 2e\mu\,\frac{\kappa+1}{r^3} + \frac{e^2\mu^2}{r^4} \quad .$$

In dimensionless coordinates, $y = \frac{r}{|e\mu|} = r\frac{2m}{g\alpha} = \frac{r}{r_0}$, we get precise- the equation which we have seen was exactly soluble:

$$\left\{ \frac{d^2}{dy^2} + \lambda^2 - \left[\frac{\kappa(\kappa+1)}{y^2} + \varepsilon\frac{2(\check{\kappa}+1)}{y^3} + \frac{1}{y^4} \right] \right\} \psi_1 = 0. \; \lambda^2 = |e\mu|^2 E^2. \quad (1)$$

It follows from our Theorem in Sect. 6, that for $\varepsilon = -1$, eq. (1) has exactly $(\kappa + 1)$ discrete solutions. The lowest one, $\kappa = 1$, has zero energy. For $\kappa = 2$, we get two solutions with $\lambda = 0$, and a solution $\lambda = i$ which corresponds to a complex energy with real part zero and imaginary part equal to $1/e\mu$.

It is remarkable that a massless particle (moving thus with the velocity of light) can form a bound-state with another charged particle, if it has a magnetic moment. The non-localizability of massless particles refer to free particles, if there are strong forces one can localize massless particles (also massive particles) to as small a dimension as one wishes - there is no contradiction with the uncertainty principle.

If the neutrino has a mass and one introduces the magnetic coupling in a non-minimal way, then the previous conclusions still hold. The zero energy solutions can be given explicity also in this case.[9]

The radial wave functions for the zero-energy solutions are given by

$$\begin{pmatrix} g(r) \\ if(r) \end{pmatrix} = \begin{pmatrix} C' \, r^{-(\kappa+1)} \, e^{\varepsilon r_0/r} \\ C \, r^{\kappa-1} \, e^{-\varepsilon r_0/r} \end{pmatrix}.$$

Let us take a definite sign of ε, say $\varepsilon = -1$, then for normalizable solutions C must vanish and κ must be positive. Hence we get normalizable solutions of the form

$$\begin{pmatrix} C' \, r^{-(|\kappa|+1)} \, e^{-r_0/r} \\ 0 \end{pmatrix}.$$

Parity of these states are $(-1)^{|\kappa|}$. In spectroscopic notation, they are $P_{1/2}$, $D_{3/2}$, $F_{5/2}$, etc. For $\varepsilon = +1$, on the other hand, only the states $S_{1/2}$, $P_{3/2}$, $D_{5/2}$, etc. occur, $\kappa < 0$, parity $-(-1)^{|\kappa|}$.

Combining these with the spin of the particle at the center we arrive at the states J^P:

$$\varepsilon = -1 : \quad 0^-, 1^- ; 1^+, 2^+ ; 2^-, 3^- ; \ldots$$

$$\varepsilon = +1 : \quad 0^+, 1^+ ; 1^-, 2^- ; 2^+, 3^+ ; \ldots$$

If the system has an additional negative intrinsic parity as in (e^+e^-) or $(e^-\nu)$, we have to switch all the parities. To the interaction (1) we must still add the spin-spin interaction.

It is remarkable that we get these sequences of J^P - states precisely as observed in the meson sequences:

(i) $\pi(0^-)$, $\rho(1^-)$; $A_1(1^+)$, $f(2^+)$; $A_3(2^-)$, $\omega(3^-)$; $h(4^+)$...

(ii) $\eta(0^-)$, $\omega(1^-)$; $B(1^+)$, $A_2(2^+)$;

(iii) $\eta'(0^-)$, $\phi(1^-)$; . . .

(iv) $\delta(0^+)$, $D(1^+)$; $\rho'(1^-)$, . . .

(v) $S^*(0^+)$, . . .

(See also part II).

Note: The $(i\vec{\alpha}\cdot\vec{E})$ term in eq. (7), Sect. 4, gives an additional $1/r^3$ - term to the potential in (1).

10. Other Relativistic Two-body Equations with Dipole Interactions

There are many forms of covariant two-body equations derived from field theory in various approximations.

A general one-time two-fermion equation with form factors has been applied to the problem of magnetic interactions.[11] The 16-component equation for 2 spin-½ particles can be reduced to two 8-component equations. In the static limit we recover the previous results, but a very general treatment of coupled radial equations is still an outstanding problem.

A phenomenological covariant 2-body equation including the spins and relative motion of both of the particles has recently been discussed by Piron and Reuse.[12]

The radial equations in this model also contain potentials with $1/r$, $1/r^2$, $1/r^3$, and $1/r^4$ terms. For e^-e^+ (or $p\bar{p}$) type of systems, there is no deep potential well in the $^1S_0$, $^1P_1$, $^1D_2$ states (singlet S = 0), but there is in the $^3P_1$, $^3D_2$; $^3P_0$, . . . states, and in the superpositions of $^3S_1 + {}^3D_1$, states. . .

II. NUCLEAR AND HADRON MODELS

The general idea and the background of applying the magnetic interactions to hadrons and nuclei is described in Appendix I. In Appendix II we give an assignment of hadron multiplets in terms of magnetically bound leptons. More details on this assignments has been already given in the literature.[13] Here we discuss some specific quantitative results which show the distinctive features of magnetic forces and may also provide some crucial tests for this theory.

1. Characteristics of Magnetic Resonances for Hadron Physics

A) The magnetic resonances in the deep potential well have

masses larger than the masses of the constituents. In most composite models of hadrons and nuclei the binding energy is negative. We have seen that here we can have very massive narrow magnetic resonances between very light, and even between massless particles. This is a new perspective, and quite different then the general intuition concerning composite particles.

B) The characteristic energies in all models comes out to be

$$\frac{m}{\alpha} = 70 \text{ MeV} \text{ , for the normal magnetic moment,}$$

and

$$\frac{m}{\alpha^2} = 9.5 \text{ GeV} \text{ , for the anomalous magnetic moment,}$$

where m is the electron mass. It is empirically known that masses of low lying particles are integer or half-integer multiples of the first number. The second mass value is in the region of ψ and γ - particles.

C) The magnetic potential is energy and angular momentum dependent. It is approximately harmonic at the dip of the potential, to confine a finite number of levels for each angular momentum, but then turns over. It has thus common features with the oscillator models of phenomenological spectroscopy but does not increase linearly to infinity as the phenomenological quark confining potentials. The latter leads to undesirable large van-der Waals forces between two hadrons.[14]

2. High Energy Cross-Sections

An important property of the magnetic interactions is the logarithmic increase of total elastic cross section with energy. In first and second Born approximation one finds[15]

$$\sigma \sim \sigma_o \ln E + \sigma_1 \frac{\ln^2 E}{E} \quad . \tag{1}$$

In the potential picture a form factor is necessary to regularize the singular potential at r = 0, in order to obtain this result. In covariant perturbation theory for the same reason we integrate the differential cross-section from some t_{min} to t_{max}, to avoid the infinite total cross, as in the Coulomb case.

The differential cross-section due to charge-dipole coupling in the lowest order is given by[16]

$$\frac{d\sigma}{dt} \simeq \pi \frac{\alpha\mu^2}{m^2} \frac{(1 - y)}{ys} \quad , \tag{2}$$

where y is the ratio fo the final energy E´ to initial energy E of the particle in the Laboratory frame. The total cross-section then gives the first term in eq. (1). These results refer to the magnetic interactions between the stable constituents of the hadrons. In order to obtain lepton-hadron, or hadron-hadron scatterings, we have to sum over all constituents (in the impulse approximation), and further include, the exchange and rearrangement of constituents.

Phenomenologically, however, high energy cross-sections are fitted with the two terms given in eq. (1). We may thus have a dynamical theory for the so called two-component Pomeron exchange models.[15]

3. Strong Spin-Dependent Forces and Narrow Resonances in pp, $p\bar{p}$ and np Experiments

Recent experiments with polarized nucleons on polarized targets have indicated rather startling and unexpected spin dependence in pp and np scatterings,[17, 18] as well as possible narrow resonances in these channels.[19] These results do not seem to be explainable in the standard quark models,[20] but follow quite naturally from our considerations on magnetic interactions.

In the pp - case, at 12 GeV/c the experiment gives

$$\frac{\frac{d\sigma}{dt}\,pp\,(\uparrow\uparrow)}{\frac{d\sigma}{dt}\,pp\,(\uparrow\downarrow)} \sim 4-5, \text{ at } P_\perp{}^2 \sim 5\ (\text{GeV/c})^2 \quad ,$$

whereas in np - case, the result is just the opposite (even at lower $P_\perp{}^2$)

$$\frac{\frac{d\sigma}{dt}\,np\,(\uparrow\downarrow)}{\frac{d\sigma}{dt}\,np\,(\uparrow\uparrow)} \sim \quad 4-5$$

Because the magnetic moment of the neutron is negative, we bring immediately the two above equations into one.

4. Mass Formulas for Hadrons and Nuclei

We can give a dynamical explanation to the empirical mass formula of Sakharov and Zeldovich.[21] Accordingly, the masses of hadrons can be expressed as (in our version)

$$M = m_o + |S|\, m_\mu + \lambda \sum_{i,j} \langle \vec{\mu}_i \cdot \vec{\mu}_j \rangle \quad .$$

Here S is the strangeness and m_μ the mass of the muon. This already incorporates (i) our interpretation of the strangeness as the number of muons (see Appendices), and (ii) the nearly zero energy bound states with spin-orbit forces alone (see I. 9) - hence the total mass is the sum of (essentially) muons. The last term expresses spin-spin interactions. Sakharov and Zeldovich evaluate this last term for quarks, and modify the magnetic moments of strange quarks.

From the dynamical model we discussed in part I, we can considerable improve such a mass formula.

If we go over to nuclei, the spin-orbit zero-energy solutions imply immediately an additive mass formula for nucleons, to which we have to add the spin-spin and exchange potentials. For example for deuteron, we have

$$M = m_p + m_n + \lambda < \frac{\mu_p \cdot \mu_n}{r^3} > + \text{Exchange term.}$$

The first two terms constitute an exact mass formula under spin-orbit forces only (thus giving a mass with an accuracy of .2% from first principles).

Quantitative mass calculations will be presented elsewhere.

5. Regge Poles only

Another remarkable feature of magnetic potentials discussed in Part I is that the scattering amplitude can be shown to be analytic in the whole of the angular momentum plane.[22] This comes simply because of the strong repulsive nature of the potential at $r = 0$, where the behavior of the amplitude becomes independent of the angular momentum barrier terms. Consequently, the scattering amplitude is a sum of Regge pole terms without cuts and background integrals. This agrees with the phenomenological analysis of the amplitudes in terms of the pole terms alone. This general feature should persist essentially if we go from constituent amplitudes to the hadron-hadron amplitudes using impulse approximation. It would be interesting to generalize this result to more accurate relativistic models.

6. Interpretation of Isospin, Finite Groups and Exchange Forces

The isospin concept first originated in the neutron-proton system. It means that strong interactions are the same in the states pp, nn and $\frac{1}{2}$[pn + np], (I = 1), and have another value in the state $\frac{1}{2}$ (pn - np), (I = 0). In our interpretation of the n as the $(pe\bar{\nu})$-state, the above separation simply means the symmetrization or antisymmetrization of the bare p - p system relative to the exchange of $(e\bar{\nu})$. This is the exact analog of the symmetric and antisymmetric states (σ_g and σ_μ) in H_2^+ - ion relative to the e-exchange.

The third component of the iso-spin simply counts the absolutely stable particles:

$$I_3 = \frac{N_p}{2} + \frac{1}{2}\left(N_{e+} - N_{e-} + N_\nu - N_{\bar{\nu}} \right) ,$$

whereas the isospin raising and lowering indices exchange $(e\bar{\nu})$ between two interacting hadrons. This agrees also precisely with the isospin multiplets and lowering, raising operators in the multiplets

Tables in Appendix II.

In fact we can introduce isospin multiplets in molecular physics in the same way. For example, pp, HH, $\frac{1}{2}$ (pH ± Hp) where H is an H-atom, form the I = 1 and I = 0 states.

For two pions the pure isospin states are those which have definite symmetry properties with respect to all perturbations of electrons and neutrinos: For example, $e^+\nu\, e^-\bar{\nu}$ is $\pi^+\pi^-$; $e^+e^-\nu\bar{\nu}$ is $\pi^o\pi^o$; $\bar{\nu}\nu e^-e^+$ is again $\pi^o\pi^o$; $\bar{\nu}e^-e^+\nu$ is $\pi^-\pi^+$. This gives for the completely symmetric states under permutations

$$|\pi\pi>_{symm.} = 2|\pi^o\pi^o> + |\pi^+\pi^-> + |\pi^-\pi^+>$$

which is a member of the I = 2 states, etc.

We are thus led to the conclusion that finite subgroups of the rotation group (e. g. tetrahedral, octahedral, icosahedral,. . .) are sufficient to derive isospin and isospin selection rules. This fact has been known long time ago.[23] We now have a physical explanation and a concrete basis for this result. An immediate support comes from the fact that Isospin values observed in nature for hadrons are not very high, namely those related to the representations of finite subgroups of SU(2). If the group were indeed the abstract Lie group SU(2), all iso-spin values up to infinity should occur. Moreover higher I-values, indicate in our picture higher exchange operations, and higher number of lepton-pair ($e\bar{\nu}$) constituents.

Thus if iso-spin and hypercharge, hence all the operations of SU(3), (and further SU(4)) have a simple physical interpretation in terms of the constituents, exchange of lepton pairs, creation and annihilation (e. g. excitation) of μ, etc. then the need for gauging this group into a Yang-Mills theory disappears[23] (see also Appendix I). We should remark that the imposition of a symmetry group, like SU(3), abstractly, does not explain the quantum numbers, it assumes them. The explanation must be based on a dynamical picture of the structure of hadrons, which we have done here.

7. Further Results

The following applications have already been discussed and we have nothing more to add at this time.[13]

a) K^o, $\bar{K}^o$ mixing, $K_L - K_S$ mass difference and CP-violation: The mixture of K^o, $\bar{K}^o$ is physically interpreted as the exchange of $(\nu\bar{\nu})$ between e^- and e^+ in a magnetic double-well potential. There is a complete analogy with the NH_3 - mass problem, which allows one to determine the $K_L - K_S$ mass difference. A mechanism for CP-violation emerges.

b) Barrier penetration: The existence of the potential barrier (see Figs. 4 and 6) implies that the scattering between for example, e^+e^- or e^-p, will be given by Coulomb scattering, except at definite energies near the resonance energy in the potential well where the phenomenon of "resonance penetration"[24] takes place and we get a small bump superimposed on the background amplitude. In fact, if the Born approximation is valid, as it seems to be the case experimentally, these small bumps are the only manifestations of the deep potential well. Indeed experimentally all the resonances in the (e^+e^-) - systems are observed as small bumps on the Born-amplitude cross section. Now by definition the effective potential from field theory is so defined that its Born approximation coincides with the lowest order Feynman diagram[25].

8. A Nuclear Model

In this final Section we propose a new nuclear model based on the absolutely stable constituents p, e, $\bar{\nu}$. We consider a rather rigid closed-packed lattice of A protons. In this lattice N lepton pairs $(e^-\bar{\nu})$ are freely moving. Now if a pair $(e^-\bar{\nu})$ is attached to a proton it will look like a neutron according to our model of the neutron. If that pair $(e^-\nu)$ jumps to another neutron, it will have the same effect as the motion of a neutron or that of a proton in the opposite direction. Thus the nucleons are not moving but only the light lepton pairs. Such a simple model has a number of remarkable properties:

(i) It reconciles the two contradictory features of successful nuclear models: the liquid drop model or the close-packed structure of nucleons on the one hand, and the shell-model of the freely moving nucleons on the other hand.

(ii) It accounts automatically for the exchange forces, spin-dependent forces, isotopic spin.

(iii) It incorporates as approximations the meson theory of nuclear forces, the vector dominances, and Regge pole models.

The simplest case, the deuteron, has already been mentioned, it would be interesting to apply the magnetic forces and exchange of leptons to more complicated 3 or 4-body problems.

Note added in Proof: Two significant developments have occured since the writing of this report:

(i) By regrouping the leptons and baryons in quartets of SU(4) group which in turn have SU(3) triplets in two different ways, it was possible to show that our multiplets in Figs. 8,9 and 10 are in one-to-one correspondance with the standard SU(3) and SU(4) group analysis and the integrally charged quark model of Han and Nambu with "colour" degree of freedom.[33]

(ii) Dynamically the resonance problem describing the pion $(e\bar{\nu})$ and the muon $(e\nu\bar{\nu})$ have been quantitatively studied. A neutrino magnetic moment of the order of 10^{-10} (well within the experimental limit) does reproduce the lifetimes of these resonances, i.e. a calculation of the Fermi coupling constant from first principles.[34]

Appendix I. The Idea of Stable Particles Model

§1. Around 1932, just before the discovery of the unstable neutron, only three particles were thought to be fundamental: proton, electron and the hypothetical neutrino (to account for the missing energy in β-decay). The fundamental particles were naturally taken to be absolutely stable. The antiparticle to the electron, position, was also predicted by Dirac equation and just about to be discovered. There was a great reluctance to increase the number of fundamental objects. Now after about 50 years, these three particles, and their antiparticles, are still the only absolutely stable particles. In this sense, the world of physics has remained stable, a remarkable fact considering the tremendous proliferation of the unstable "particles". We should of course add the field of photons (or more generally the electromagnetic field) to the list of stable entities (Table I).

Table I: ABSOLUTELY STABLE ENTITIES (1932-1980)

Particles	p	e^-	ν	γ (or EM-Field)
Antiparticles	$\bar{p}$	e^+	$\bar{\nu}$	γ

Although it has been recently speculated that proton might be unstable, the speculated lifetime, $\sim 10^{32}$years, is 20 orders of magnitude longer than the age of universe, hence every proton since the "beginning" is essentially still here and has plenty of time to be a fundamental building block of matter.

§2. All unstable "particles" discovered beginning with the neutron eventually decay into the above absolutely stable particles: $n\to pe^-\bar{\nu}$, $\mu\to \nu\bar{\nu}$, $\pi\to\mu\nu\to e\nu\bar{\nu}\nu$, etc. They are, therefore, resonance poles in the appropriate channels of the S-Matrix between absolutely stable particles. (Fig. 7)

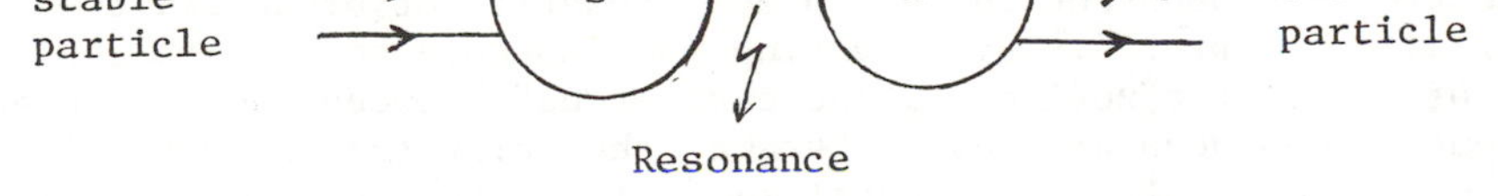

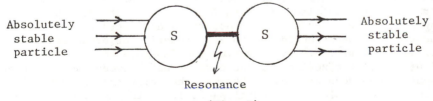

(Fig. 7)

We are thus led to the

Proposition: Only absolutely stable particles can be the truly elementary constituents of matter.

For if these entities are really industructable they must be industructable when forming composite objects. In other words, their "virtues" (Newton), or their "quantum numbers", as we would say today, must be indestructable. Hence the conservation laws they generate are infrangible. We shall see that this is the case.

From the point of simplicity the most economical model of describing the unstable resonances and their quantum numbers is to consider them to be composed of absolutely stable particles into which they disintegrate, instead of inventing new unobserved (or unobservable) objects. Even if we did not know what forces act between the absolutely stable particles, it would be simpler to infer these forces, rather than to infer the forces between the hypothetical and vastly numerous constituents. Fortunately, there is one class of known and familiar force which, acting between the absolutely stable particles, shows enough strength and richness, to provide the essential features of leptonic, hadronic and nuclear states. These are the magnetic forces.

§3. Before the discussion of these dynamical aspects we shall review how and why the basic entities of matter changed drastically from the few absolutely stable particles, to the multitude of quarks (6) their coloured duplicates (times 3), leptons (6), gluons (8?). Higgs bosons (n?).

In the pre-Yakawa period of particle physics, the absolutely stable particles were indeed considered to be the building blocks of nuclei as well as providing the necessary exchange forces. Thus, Pauli's model of the neutron considered the neutron to be a bound-state of ($pe^-\nu$) into which it decays. Properties of nuclei and nuclear forces were explained by the exchange of $e\bar{\nu}$, e^+e^-, $\nu\bar{\nu}$ between the nucleons by Tamm and Ivanenko , Wick , von Weiszäcker , Wentzel , Gamow and Teller[26], etc.

These theories were not followed further and somewhat forgotten, because no forces were known between the stable particles which would provide a deep enough potential well to hold for example ($pe\nu$) into a semi-stable neutron and also account for its instability. To explain the latter Fermi[27] introduced the notion of "creation of the electron and the neutrino" at the surface of the neutron, thereby giving a new name to decay processes, the weak interactions. Later the neutron has been considered to be a constituent of nuclei together with the proton. By elevating the unstable neutron to the level of proton the notion of "nuclear forces" between the proton and the neutron has been created.[28] Yukawa[29] then elevated the unstable pion to the level of a fundamental particle as the agent of nuclear forces.

Thus a new period of particle physics began when more and more unstable particles were treated as fundamental particles or fields. These efforts are not wrong, but they mask the basic issue of fundamental objects and fundamental forces, by using all sorts of effective forces between these composite objects. Thereby we introduce apparently new distinct forces, e.g. strong and weak interactions. It is analogous to the introduction of chemical forces between neutral atoms as distinct from the Coulomb forces inside the atom. We now know of course that the forces between atoms and within the atoms are the same and already unified. I think this is one of the strongest arguments against treating any unstable particle as a fundamental field, and this analogy will persist in our construction of an already unified theory for hadrons.

§4. Later on more and more unstable particles have been discovered ρ $\mu,\pi,K,\Lambda,\Sigma,\eta,\Xi,\Omega\cdots$, and others which are truly short lived resonances: ,Δ, K^*,---etc., and the idea of absolutely stable constituents was even further removed.

We can identify the following periods in the developments in particle physics over the last quarter of century:

Period I. Recognition of empirical regularities of multiplets of unstable states grouped according to certain quantum numbers. A successful example is the empirical forms of octets of mesons, baryons and leptons.[30]

Period II. Recognition of the group structure of multiplets, e.g. the SU(3)-group structure of the above multiplets.[31] There is a general mathematical result that every group has some fundamental representations such that every other representation (multiplet) can be obtained from the product of these fundamental representations. In the case of SU(3), there are two such fundamental representations both of dimension 3, called 3 and $\bar{3}$. In the next period this mathematical fact has been literally given a physical meaning:

Period III. Assumption that the fundamental representations do physically exist and compose to give the observed multiplets.[32] The objects 3, $\bar{3}$ have been called quarks, and the ensuing model, the "naive quark model", to indicate the candidness of the step.

Period IV. If q and $\bar{q}$ are really the building blocks, then one should attempt a field theory with these objects.

Period V. There should then be fundamental forces between these q, $\bar{q}$ constituents which lead in analogy with electrodynamics to the gauge theories in which gluons are exchanged between these constituents.

Period VI. It is again a mathematical result that the fundamental representations, q, $\bar{q}$, obey the Bose statistics, but carry spin ½. This leads to spin statistics difficulties and a new degree of freedom, colour, has been introduced. In the final period of developments the original SU(3)-group (of what is called flavor) has been replaced by another SU(3)-colour group with respect which the non-abelian gauge group of gluons is formulated. Furthermore the mechanism by which observed particles acquire masses is accomplished by a set of so called Higgs mesons.

§5. If atomic and nuclear physics had followed the same steps we would now have a quark-gauge theory of atomic and nuclear phenomena. As a simple example, let me consider the quantum theory of the H-atom:

i) The recognition of n^2-multiplets of the levels of H-atom. ($2n^2$ with spin).

ii) Recognition of the group structure of these multiplets. For non relativistic atom this group is SO(4). It has two fundamental representations of dimension 4; 4 and $\bar{4}$.

iii) Naive quark model in which we think the multiplets to be built from 4 and $\bar{4}$. This is indeed possible. In this case these "quarks" come in really as normal modes of the atomic matter. The Hamiltonian of the atom $H = \frac{1}{2m}p^2 + \frac{\alpha}{r}$ can be reexpressed by suitable transformations in the form of normal modes,

$$N = \sum_{i=1}^{4} a_i^+ a_i,$$

In fact N gives the principal quantum number n of the multiplet, and the energy is a function of n (mass formula), and increasing multiplets have increasing number of "quarks". Again, a, a^+ satisfy the Bose-statistics, but carry spin ½.

In this case, we note that the dynamical force $\frac{\alpha}{r}$ is already included, the normal modes diagonalize this interaction, we do not need any additional "gluon" forces, the normal modes are additive. But if we did not know the Bohr picture of the atom with two constituents, p and e and the Coulomb force, we might have taken the 4 quarks (normal modes) as the true constituents, introduce a colour, to resolve spin-statistics conflict, and introduce new forces, gluons .

It seems, therefore, that there should be underlying the quark picture of the particle multiplets, also a Bohr-picture with stable constituents and a unified force. This is what we are aiming to uncover.

§6. It has been shown recently in a large number of detailed mathematical models that a remarkable effect of magnetic interactions at short distances has been overlooked.[13]

Up to now the magnetic moment interaction terms, say between electrons and protons, have been treated in perturbation theory. Typically one solves an unperturbed Coulomb problem and takes the matrix elements of the magnetic potentials between such Coulomb states. A myth has been prevailing that magnetic moment forces are always small and constitute inessential complications to the other, Coulomb or nuclear, forces. However, if these magnetic terms are treated exactly and non-perturbatively, one obtains an essentially new and unexpected behavior at short distances (or high energies), in particular the emergence of massive narrow resonances which have the properties of nuclear and hadronic states. Because this result might appear to be incredulous, we give a simple model in Appendix A, showing the existence of narrow, massive resonances.

§7. After this brief discussion of the dynamical problem we now address ourselves to the kinematical problem of building up the multitude of hadronic states and their quantum numbers from just three stable particles p,e,ν and their anti-particles. Also this might appear at first to be an impossible task. We first construct the neutron, as the magnetic bound state of (peν). Here the ν helps to compensate the large magnetic moment of the electron. The only perhaps new property assigned to the stable particles is the following:

Assumption: There exists a 4-component neutrino with an anomalous magnetic moment. Then such a neutrino can be shown to be able to form bound states with other particles. The neutrinos coming from the decay of π, K, ...are polarized and can be approximated by two-component neutrinos.

Next we make the π-mesons from electrons and neutrinos. The protons and neutrons in turn form nuclei. The ($e\nu$) cloud of the neutron can be exchanged between proton and the neutron giving rise to approximate meson exchanges as well as to the concept of isospin (see §8).

Since the muon (μ) decays into ($e\,\nu\,\bar{\nu}$) we may think the muon to be a magnetic bound state of these three particles or a kind of "magnetic excited state" of the electron, in which ($\nu\bar{\nu}$) plays the role of the photon in electric excited states. The muon in turn participates in forming new resonances with the previous particles (p,e,ν). In fact the muon brings in the strangeness quantum number, muon number being precisely equal to the strangeness quantum number S.

We may continue this process with other leptons coming in, in turn, ν_μ, τ, ν_τ, each time a new quantum number being introduced.

Appendix B shows the resulting multiplet structures.

§8. The intuitive physical meaning of the internal quantum numbers is shown in Table II.

Table II. Internal quantum numbers

Number of protons is conserved	Baryon number B
Number of electron and neutrinos is conserved	Lepton number L
Number of p and e is conserved	Charge Q
Number of μ is conserved in "strong" interactions (before μ has time to decay)	Strangeness S
($e\nu$)'s are exchanged between protons (symmetrization of states relative to $e\nu$-exchange)	Isospin $\vec{I}$

With respect to Table II, we might add, that stable particles are only produced in particle-antiparticle pairs. Furthermore the concept of isospin is in exact analogy with the symmetrization of, for example, H_2^+-molecule states with respect to the electron exchange between the two protons.

Having obtained a simple interpretation of the mysterious internal quantum numbers in terms of the known structural properties of hadrons and leptons we can state that "the only fundamental quantum numbers are those of space-time symmetry (mass,spin, parity), and the identity of absolutely stable particles". These symmetries are infrangible. All other derive from the composite structure.

The internal "colour" degree of freedom never arises; there is no spin-statistics difficulty with the stable particle constituents; they are fermions. Nor is there any quark confinement problem.

§9. The picture we have presented is an already unified theory, unified moreover by electromagnetism which then appears as the only force field also in particle physics. The strong interactions are due to (1)magnetic forces,(2)rearrangement and pair production of stable constituents. Weak interactions come from barrier penetration of resonances with or without μ-decay, resonance scattering and resonance penetration.

It is surprising how the high energy phenomenology fits into this simple framework. The rules of particle physics are no longer in darkness, but can all be given an intuitive explanation, so that high energy physics appears no longer as a new abstract field, but a continuation of atomic and nuclear physics with the only inclusion of the neutrino and the magnetic forces. In fact, a simple reflection shows that many models of the last half century, the meson theory of nuclear forces, the vector dominance model(VDM), Regge pole exchange models, dual models, etc. all appear as special approximations of the stable particle model, so they can be unified and, in principle, derived from this model.

§10. Naturally one would like to have crucial tests of the stable particle model. On the theoretical side besides the qualitative agreement with many phenomena, the most important test would be the precise calculation of absolute masses of pion, neutron, etc. because we have in principle a known dynamics. No other theory is able to do this even in principle. On the other side, we may cite the recent unexpected large spin effects[17] and narrow resonances in pp,$p\bar{p}$ and np-scattering. We would predict similar effects in high energy polarized e^+-e^- scattering.[18] The model predicts further narrow resonances in, e.g. e-e scattering of charge Q=2, and new type of hadrons as in (ep)systems - without the neutrino -,probably very high mass hadrons but with electronic magnetic moments i.e. of the order of 4000 nuclear magnetons! (A candidate might be the peculiar centauro events recently found in cosmic rays).

APPENDIX II. Hadron Multiplets

We show here the hadron multiplets in the well-known SU(3)-form. It is easy to generalize it to the SU(4)-form.

Fig. 8 show the pseudoscaler mesons. The vector mesons are the same, the $J=1^-$ resonance states of the same constituents, etc.

Fig. 9 and 10 show the baryon octet and decouplets. We add the following remarks:

(i) To each hadron one can add pairs like $(e^+e^-, \nu\bar{\nu}, \ldots)$ without changing the quantum numbers.

(ii) Every baryon eventually decays into a proton (B=1), so proton and the surrounding meson cloud $(\ell\bar{\ell})$ is the general picture of all baryons. The baryon octet looks more symmetrical group theoretically if we rewrite then in the $(\Lambda\ell\bar{\ell})$- representation - they every member of the octet has three constituents $(\Lambda\ell\bar{\ell})$.

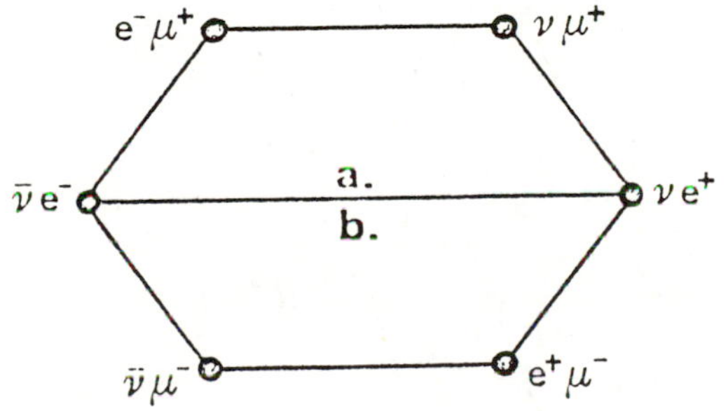

$$a = \frac{1}{\sqrt{2}}(\bar{\nu}\nu - e^+e^-),\ b = \frac{1}{\sqrt{6}}(\bar{\nu}\nu + e^+e^- - 2\mu^+\mu^-)$$

FIGURE 8. The Meson Octet

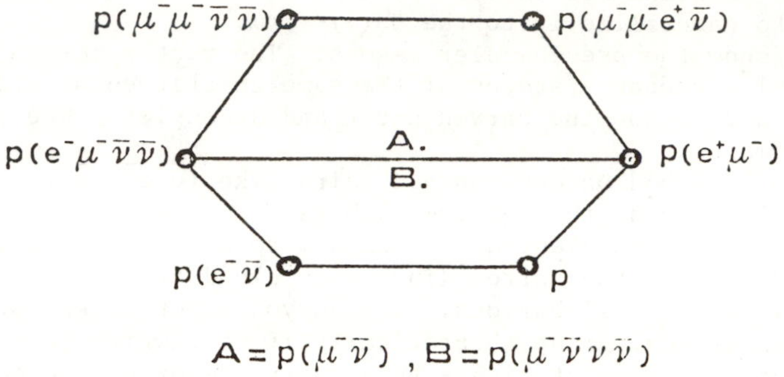

FIGURE 9. The Baryon Octet

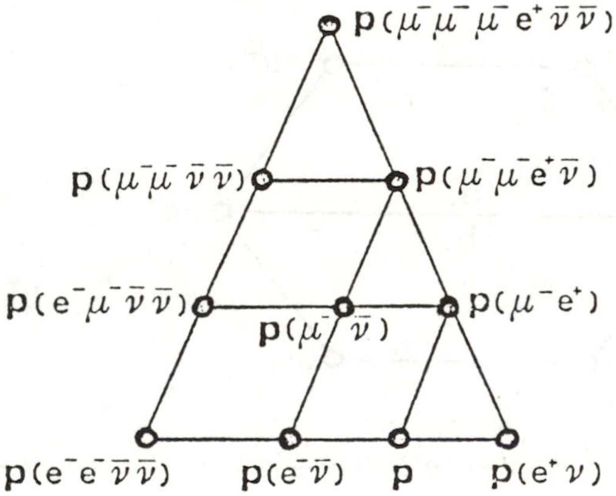

FIGURE 10. The Baryon Decouplet. The nearly linear mass formula of about the μ mass is a consequence of nearly zero-energy bound states in the magnetic potential well.

ACKNOWLEDGMENT

It is a pleasure to acknowledge many stimulating discussions and fruitful collaboration with J. Kraus, M. Berrondo, G. Strobel, R. Rączka, A. Skorupski, A. Hacinliyan, H. Beker, A. Baran, N. Unal, H. Salecker, T. Anders, A. D. Krisch, C. Piron, F. Reuse, H. Behncke, H. Aly, L. Licht.

References

1) C. Störmer, Zeit. f. Astrophys. 1, 237 (1930).
S. Chapman and J. Bartels, Geomagnetism, Vol. 2, 833 (Oxford Clarendon Press, 1940).
M. S. Vallarta, Obra Cientifica, (Collected Papers), Univ. of Mexico, 1978.
J. McConnell and E. Schrödinger, Proc. Roy. Irish Acad. 49A, 259 (1944).
2) A. J. Dragt and J. M. Finn, J. Geophys. Res. 81, 2327 (1976).
3) J. Ashkin and T.-Y. Wu, Phys. Rev. 73, 973 (1948).
4) A. Messiah, Quantum Mechanics, Vol. 2, p. 541 (North Holland, 1962).
5) A. de Shalit and H. Feshbach, Theoretical Nuclear Physics, Vol. 1, p. 13 (John Wiley, 1974).
6) A. O. Barut, M. Berrondo and G. Garcia-Calderón, J. Math. Phys. 21, 1851 (1980).
7) A. Skorupski and A. Senatorski (to be published).
8) A. O. Barut and J. Kraus, J. Math. Phys. 17, 506 (1976).
9) A. O. Barut, J. Math. Phys. 21, 568 (1980).
10) A. O. Barut, A. Baran and N. Unal, Radial Equations for Dirac particle with anomalous magnetic moment in a dipole field (to be published).
11) A. O. Barut and R. Rączka, Acta Phys. Polon. B10, 687 (1979).
12) C. Piron and F. Reuse, Helv. Phys. Acta 51, 157 (1978).
F. Reuse, Helv. Phys. Acta 51, 157 (1978); Univ. of Geneva preprints (1980).
13) A. O. Barut, Surveys in High Energy Physics, 1, 131-140 (1980).
14) S. Matsuyama and H. Miyazawa, Progr. Theor. Phys. 61, 942 (1979).
15) A. O. Barut and A. Hacinliyan, Lett. N.C. (1980).
16) A. O. Barut, The magnetic moment of the Neutrino, Proceedings of the Neutrino Mass Conference, Wisconsin 1980, AIP Conference Proceedings (1980) and references therein.
17) I. P. Auer et al, Phys. Lett. 67B, 113 (1977); Phys. Rev. Lett. 41, 354 (1978).
18) K. Hidaka and A. Yokosawa, Surveys in High Energy Physics 1, 141 (1980).
19) D. G. Crabb et al, Phys. Rev. Lett. 43, 983 (1979).
20) A. D. Krisch, Scientific American, 240, May 1979, p. 68; and Proc. Orbis Scientia 1979.
21) Ya. B. Zeldovich and A. D. Sakharov, Sov. J. Nucl. Phys. 4, 283 (1967).
22) E. Predazzi and T. Regge, Nuovo Cim. 24, 518 (1962).
23) K. M. Case, P. Karplus and C. N. Yang, Phys. Rev. 101, 874 (1956).
24) R. W. Gurney, Nature, 123, 565 (1929).
25) A. I. Akhiezer and V. B. Berestetshii, Quantumelectrodynamics, (Interscience 1965).

26) W. Pauli, Letter to the Radioactivity Meeting, 4 Dec. 1930; reprinted in Collected Scientific Papers, Vol. 2, Interscience (1964), p. 1313; Handbook der Physik, 1933 Edition, Band 24/1, p. 233.
D. Ivanenko and I. Tamm, Nature, 133, 981(1934) [$e\bar{\nu}$-exchange]
G. Wick, Atti Acad. Lincei 21, 170 (1935) [anomalous magnetic moment of nucleons from ($e\bar{\nu}$- exchange)]
C. von Weiszäcker, Z.f. Phys. 102,572-602 (1936) [spin-dependence of nuclear forces]
G. Wentzel, Helv. Phys. Acta, 10, 107 (1936)
G. Gamow and E. Teller, Phys. Rev. 51, 289 (1937) [e^+e^-,$\nu\bar{\nu}$ exchange]

27) E. Fermi, Z.f. Phys. 88, 161 (1934) [Engl. transl. Am. J. Phys 36, 1150 (1968)]

28) W. Heisenberg, Z.F. Phys. 77, 1(1932); 156(1932); 80, 587 (1932); 96, 473(1935).

29) H. Yukawa, Proc. phys. math. Soc. (Japan) 17, 48 (1935).

30) A. O. Barut, Nuovo Cim. 10, 1146(1958)

31) M. Gell-Mann, Phys. Rev. 125, 1066 (1962)
Y. Ne'eman, Nucl. Phys. 26, 222 (1961)

32) G. Zweig, Cern preprint, TH401, 1964 (unpublished) M. Gell-Mann and Y. Ne'eman. The Eight-fold way, W. A. Benjamin,N.Y. (1969)

33) A.O. Barut and S.A. Basri, "Leptons and bare baryons as integrally charged "quarks" (subm.for publication); and "Theory of fundamental Particles based on Clifford Algebra C_7." (subm.for publication)

34) A.O. Barut and G.L.Strobel, "A calculation of the Fermi coupling constant" (subm. for publication); "Neutrino magnetic moment and the decay of the muon" (submit. for publication).

LATTICE GAUGE THEORIES: AN INTRODUCTION*

Laurence Jacobs
Instituto de Física, UNAM, México 20, D.F.

I. PRELIMINARIES

In the course of these four lectures I shall try to introduce the lattice formulation of Quantum Field Theory as proposed by Wilson[1] and Polyakov[2].

Since the subject is vast, no decent comprehensive overview of the current state of the theory can be given in so short a time. I have therefore chosen to proceed mainly by dealing with examples rather than by attacking the general theory. The examples are representative and serve to illustrate the concepts involved as well as some of the main techniques which have been developed to analyze these theories.

The first two lectures are based mainly on material taken from the excellent reviews by Kogut and Wilson[3] and Kogut[4], as well as from which I gave at the University of Salamanca, Spain, in the summer of 1979[5]. The reader who wishes to learn more about this field is urged to consult these references.

The course is organized as follows. After a very brief discussion of the relationship between Statistical Mechanics and Feynman's formulation of Quantum Mechanics[6], the subject matter is introduced with a description of the ordinary, two-dimensional Ising spin system. This model is analyzed in the symmetric formulation as well as in the Hamiltonian approach of Kogut and Susskind[7] as discussed by Fradkin and Susskind[8]. Wegner's generalization[9] of the Ising model to implement a local (gauge) invariance follows. An analysis of models with a discrete gauge invariance group, Z(N), with particular emphasis on Z(2), comprises the last portion of the course.

As remarked earlier, these notes do not constitute a course in Lattice Gauge Theories; rather, they are intended as a motivation and introduction to the existing literature on these powerful and promising new techniques in high-energy physics.

II. INTRODUCTION

The reasons for the interest in Lattice Gauge Theory among particle physicists are manifold. Most important among these, however, is the fact that the formulation of gauge theories —widely believed to be the most viable candidates for the description of elementary— particle dynamics-on a space-time lattice leads to the solution of a variety of problems which had hitherto remained obscured by the maladies and intricacies inherent in the continuum formulation. As we shall soon see, moreover, once the 3+1-dimensional field theory has been appropriately put on the lattice, it becomes essentially a problem of Statistical Mechanics, enabling one to use the powerful methods developed in that field to analyze problems of interest in Quantum Field Theory. On the

ISSN:0094-243X/81/710109-26$1.50

purely technical side, the lattice provides a gauge-invariant, non-perturbative, ultraviolet cutoff which serves to define the theory in a correct fashion.

The connection between 3+1-dimensional Field Theory and 4-dimensional Statistical Mechanics can be seen best through the use of the Transfer Matrix.

To see this connection consider a one-dimensional harmonic oscillator[6,10] described by the Lagrangian

$$L = \frac{1}{2}(\dot{x}^2 - \omega^2 x^2) \tag{2.1}$$

Going now to an imaginary time lattice of spacing a, the amplitude for going from (X_a, t_a) to (X_b, t_b) is given by the path integral[6]

$$Z = \sum_{\text{PATHS}} \exp\left[-\frac{1}{\hbar} S_P\right] \tag{2.2}$$

where $S_P = \int_{t_a}^{t_b} L\, dt$ corresponds to the (rotated) action for a particular world line joining the two points.

On the lattice, we can write

$$S_P(a) = a \sum_j \frac{1}{2}\left[\left(\frac{x_{j+1} - x_j}{a}\right)^2 + \omega^2 x_j^2\right] \tag{2.3}$$

and the path integral becomes

$$Z = \int_{-\infty}^{\infty} \prod_i dx_i \exp\left[-\frac{1}{\hbar} S\right] \tag{2.4}$$

Notice that eqs. (2.3) and (2.4) provide the description of a one-dimensional statistical mechanics problem with Z being the corresponding nearest-neighbor interaction partition function and the important correspondence $\hbar \leftrightarrow \Theta$ (Temperature). This relation is completely rigorous and general: just as Θ is a measure of the fluctuations of a thermodynamical system, $\hbar$ is a measure of the quantum-mechanical fluctuations; the classical path ($\hbar \to 0$) corresponds to a frozen ($\Theta \to 0$) lattice.

The local nature of eq. (2.3) allows us to write

$$Z = \prod_i \int dx_i\, T_{x_{i+1} x_i} \tag{2.5}$$

with

$$T_{xx'} = \exp -\frac{1}{2}\left[\frac{1}{a}(x - x')^2 + a\omega^2 x^2\right], \tag{2.6}$$

and think of $T_{x,x'}$ as the matrix elements of some operator $\hat{T}$- the transfer matrix. To see that, set up a Hilbert space with the operators x and p :

$$\hat{x}|x\rangle = x|x\rangle$$
$$i\,[\hat{p},\hat{x}] = \hbar \tag{2.7}$$

and the properties

$$\langle x'|x\rangle = \delta(x-x')$$
$$\int |x\rangle\, dx\, \langle x| = 1 \tag{2.8}$$
$$\exp(-i\hat{p}a)|x\rangle = |x+a\rangle$$

Define now $\hat{T}$ as

$$\langle x'|\,\hat{T}\,|x\rangle = T_{xx'} \tag{2.9}$$

One recognizes $\hat{T}$ as the usual time evolution operator of quantum mechanics (for imaginary time).

With these definitions,

$$\begin{aligned} Z &= \int \prod_i dx_i \,\langle x_{i+1}|\hat{T}|x_i\rangle \\ &= \int \langle x_b|\hat{T}|x_{N-1}\rangle\, dx_{N-1}\, \langle x_{N-1}|\hat{T} \cdots \\ &\qquad \cdots \langle x_2|\hat{T}|x_1\rangle dx_1 \langle x_1|\hat{T}|x_a\rangle \\ &= \langle x_b|\hat{T}|x_a\rangle \end{aligned} \tag{2.10}$$

Here $(N-1).a = t_b - t_a$. Imposing periodic boundary conditions and summing over all initial positions one obtains the familiar result,

$$Z = \mathrm{tr}\,(\hat{T}^N) \tag{2.11}$$

Green's functions are obtained by inserting polynomials of the X_i into eq. (2.4) or, equivalently, by inserting $\hat{x}$'s in the appropriate places in eq. (2.11).

To construct $\hat{T}$ in terms of $\hat{p}$ and $\hat{x}$ is now straightforward. By using eqs. (2.7) and (2.8) and the definition of $T_{xx'}$ given in eq. (2.6) one readily obtains,

$$\hat{T} = \exp\left(-\frac{a}{4\hbar^2}\omega^2\hat{x}^2\right)\exp\left(-\frac{a}{2\hbar}\hat{p}^2\right)\exp\left(-\frac{a}{4\hbar^2}\omega^2\hat{x}^2\right). \qquad (2.11)$$

Using now the Baker-Hausdorff formula it follows that, as $a \to 0$,

$$\hat{T} = \exp\left[-\frac{a}{\hbar}\hat{H} + O(a^2)\right] \qquad (2.12)$$

with $\hat{H} = \frac{1}{2}(\hat{p}^2 + \omega^2\hat{x}^2).$

One recognizes eq. (2.12) as the expression for the time-development operator for the simple harmonic oscillator. In terms of $\hat{T}$, one can study the time-evolution of the wave function, $\psi(x,t)$,

$$\begin{aligned}\psi(x',t') &= \int dx\, Z(x',t'|x,t)\,\psi(x,t)\\ &= \int dx\, \langle x'|\hat{T}|x\rangle\,\psi(x,t)\\ &= \int dx\, \delta(x-x')\left[1 - \frac{a}{\hbar}\hat{H} + O(a^2)\right]\psi(x,t)\end{aligned}$$

or,

$$\frac{\hbar}{a}\left[\psi(x',t') - \psi(x',t'+a)\right] = \hat{H}\,\psi(x',t) \qquad (2.13)$$

which becomes the (Euclidean) Schrödinger equation in the limit as $a \to 0$.

The generalization of this procedure to the case of a 3+1-dimensional field theory[11] establishes the claimed connection with 4-dimensional statistical mechanics. When this procedure is carried out, the following correspondence is found:

Field Theory	Statistical Mechanics
Coupling constant	Temperature
Vacuum energy density	Free energy density
Propagator	Correlation function
Mass gap	Inverse correlation length
Ground-state expectation values	Ensemble averages

Using this identity in eq. (3.2) we find

$$Z_H(\beta) = \sum_{\sigma_N} \cdots \sum_{\sigma_1} \prod_{\langle ij \rangle} [\cosh\beta + \sigma_i\sigma_j \sinh\beta]$$

$$= (\cosh\beta)^{2N} \sum_{\sigma_N} \cdots \sum_{\sigma_1} \prod_{\langle ij \rangle} [1 + \sigma_i\sigma_j \tanh\beta] \qquad (3.4)$$

Our goal now is to expand $(\cosh\beta)^{-2N} Z_H(\beta)$ in powers of Tanh β. Since $\sum_{\sigma=\pm1} \sigma = 0$ and $\sum_{\sigma=\pm1} 1 = 2$, the expansion's only contributions come from terms in which all σ_i's appear squared. This suggests a useful graphical expansion. If we represent $\sigma_i\sigma_j \tanh\beta \equiv U_{ij}$ as a link joining sites i and j, this condition is equivalent to the statement that only graphs with the U's forming closed paths will contribute to the sum. For example,

$$= U_{ij} U_{jk} U_{kl} U_{li}$$

$$= (\tanh\beta)^4 (\sigma_i\sigma_j)(\sigma_j\sigma_k)(\sigma_k\sigma_l)(\sigma_l\sigma_i)$$

$$= (\tanh\beta)^4 \qquad (3.5)$$

The next contributions are

$$= (\tanh\beta)^6 \qquad (3.6)$$

and

$$= (\tanh\beta)^8 \qquad (3.7)$$

The number of ways each graph can appear can be found quite easily. One finds the multiplicity to be 2^N N for (3.5), 2^{N+1} N for (3.6) and 2^{N-1} N (N-5) for (3.7). Systematic rules have been found[12] which allow pushing the expansion to high orders.

Collecting these results one obtains

In what follows, we shall take advantage of these connections as we analyze specific examples.

The first problem to be discussed is that of the two-dimensional Ising model. The model does not have any wide applicability, but owing to its simplicity, it will allow us to develop very general methods, some of which we will then apply to the important case of gauge systems in four dimensions.

III. THE TWO-DIMENSIONAL ISING MODEL

The Ising model can be solved exactly in two dimensions. Nevertheless, for the reasons given above, it is instructive to apply here the approximate methods which are needed in more complicated situations.

As a first step in the analysis of the model, we will compute approximate expressions for the partition function which are valid for weak and strong coupling. We will then see that there exists a profound relationship between these two series from which much can be deduced. Next, the Hamiltonian theory will be written down and analyzed in a similar fashion. Here we will again notice that the high- and low-temperature regions are related. However, unlike the previous situation where the transformation is dynamical in origin, in this case it will emerge as a purely geometrical relationship between two formulations of the same lattice system. This is a particular case of the so-called duality transformation. Lastly, we will solve the model exactly and discuss the relationship between the exact solution and what we learned from the approximate calculations. The relevance of phase transitions for the definition of a local, relativistic field theory as the limiting case of the lattice model is clearly seen in this example.

The action for the model is given by (N sites)

$$S_k = -k \sum_{\langle ij \rangle} \sigma_i \sigma_j \tag{3.1}$$

where $\langle ij \rangle$ denotes summation over nearest-neighbor sites on the lattice, the "spins" σ_i take the values ± 1 and k is a constant which measures the interaction strength.

In terms of S_k, the partition function is given by

$$Z_k(\beta) = \sum_{\sigma_N=\pm 1} \cdots \sum_{\sigma_1=\pm 1} \exp\left(-\beta S_k\right) \tag{3.2}$$

Since the effective coupling strength appearing in eq. (3.2) is βk, without loss of generality we set $k = 1$ and use β as a parameter. For small β, (3.2) can be expanded in a power series. We do this by first recalling that, for $\sigma = \pm 1$,

$$\exp(\beta\sigma) = \cosh \beta + \sigma \sinh \beta . \tag{3.3}$$

$$Z_H(\beta) = (\cosh\beta)^{2N} 2^N \left\{ 1 + N(\tanh\beta)^4 + 2N(\tanh\beta)^6 + \frac{1}{2}N(N-5)(\tanh\beta)^8 + O\left[(\tanh\beta)^{10}\right] \right\} \tag{3.8}$$

For large β (low temperature) the lowest state has all spins aligned. The successive contributions to $S(\beta)$ come from flipping spins. A suggestive graphical description of the first three contributions consists of again drawing links but now appearing only between sites of different spin. Denoting a flipped spin by a circle we have

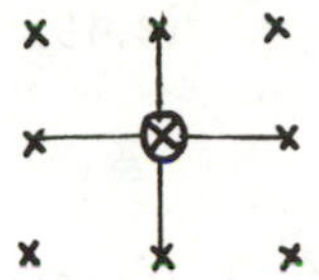

$$\longrightarrow \quad \exp 2(N-4)\beta \tag{3.9}$$

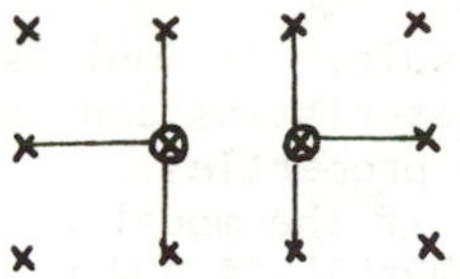

$$\longrightarrow \quad \exp 2(N-6)\beta \tag{3.10}$$

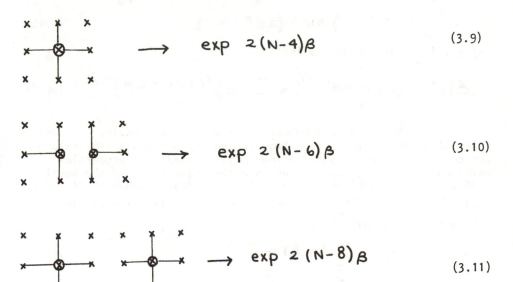

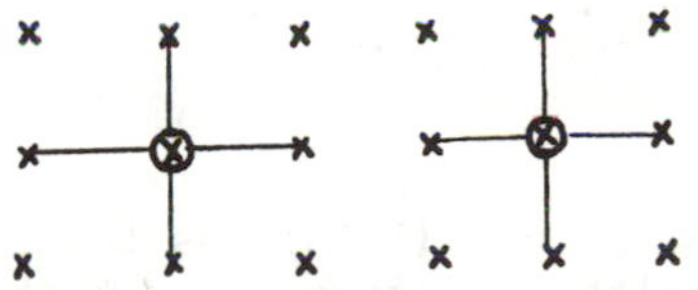

$$\longrightarrow \quad \exp 2(N-8)\beta \tag{3.11}$$

In this way one obtains

$$Z_L(\beta) = e^{2N\beta}\left[1 + Ne^{-8\beta} + 2Ne^{-12\beta} + \frac{1}{2}N(N-5)e^{-16\beta} + O(e^{-20\beta})\right] \tag{3.12}$$

The similarity between eqs. (3.8) and (3.12) is obvious. Defining a new coupling

$$\beta^* = -\frac{1}{2}\ln(\tanh\beta) \tag{3.13}$$

One finds

$$Z_L(\beta^*) = e^{2N\beta^*}\left[1 + N(\tanh\beta^*)^4 + 2N(\tanh\beta^*)^6 + \tfrac{1}{2}N(N-5)(\tanh\beta^*)^8 + \cdots\right]$$

so that

$$Z_L(\beta^*)(e^{2\beta^*})^{-N} = Z_H(\beta)2^{-N}(\cosh^2\beta)^{-N} \tag{3.14}$$

Eq. (3.13) can be also written as

$$\sinh(2\beta)\sinh(2\beta^*) = 1 \tag{3.15}$$

Using this result in eq. (3.14) leads to

$$Z(\beta^*)/(\sinh 2\beta^*)^{N/2} = Z(\beta)/(\sinh 2\beta)^{N/2} \tag{3.16}$$

Eq. (3.16) is a very powerful and unexpected result. It implies that all of the system's high-(low) temperature characteristics can be deduced from the corresponding low-(high) temperature properties by using eq. (3.13). In particular, one can argue that, if the model has a unique phase transition, it must take place at a temperature where $\beta = \beta^*$. From (3.15) this implies that $\sinh^2(2\beta_c) = 1$, giving the famous Kramers-Wannier[13] result:

$$\beta_c = \frac{1}{2}\ln(1+\sqrt{2}). \tag{3.17}$$

The power of the duality transformation, as eq. (3.13) has been called, is apparent when one considers the wealth of information which can be obtained from it. In particular, the exact result, eq. (3.14) follows from a few trivial terms in a perturbative expansion. As we shall see below, this transformation is a special case of a general geometric map. In this particular situation, the system is mapped into itself under duality; in more general cases, the duality transformation maps a given model into a different one. Here its use is also apparent in that it may be possible to map a complicated system into one which is simpler or, in some instances, even solvable.

The next method we will use to analyze the model proceeds by the construction of the transfer matrix. The Hamiltonian is then identified and its spectrum studied[8]. The geometric nature of the duality transformation will be obvious in this approach.

Since the construction of a Hamiltonian requires a highly anisotropic lattice in which the spacing in the time direction is taken to zero, the starting action differs from eq. (3.1) in the appearance of an extra coupling parameter. Denoting unit vectors in the time and space directions by t and x respectively, the action is written as

$$S(\beta_x,\beta_\tau) = -\beta_\tau \sum_n \sigma_n \sigma_{n+\hat{\tau}} - \beta_x \sum_n \sigma_n \sigma_{n+\hat{x}}$$
$$\sigma_N = \pm 1 \tag{3.18}$$

For the present purpose, it is convenient to add a constant term to the action and rewrite it in the form

$$S(\beta_x,\beta_\tau) = \tfrac{1}{2}\beta_\tau \sum_n [\sigma_{n+\hat{\tau}} - \sigma_n]^2 - \beta_x \sum_n \sigma_n \sigma_{n+\hat{x}} \tag{3.19}$$

It is useful to label rows by alternating the notation for a spin at site n. Calling these successive spins S(n) and σ(n) i.e., such that s(n + t) = σ(n), etc. we get

$$S(\beta_x,\beta_\tau) = \sum_\ell \mathcal{L}(\ell+1,\ell) \tag{3.20}$$

with

$$\mathcal{L} = \tfrac{1}{2}\beta_\tau \sum_m^M [\sigma(m) - s(m)]^2 - \tfrac{1}{2}\beta_x \sum_m^M [\sigma(m+1)\sigma(m) + s(m+1)s(m)], \tag{3.21}$$

M is the number of sites along $\hat{x}$. Since, for each row there are 2^M configurations, the transfer matrix, T, is a $2^M \times 2^M$ matrix. Along a diagonal of T, s(m) = σ(m) for all m, so (3.21) gives

$$\mathcal{L}(\text{no flips}) = -\beta_x \sum_m \sigma(m+1)\sigma(m) \tag{3.22}$$

giving the 0-flip contribution

$$T_0 = \exp\left[\beta_x \sum_m \sigma(m+1)\sigma(m)\right] \tag{3.23}$$

The n-flip contribution (that is, the one coming for n spins such that s(m) = - σ(m)) is easily obtained:

$$T_n = \exp(-2n\beta_\tau)\exp\Big[\tfrac{1}{2}\beta_x \sum_m [\sigma(m+1)\sigma(m) + s(m+1)s(m)]\Big] . \tag{3.24}$$

To find the Hamiltonian, such that

$$\hat{T} = \exp(-\tau\hat{H}) \approx 1 - \tau\hat{H}$$

as $\tau \to 0$, first note that, if

$$\hat{T}_0 \approx 1 - \tau\hat{H}_0 \tag{3.25}$$

then (3.22) implies $\beta_x \sim \tau$. Similarly $\hat{T}_1 \approx -\tau\hat{H}_1$ leads to $\exp(-2\beta_\tau) \sim \tau$, so we must have $\beta_x \sim \exp(-2\beta_\tau)$. Calling the proportionality constant λ we have that

$$\beta_x = \lambda \exp(-2\beta_\tau) \tag{3.26}$$

Identifying

$$\tau = \exp(-2\beta_\tau)$$
$$\beta_x = \lambda\tau \tag{3.27}$$

so, as $\tau \to 0$, $\beta_\tau \to \infty$ and $\hat{T}_n \longrightarrow \exp(-2n\beta_\tau)$. To write $\hat{H}$ as an operator we need spin-flip operators. Letting spin up be $\binom{1}{0}$ and spin down be $\binom{0}{1}$, we can do this by putting a Pauli matrix $\hat{\sigma}_3$ at each site,

$$\hat{\sigma}_3 = \begin{pmatrix} 1 & 0 \\ 0 & -1 \end{pmatrix} \qquad \begin{aligned} \hat{\sigma}_3(\uparrow) &= +(\uparrow) \\ \hat{\sigma}_3(\downarrow) &= -(\downarrow) \end{aligned} \tag{3.28}$$

then $\hat{\sigma}_1$, is a spin-flip operator in this basis:

$$\hat{\sigma}_1 = \begin{pmatrix} 0 & 1 \\ 1 & 0 \end{pmatrix} \qquad \begin{aligned} \hat{\sigma}_1(\uparrow) &= (\downarrow) \\ \hat{\sigma}_1(\downarrow) &= (\uparrow) \end{aligned} \tag{3.29}$$

so $\hat{T}$ becomes

$$\hat{T} = 1 + \tau\left[\lambda\sum_m \hat{\sigma}_3(m)\hat{\sigma}_3(m+1) + \sum_m \hat{\sigma}_1(m)\right] + \tau^2\sum_{m\neq n}\hat{\sigma}_1(n)\hat{\sigma}_1(m) + \ldots \tag{3.30}$$

Eq. (3.25) then tells us that

$$\hat{H} = -\sum_m \hat{\sigma}_1(m) - \lambda\sum_m \hat{\sigma}_3(m)\hat{\sigma}_3(m+1)\,, \tag{3.31}$$

which is a one-dimensional quantum Hamiltonian for an Ising model in a transverse magnetic field.

Next we analyze (3.31) perturbatively in powers of λ. One result which will follow from this is that

$$1/\lambda \sim \text{(temperature)}\,. \tag{3.32}$$

Instead of (3.31) we will use

$$\hat{H} = \sum_m \left[1 - \hat{\sigma}_1(m)\right] - \lambda\sum_m \hat{\sigma}_3(m)\hat{\sigma}_3(m+1) \tag{3.33}$$

(this will make the ground-state energy vanish for $\lambda = 0$). Writing $\hat{H} = \hat{H}_o + \lambda\hat{V}$, the minimization of $\hat{H}_o$ at $\lambda = 0$ requires that the lowest state, $|0\rangle$ be such that

$$\hat{\sigma}_1(n)|0\rangle = |0\rangle \tag{3.34}$$

for all n. If we call $|0\rangle$ the state with all spins up, a basis for (3.34) will be $(\uparrow) = \binom{1}{1}$ for spin up and $(\downarrow) = \binom{1}{-1}$ for spin down. In this basis $\hat{\sigma}_3$ will act as a spin-flip operator: $\hat{\sigma}_3\uparrow = \downarrow$, $\hat{\sigma}_3\downarrow = \uparrow$

Hence, the perturbation, $\hat{V}$, flips spins on adjacent sites. The expansion in λ is therefore an expansion in the number of flipped spins. The first excited state above $|0\rangle$ contains one spin which is flipped (as $\lambda\to 0$), the N-fold degeneracy of this state (it can occur at any of the N sites) is removed by introducing a zero-momentum normalized state,

$$|-1\rangle = \frac{1}{\sqrt{N}}\sum_n \hat{\sigma}_3(n)|0\rangle\,. \tag{3.35}$$

The expansion for the energy of a state $|s\rangle$ is $(\hat{H}_o|s\rangle = \epsilon_o|s\rangle)$,

$$E_s = \sum_{j=0} \epsilon_j \lambda^j \tag{3.36}$$

where

$$\epsilon_1 = \langle s|\hat{V}|s\rangle$$
$$\epsilon_2 = \langle s|\hat{V}g\hat{V}|s\rangle$$
$$\epsilon_3 = \langle s|\hat{V}g\hat{V}g\hat{V}|s\rangle - \langle s|\hat{V}|s\rangle\langle s|\hat{V}g^2\hat{V}|s\rangle\,, \tag{3.37}$$

etc.
with the resolvent g given by

$$g = (1 - |a\rangle\langle a|)/(E_o - E_a)\,. \tag{3.38}$$

Denoting by a vertical line a spin flipped at site n, the terms in (3.37) can be obtained by evaluating graphs. The zeroth-order energy is $\epsilon_o = 2$, since it has one spin flipped. Since $\hat{V}$ flips spins to the left or to the right of this site,

$\epsilon_1 = \langle -1|\hat{V}|-1\rangle =$ 

$$= -\frac{1}{N}\sum_{n,\ell} \langle 0|\hat{\sigma}_3(n)\sum_m \hat{\sigma}_3(m)\hat{\sigma}_3(m+1)\hat{\sigma}_3(\ell)|0\rangle\,. \tag{3.39}$$

The only terms which contribute to the sums are n = m and $m+1 = \ell$ or $\ell = m$ and $m+1 = n$. Since $\hat{\sigma}_3^2 = \mathbb{1}$, each term gives a factor N so

$$\epsilon_1 = -2\,. \tag{3.40}$$

All other terms in eq. (3.39) will flip at least one spin when acting on $|0\rangle$, so they vanish when contracted with $\langle 0|$.

To order λ^2, there are two graphs:

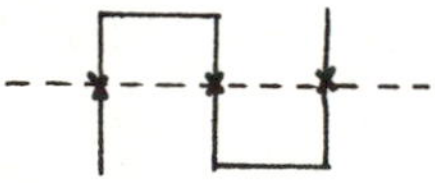 $\rightarrow \quad -2\lambda^2/4$ (3.41)

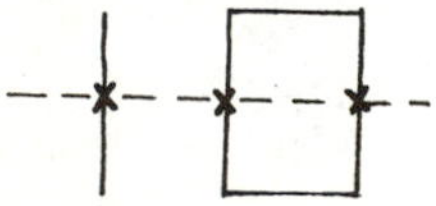 $\rightarrow \quad -(N-2)\lambda^2/4$ (3.42)

The dashed line represents a sum over intermediate states: three flipped spins, each giving two units of energy. The denominators are the energy difference between the initial state (2) and the intermediate state (2 x 3).

There are four graphs to order λ^3 which contribute to $\langle -1 | \hat{V} g \hat{V} g \hat{V} | -1 \rangle$:

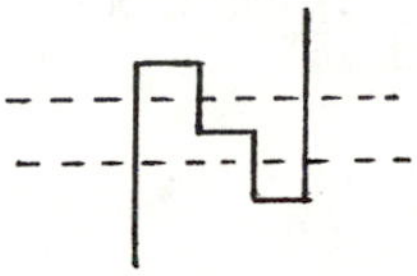 $\rightarrow \quad -2\,\lambda^3/16$ (3.43)

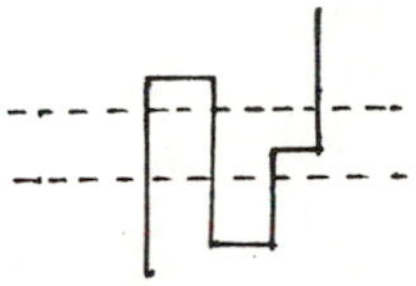 $\rightarrow \quad -2\,\lambda^3/16$ (3.44)

 $\rightarrow \quad -2(N-3)\,\lambda^3/16$ (3.45)

 $\rightarrow \quad -2\,\lambda^3/16$ (3.46)

The rest of ϵ_3 comes from the term $\langle -1 | \hat{V} | -1 \rangle \times \langle -1 | \hat{V} g^2 \hat{V} | -1 \rangle$, which can be obtained from the order λ and λ^2 graphs, the only difference coming from the energy denominators which are now $(2-6)\times(2-6)=16$,

$$\langle -1 | \hat{V} | -1 \rangle \langle -1 | \hat{V} g^2 \hat{V} | -1 \rangle = (-2\lambda)\left(\frac{2}{16} + \frac{N-2}{16}\right)\lambda^2 , \qquad (3.47)$$

which could be represented by

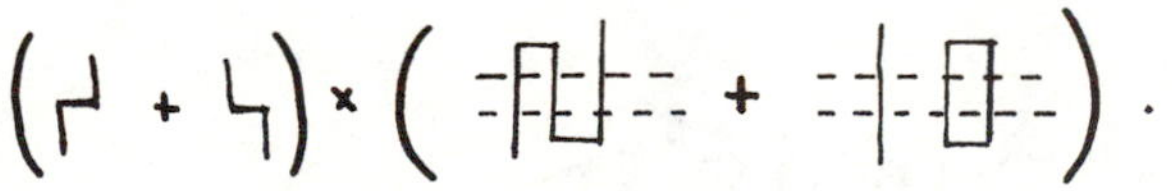 . (3.48)

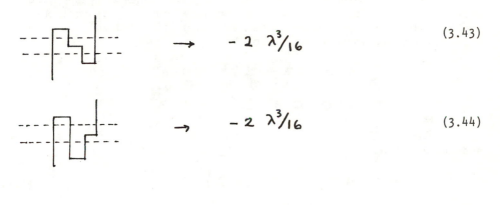

To obtain the mass gap (the energy difference between the first excited state and the ground state), we need the ground-state energy to $O(\lambda^3)$. To do this, we use eqs. (3.37) with $|s\rangle = |0\rangle$, $\epsilon_0 = 0$

trivially, $\epsilon_1 = \langle 0|\hat{V}|0\rangle$ is also zero, since $\hat{V}|0\rangle$ has a flipped spin.

The first non-zero term comes from the graph

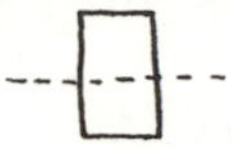

$$\rightarrow \quad -N\lambda^2/4 \tag{3.49}$$

There are obviously no λ^3 contributions, since only applying $\hat{V}$ an even number of times on $|0\rangle$ can produce something proportional to $|0\rangle$ (V must flip up as many spins as it has flipped down).

Collecting these results, to $\Theta(\lambda^3)$:

$$\begin{aligned} E_{-1} &= 2 - 2\lambda - \frac{N}{4}\lambda^2 + 0\cdot\lambda^3 \\ E_0 &= 0 + 0\cdot\lambda - \frac{N}{4}\lambda^2 + 0\cdot\lambda^3 \end{aligned} \tag{3.50}$$

and the mass gap, $G(\lambda) = E_{-1}(\lambda) - E_0(\lambda)$ is

$$G(\lambda) = 2(1-\lambda) \tag{3.51}$$

If one pushes the calculation to higher orders, the cancellations which took place at orders λ^2 and λ^3 are seen to occur at all higher orders, so that eq. (3.51) is actually exact. We must find that the gap vanishes at $\lambda = 1$. To study the large-λ region, one could proceed to a $1/\lambda$ expansion in the usual way by looking at the operator $1/\lambda \cdot H$. As we shall soon see, however, this is really not necessary, due to the self-duality of the model.

Consider the transformation $D[n] = n + 1/2$:

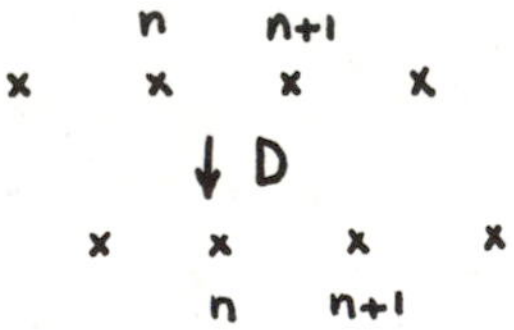

To define a Hamiltonian on the dual lattice we must define appropriate spin operators there. Looking at the structure of H given in eq. (3.31) one can see that the trick is to define a set of operators, $\hat{\alpha}_1$ and $\hat{\alpha}_3$, which satisfy the same algebra as the $\hat{\sigma}$'s but which interchange the role of the spin-flip term in $\hat{H}$ to the interaction term in $\hat{H}_D$. Specifically, we want to construct $\hat{\alpha}$'s such that

$$\hat{\alpha}_1(n) = \hat{\sigma}_3(n+1)\,\hat{\sigma}_3(n) \tag{3.52}$$

$$\hat{\alpha}_1(n+1)\,\hat{\alpha}_3(n) = \hat{\sigma}_1(n) \tag{3.53}$$

with an algebra

$$\begin{aligned} &\{\hat{\alpha}_1(n),\, \hat{\alpha}_3(n)\} = 0 \\ &\hat{\alpha}_1^2 = \hat{\alpha}_3^2 = \mathbb{1} \\ &[\hat{\alpha}_1(n),\, \hat{\alpha}_3(m)] = 0 \qquad n \neq m \end{aligned} \tag{3.54}$$

This can be achieved by defining $\hat{\alpha}_1$ through eq. (3.52) and

$$\hat{\alpha}_3(n) = \prod_{m<n} \hat{\sigma}_1(m) \tag{3.55}$$

as can be directly verified. Using the $\hat{\alpha}$'s we get the dual Hamiltonian,

$$\begin{aligned} \hat{H}_D &= -\sum_n \hat{\alpha}_3(n+1)\,\hat{\alpha}_3(n) - \lambda \sum_n \hat{\alpha}_1(n) \\ &= \lambda\left[-\sum_n \left(\hat{\alpha}_1(n) + \frac{1}{\lambda}\,\hat{\alpha}_3(n+1)\,\hat{\alpha}_3(n)\right)\right]. \end{aligned} \tag{3.56}$$

But the $\hat{\alpha}$'s satisfy the same algebra as the $\hat{\sigma}$'s, so (3.56) has the same operator structure as (3.31), implying that

$$\hat{H}(\lambda) = \lambda\,\hat{H}(1/\lambda) \tag{3.57}$$

the promised self-duality relation. From eq. (3.57) we deduce that $G(\lambda) = \lambda G(1/\lambda)$, so if $G(\lambda_c) = 0$ so should $G(1/\lambda_c) = 0$, leading to the conclusion that $\lambda_c = 1$; confirming our previous analysis. The fact that the λ expansion is valid at $\lambda = 1$ substantiates our assertion that (3.51) is exact. In fact, duality leads to

$$G(\lambda) = 2\,|1-\lambda| \tag{3.58}$$

As a final step in our study of the Ising model in two dimensions, we shall use the Hamiltonian eq. (3.31) to solve it exactly[14]. The main point of doing this is to show explicitly how scale invariance sets in at a critical point, trivially leading to a continuum limit and defining there a relativistic field theory.

We begin with a Hamiltonian which is equivalent to (3.31)(we drop all hats from operators):

$$H = -\sum_n \sigma_3(n) - \lambda \sum_n \sigma_1(n+1)\,\sigma_1(n) \tag{3.59}$$

As we shall see, eq. (3.59) is really an obscure way of writing a theory for a free, massless Majorana fermion. To see this, recall the Jordan-Wigner transformation[15].

Define raising and lowering operators $\sigma^+(n)$ and $\sigma^-(n)$ as

$$\sigma^{\pm}(n) = \frac{1}{2}\left[\sigma_1(n) \pm i\,\sigma_2(n)\right], \tag{3.60}$$

with n = -N, -N+1,..., N. Fermionic operators can then be constructed. They are given by

$$C(n) = \prod_{\ell=-N}^{n-1} \exp\left[i\pi\sigma^+(\ell)\sigma^-(\ell)\right]\sigma^-(n)$$

$$C^\dagger(n) = \prod_{\ell=-N}^{n-1} \exp\left[-i\pi\sigma^+(\ell)\sigma^-(\ell)\right] \tag{3.61}$$

Using the Pauli spin algebra, it is not difficult to verify the Fermionic nature of C and C^+:

$$\{C(n), C^\dagger(m)\} = \delta_{nm},$$

$$\{C(n), C(m)\} = \{C^\dagger(n), C^\dagger(m)\} = 0 \tag{3.62}$$

Moreover, one easily obtains:

$$C(n) = \prod_{\ell=-N}^{n-1}\left[-\sigma_3(\ell)\right]\sigma^-(n),$$

$$C^\dagger(n) = \sigma^+(n)\prod_{\ell=-N}^{n-1}\left[-\sigma_3(\ell)\right]; \tag{3.63}$$

from which follows that

$$\begin{aligned}\sigma_3(n) &= 2\,\sigma^+(n)\sigma^-(n) - 1\\ &= 2\,C^\dagger(n)C(n) - 1\end{aligned}$$

$$\begin{aligned}C(n)C^\dagger(n+1) &= -\sigma^-(n)\sigma^+(n+1)\\ C^\dagger(n)C^\dagger(n+1) &= \sigma^+(n)\sigma^+(n+1)\\ C(n)C^\dagger(n+1) &= -\sigma^-(n)\sigma^-(n+1)\end{aligned} \tag{3.65}$$

From these one finds

$$\sigma_1(n)\sigma_1(n+1) = \left[C^\dagger(n) - C(n)\right]\left[C^\dagger(n+1) - C(n+1)\right] . \tag{3.66}$$

These relations enable us to write H in terms of C and $C^\dagger$:

$$H = -2\sum_n C^\dagger(n)C(n) - \\ -\lambda\sum_n \left[C^\dagger(n) - C(n)\right]\left[C^\dagger(n+1) - C(n+1)\right] . \tag{3.67}$$

But now H is a quadratic form, so the solvability of the model becomes apparent.

It is useful to write H in momentum space. Introduce the operators

$$a_k = (2N+1)^{-1/2} \sum_{n=-N}^{N} e^{ikn}\, C(n)$$

$$k = 0, \pm 2\pi/(2N+1), \ldots, \pm 2N\pi/(2N+1) . \tag{3.68}$$

It is straightforward to check that

$$\{a_k^\dagger, a_{k'}\} = \delta_{kk'} \tag{3.69}$$

$$\{a_k^\dagger, a_{k'}^\dagger\} = \{a_k, a_{k'}\} = 0$$

by using the completeness relations for exp -i k (n-m) . Inverting eq. (3.68) allows us to write H in terms of a_k and $a_k^\dagger$ as

$$H = -2\sum_k a_k^\dagger a_k - \lambda\sum_k \left(e^{-ik} a_k^\dagger a_{-k}^\dagger + 2(\cos k) a_k^\dagger a_k - e^{ik} a_k a_{-k}\right)$$

$$= -2\sum_k (1+\lambda\cos k) a_k^\dagger a_k - \lambda\sum_k \left(e^{-ik} a_k^\dagger a_{-k}^\dagger - e^{ik} a_k a_{-k}\right)$$

$$= -2\sum_{k>0} (1+\lambda\cos k)(a_k^\dagger a_k + a_{-k}^\dagger a_{-k}) + 2i\lambda\sum_{k>0} (\sin k)(a_k^\dagger a_{-k}^\dagger + a_k a_{-k}) \tag{3.70}$$

To simplify this operator, perform a canonical transformation to a new set of operators $\eta_k, \eta_k^\dagger$:

$$\eta_k = u_k a_k + i v_k a^\dagger_{-k} \tag{3.71}$$

with u, v , real and even functions of k.
To satisfy

$$\{\eta_k , \eta^+_{k'}\} = \delta_{kk'}$$

$$\{\eta^+_k, \eta^+_{k'}\} = \{\eta_k, \eta_{k'}\} = 0 , \tag{3.72}$$

the condition $u_k^2 + v_k^2 = 1$ must be met. Writing $u_k = \cos \theta_k$, $v_k = \sin \theta_k$ and inverting (3.71) we obtain

$$\begin{aligned} H = \sum_{k>0} & [-2(1+\lambda \cos k) \cos 2\theta_k \\ & +2\lambda(\sin k) \sin 2\theta_k](\eta^\dagger_k \eta_k + \eta^+_{-k}\eta_{-k}) \\ & + \sum_{k>0} [4i(1+\lambda \cos k) \sin 2\theta_k \\ & + 4i\lambda(\sin k) \cos 2\theta_k](\eta^\dagger_k \eta^\dagger_{-k} + \eta_k \eta_{-k}) . \end{aligned} \tag{3.72}$$

H can be diagonalized if

$$(1+\lambda \cos k) \sin 2\theta_k = -\lambda(\sin k) \cos 2\theta_k \tag{3.73}$$

or

$$\tan 2\theta_k = -\lambda \sin k/(1+\lambda \cos k) . \tag{3.74}$$

This defines θ_k. Using eq. (3.74) in H,

$$H = 2\sum_k \Lambda_k \eta^\dagger_k \eta_k \tag{3.75}$$

(up to irrelevant constants), with

$$\Lambda_k = 2(1+2\lambda \cos k + \lambda^2)^{1/2} \tag{3.76}$$

The minima of Λ_k occur for $k = \pm\pi$ where

$$\Lambda_{\pm\pi} = 2\,|1-\lambda| \tag{3.77}$$

this time proving that eq. (3.51) is exact.

Up to now we have been using lattice units (in which the lattice spacing is unity). To restore physical units we write (lattice spacing = a)

$$k = \pi + ap \tag{3.78}$$

where we now measure momenta from the minimum value of Λ_k and p has appropriate dimensions. For the energy, we must similarly define

$$E(p) = \Lambda_k / 2a \,. \tag{3.79}$$

Since we want p to be finite as $a \to 0$, k must tend to π in the limit. From eq. (3.76),

$$\frac{1}{2}\Lambda_k^2 \approx (1-\lambda)^2 + \lambda\,(ap)^2 \,, \quad a \approx 0 \,, \tag{3.80}$$

and hence

$$E(p) = \left[\frac{(1-\lambda)^2}{a^2} + \lambda p^2 \right]^{1/2} \tag{3.81}$$

is the energy-momentum relation. We see that, away from the critical point, $\lambda = 1$, E(p) has no limit as $a \to 0$. On the other hand, at $\lambda = 1$ all traces of the lattice disappear and we find

$$E(p) = |p| \,; \tag{3.82}$$

the correct relativistic spectrum of a massless particle.

IV. GAUGE SYSTEMS

The generalization of spin systems to theories with a local symmetry was accomplished by Wegner[9] in 1971. In an article which was, until recently, seldom quoted, he invented lattice gauge theories. His motivation had nothing to do with particle physics, rather, he wondered whether a system which could not magnetize could still have a non-trivial phase structure. He was led to this question by the behavior of the planar model, a model like those we have been discussing except for the fact that, unlike the Ising case, where spins take values on Z_2 they take values on O(2) in this case. It was known that this system could not magnetize, but there were also indications[16] that it had a phase transition at finite temperature. Wegner realized that, by constructing a model with a local symmetry, no local operator in the theory could develop a non-zero expectation value. He then showed, by constructing appropriate non-local order parameters, that these new theories could indeed undergo phase transitions.

We will discuss Wegner's model in some detail and then briefly mention some of the most relevant results in these models and their importance for particle theory.

The main idea behind the generalization of the global (up $\leftrightarrow$ down) symmetry of the Ising model to a local one, is that here "spins" are defined on the links joining adjacent sites, instead of on the sites themselves. Gauge transformations, in a way which we shall specify shortly, are performed at sites.

Consider a D-dimensional hypercubical lattice (in practice, we will generally take D = 4) and label the links by a site n and a unit vector $\hat{\mu}$. A given link is then either $(n, \hat{\mu})$ or $(n + \hat{\mu}, -\hat{\mu})$. For a gauge group G, spins will be assigned an element, $U_{n\hat{\mu}}$ of G. In the case $G = Z_2$, $U_{n\hat{\mu}} = \pm 1$.

A local gauge transformation at site n, $H(n) \in G$, acts on all spins touching n, that is, all $U_{n\hat{\mu}}$. In the case of Z_2, H(n) simply flips all spins sharing the site n (8 in D = 4). It is easy to see that the only G - invariant functions of the $U_{n\hat{\mu}}$ that one can construct must be built from closed paths of U's on the lattice.

The simplest such path is an elementary square - usually called a plaquette. In terms of plaquettes we can write down the simplest action which is G-invariant as

$$S = \sum_{n,\mu\nu} U_{n\hat{\mu}} \, U_{n+\hat{\mu},\hat{\nu}} \, U_{n+\hat{\mu}+\hat{\nu},-\hat{\mu}} \, U_{n+\hat{\nu},\hat{\nu}}$$

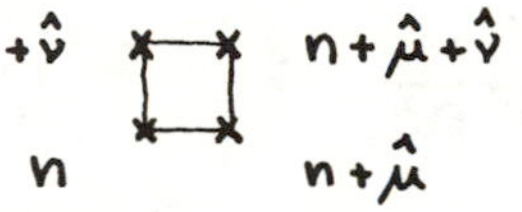

(4.1)

A more convenient notation for eq. (4.1) is to write it as

$$S(U) = \sum_{\square} s_{\square} \tag{4.2}$$

where the sum runs over all plaquettes in the lattice and the action per plaquette, $s_{\square}$, is

$$\begin{aligned} s_{\square} &= 1 - \omega \\ \omega &= U_{ij} U_{jk} U_{k\ell} U_{\ell i} \end{aligned} \tag{4.3}$$

and the indices (i,j,k,l) run around a plaquette.

The U_{ij} are oriented spins in the sense that

$$U_{ji} = U^{*}_{ij}$$

The partition function then takes the form

$$Z(\beta) = \sum_{U \in G} \exp\left[-\beta S(U)\right] . \tag{4.4}$$

When discussing the low-temperature properties of the Ising model, a useful concept which was introduced was that of a "broken-bond"; an invariant concept in that case. This concept is not locally gauge invariant and, hence, not generally useful when analyzing gauge systems. The corresponding concept in this case is that of "frustration". Consider the case of Z_2. Here $U = \pm 1$ and $S_\square$ eq. (4.3), can take the values $S_\square = 0$ or $S_\square = 2$. The second case occurs when an odd number of U's are minus one. Notice that this is gauge invariant, since a gauge transformation applied at a lattice site changes two spins, leaving the product fixed. A plaquette for which $S_\square = 2$ is called a frustrated plaquette.

We will now prove our earlier statement about the impossibility of having phase transitions with a local order parameter. This result[17] implies, in the present case, that the magnetization vanishes identically for all temperatures. This is a general theorem, but, for simplicity we shall prove it only for the case of the Z_2 model[4].

Consider the model in the presence of an external field h and compute the average "spin" $\langle U_{ij}\rangle_h$. The system will exhibit spontaneous magnetization if $\langle U_{ij}\rangle_h \neq 0$ as $h \to 0$. In this case both the global as well as the local symmetries would be broken.

Using eq. (4.4) and adding the coupling $h \quad U_{ij}$,

$$\langle U_{ij}\rangle_h = \frac{\sum_{\text{config}} U_{ij} \exp\left[\beta \sum_\square UUUU + h \sum_\square U\right]}{\sum_{\text{config}} \exp\left[\beta \sum_\square UUUU + h \sum_\square U\right]} . \tag{4.5}$$

Consider a local gauge transformation at some site n and call the set of links touching n, L. The quartic term in the action will be left invariant, but the coupling term changes by the amount

$$h \sum U \to h \sum U' - h \sum \delta U \tag{4.6}$$

where U' is the new spin and

$$\delta U_{ij} = \begin{cases} -2U_{ij} & (i,j) \in L \\ 0 & (i,j) \notin L \end{cases} \tag{4.7}$$

changing variables in (4.5), $U \to U'$

$$\langle U_{ij}\rangle_h = -\left\{\sum U'_{ij}\exp\left[\beta\sum U'U'U'U' + h\sum U' + h\sum \delta U\right\}/Z_h\right.$$

$$= \left\langle -U_{ij}\exp\left[-h\sum_{(ij)\in L}\delta U\right]\right\rangle_h \tag{4.8}$$

leading to a bound

$$\left|\langle U_{ij}\rangle_h - \langle -U_{ij}\rangle_h\right| = \left|\left\langle -U_{ij}\left[\exp\left(-h\sum_L \delta U\right)-1\right]\right\rangle_h\right|$$
$$\leq \left|e^{4Dh}-1\right|\langle U_{ij}\rangle_h \tag{4.9}$$

but $e^{4Dh}\to 1, (h\to 0)$ implying

$$\langle U_{ij}\rangle_{h\to 0} = \langle -U_{ij}\rangle_{h\to 0} \tag{4.10}$$

or

$$\langle U_{ij}\rangle_{h\to 0} = 0 \tag{4.11}$$

The theorem, stated in more general terms says that, for any operator Θ , $\langle\Theta\rangle$ = 0 unles Θ is invariant under all local gauge transformations.

So the natural question that arises is: What kind of gauge-invariant function of the U's can serve to distinguish between phases? As was noted by Wegner, the answer lies in the spatial dependence of gauge-invariant correlation functions. Consider a closed path of links on the lattice and denote this closed path by P. Define

$$W_P = \prod_{(ij)\in P} U_{ij} \tag{4.12}$$

As we have seen W_P is gauge-invariant. Moreover, its expectation value, $\langle W_P\rangle$ measures the correlations among the spins. It is clear that $\langle W_P\rangle$ falls off as the path is enlarged; however, the speed at which it does depends on the detailed dynamical properties of the system. At high temperature (strong coupling), a disordered phase is such that $\langle W_P\rangle$ falls off exponentially as the area of the minimal surface bounded by P, whereas at low temperature, the appearance of an ordered phase is signaled by a much slower fall-off, with $\langle W_P\rangle \sim \exp(-L_P)$ where L_P is the length of P.

The properties of W_P are most important when one considers the continuous groups, U(1) and SU(3), of relevance to quantum electrodynamics and quantum chromodynamics respectively. It has been argued by Wilson[1] that the dynamics of fermions in these theories is controlled by the behavior of such correlations. Concretely, the area-law signals confinement (by implying, in the appropriate limit, that the energy which binds two fermions grows linearly with their separation) whereas the perimeter law does not (energy independent of separation).

Let us proceed to the analysis of these correlation functions in the case of Z_2.

The techniques we shall use are quite similar to those of Sec. 3. The identity, eq. (3.3) reads in this case

$$\exp(\beta\, UUUU) = \cosh\beta\,(1 + UUUU \tanh\beta) \qquad (4.13)$$

then

$$\langle W_P \rangle = \frac{\sum_{\text{config}} \prod_{\square} (1 + UUUU \tanh\beta) \prod_P U}{\sum_{\text{config}} \prod_{\square} (1 + UUUU \tanh\beta)} . \qquad (4.14)$$

As we did in Sec. 3, for small β we can expand in powers of Tanh β . The same arguments apply in this case and we find that the first term in eq. (4.14) comes from tessellating the minimal surface bounded by P with plaquettes. Since the number of plaquettes is proportional to the area of such a surface, we find that

$$\begin{aligned} \langle W_P \rangle &= (\tanh\beta)^{\alpha A_P} + \cdots \\ &= \exp[\alpha A_P (\ln \tanh\beta) + \cdots] \end{aligned} \qquad (4.15)$$

Of course, considering more terms alters eq. (4.15). However, the area law remains at higher orders.

For small temperatures, the expansion is in the number of frustrated plaquettes (flipped spins in Sec. 3).

Starting from a frozen lattice, with no plaquettes frustrated, changing one of the U's will frustrate 2(D-1) plaquettes; the action of this configuration (relative to the minimum) is therefore 4(D-1) For the frozen lattice $\langle W_P \rangle = 1$ since $\prod_P U = 1$. If the changed U lies on P, this contributes a factor of -1, otherwise, the numerator in eq. (4.14) is always + 1. If L_P is the length of P and N is the number of links in the lattice,

$$\langle W_P \rangle = \frac{1 + (N - 2L_P) \exp[-4(D-1)\beta] + \cdots}{1 + N \exp[-4(D-1)\beta] + \cdots} \tag{4.16}$$

If n U's are changed, we can estimate a contribution proportional to

$$\frac{1}{n!} N^n \exp[-4n(D-1)\beta] \tag{4.17}$$

summing over n,

$$\begin{aligned}\langle W_P \rangle = &\{1 + (N-2L_P)\exp[-4(D-1)\beta] \\ &+ \tfrac{1}{2}(N-2L_P)^2 \exp[-8(D-1)\beta] + \cdots\} / \{1 + \\ &+ N\exp(-4(D-1)\beta] + \tfrac{1}{2}N^2\exp[-8(D-1)\beta] + \cdots\} \\ \approx &\exp[-(2\exp(-4(D-1)\beta))L_P] .\end{aligned} \tag{4.18}$$

Again, by considering more terms (in particular, for the n-flip contribution we assumed independence) the detailed form of (4.18) would change, but the perimeter law remains.

These results imply that the Z_2 model has a phase transition (In more than two dimensions). For D = 2 the model is trivially solvable by the following argument. The gauge invariance of the model allows us to perform gauge transformations at any site. As long as only gauge-invariant operations are considered, this freedom permits a choice of gauge. In particular, for D = 2 we can always fix the U_{ij} in some direction to be, say + 1. Calling this the time direction, eq. (4.1) reduces to ($U_{n\hat{\nu}} = 1$)

$$S_2 = \sum_n U_{n\hat{\mu}} U_{n+\hat{\nu},\hat{\mu}} , \tag{4.19}$$

which is the action for a one-dimensional ordinary Ising model. This model can be trivially shown to be disordered at all finite temperatures.

Wegner also showed that the Z_2 model is self-dual[9]. The transition takes place at the self-dual point (which in fact coincides with that of the D = 2 Ising model, eq. (3.17).

The Z_N models have received much attention in recent times as they provide an indirect way of studying the continuous case of the group U(1). Since much is known about the U(1) gauge theory, this is an ideal ground for testing the ideas and techniques of lattice gauge theory. In particular, if the lattice version of QED were to imply the confinement of electrons, its use in the non-Abelian models would be clearly suspect.

Until very recently, the only clue to the phase structure of the U(1) and SU(2) models came from the so-called Migdal recursion relations.[18] Through these relations, an approximate connection between a D-dimensional gauge theory and a D/2-dimensional spin system can be established. One finds in this way that the D = 4 U(1) model is related to the D = 2 Planar model, whereas D = 4 SU(2) is related to the O(3) Heisenberg spin model in D = 2. It is known that the planar model has a phase transition —supporting the fact that the U(1) theory describes physical electrons. On the other hand, the D = 2 Heisenberg model is rigorously known to be disordered at all finite temperatures[19] —supporting the confinement of quarks. However, these relations are only approximate and it is not known how to find corrections. In fact, they have recently found to lead to false conclusions in some particular cases,[20] so whether or not the prediction of quark confinement is correct should be investigated by other means. As for lattice QED, several recent investigations have led to the conclusion that it does have a phase transition and that, in fact, it does resemble in many ways the Planar model[21,22].

Applying Monte-Carlo techniques[23] the non-Abelian models have been studied recently[24,25,26]. The results of these investigations lend strong support to the conjecture of quark confinement, a long sought-after goal for field theorists.

A few last remarks about the Z_n gauge models. It has been suggested[27] that the Z_2 model may be exactly solvable. This possibility is supported by the impressive agreement between the perturbative calculations pioneered by Balian, Drouffe and Itzykson[28] and the non-perturbative results of Ref. 22. It has been established[21,22] that for $n \geq 4$, the Z_n theories have three phases, the first transition approaches $\beta = 1$ smoothly as $n \to \infty$ whereas, for large n, the second transition disappears with $\beta_n \sim n^2$. The U(1) limit, has therefore a single phase transition around $\beta = 1$[22]. Very high-order perturbative analysis recently performed[29] matches these results everywhere except for the middle phase, where the methods employed are not applicable. These facts, together with the self-duality of Z_3 and Z_4[30] suggests the possibility that the three models, Z_2, Z_3 and Z_4 may be exactly solvable.

[1]K.G. Wilson, Phys. Rev. D10, 2455 (1974).
[2]A.M. Polyakov, Phys. Lett. B59, 79 (1975), ibid, 82 (1975).
[3]J. Kogut and K.G. Wilson, Phys. Rep. C12, 75 (1974).
[4]J. Kogut, Rev. Mod. Phys. 51, 659 (1979).
[5]L. Jacobs, Lecture given at the University of Salamanca, Spain. (unpublished), 1979.
[6]R.P. Feynman, Rev. Mod. Phys. 20, 367 (1948).
[7]J. Kogut and L. Susskind, Phys. Rev. D11, 395 (1975).

[8] E. Fradkin and L. Susskind, Phys. Rev. D17, 2637 (1978).
[9] G. Wegner, J. Math. Phys. 12, 2259 (1971).
[10] M. Creutz, Phys. Rev. D15, 1128 (1977).
[11] See, for example, J. Kogut, Ref. 3.
[12] M. Wortis, Phase Transitions and Critical Phenomena, Vol. 3, 113 (1974). Eds. C. Domb and M.S. Green (Academic, London).
[13] H.A. Kramers and Wannier, Phys. Rev. 60, 252 (1941).
[14] The exact solution of this model was first given in a famous paper by Onsager in 1944: L. Onsager, Phys. Rev. 65, 117 (1944). It has since been solved in many different ways, at least partly due to the fantastically complicated nature of Onsager's solution. The present analysis appears in T.D. Schultz, D.C. Mattis, and E. Lieb, Rev. Mod. Phys. 36, 856 (1964). See also P. Pfeuty, Ann. Phys. (NY) 57, 79 (1970) as well as Ref. 4.
[15] P. Jordan and E. Wigner, Z. Phys. 47, 631 (1928).
[16] H.E. Stanley and T.A. Kaplan, Phys. Rev. Lett. 17, 913 (1966).
[17] S. Elitzur, Phys. Rev. D12, 3978 (1975).
[18] A.A. Migdal, Zh. Eksp. Teor. Fiz. 69, 810, 1457 (1975).
[19] N.D. Mermin and H. Wagner, Phys. Rev. Lett. 17, 1133 (1966).
[20] M. Creutz, L. Jacobs and C. Rebbi, Phys. Rev. Lett. 42, 1390 (1979).
[21] S. Elitzur, R. Pearson and J. Shigemitsu, J. Phys. Rev. D19, 3698 (1979); D. Horn, M. Weinstein and S. Yankielowicz, ibid, 4, 1013 (1980).
[22] M. Creutz, L. Jacobs and C. Rebbi, Phys. Rev. D20, 1915 (1979).
[23] M. Creutz, L. Jacobs and C. Rebbi, Phys. Rep. C (To be published); L. Jacobs and C. Rebbi, J. Comp. Phys. (To be published).
[24] K. Wilson, Cornell Univ. Rep. (1980)(unpublished).
[25] M. Creutz, Phys. Rev. D21, 2308 (1980).
[26] C. Rebbi, Phys. Rev. D (To be published. July, 1980).
[27] A.M. Polyakov, Phys. Lett. B (1979).
[28] R. Balian, J.M. Drouffe and C. Itzykson, Phys. Rev. D10, 3376 (1974); 11, 2104 (1975).
[29] J.M. Drouffe, Saclay Report (1980).
[30] Korthals-Altes, Nucl. Phys. B142, 315 (1978); T. Yonega, ibid, B144, 195 (1978).

GAUGE THEORY OF THE FUNDAMENTAL INTERACTIONS†

L. O'Raifeartaigh
Dublin Institute for Advanced Studies,
10 Burlington Road, Dublin 4, Ireland.

INTRODUCTION

During the past ten years the idea has gradually emerged that all four of the known fundamental physical interactions (strong, weak, electromagnetic and gravitational) have a common structure, namely a so-called gauge-structure. Of course, the (independent) gauge structure of electromagnetism and gravitation had been known for some time [1], but this structure was not considered to be the central feature of those interactions and did not appear to play any role in the strong and weak interactions. Now all that has changed, and the gauge-structure is considered to play a fundamental role for all four interactions.

The gauge structure manifests itself in quite different ways for the different interactions, however. In the case of gravity the gauge-structure is linked to the curvature of space-time (through Cartan's vierbein formulation of general relativity) and although it has been studied intensively in its own context [2], the gauge structure of gravitation has not yet been related to the gauge-structure of the other three interactions. In the case of electromagnetism and weak interactions, there appears to be a single gauge structure which is common to both[3]. But for the weak interactions the structure is masked at the phenomenological level because the gauge-symmetry is spontaneously broken. Finally, for the strong interactions the gauge-symmetry is unbroken. But it is again masked at the phenomenological level, this time because the gauge-group is the internal (color) symmetry group of the quarks, and this group manifests itself only in a very indirect manner.

The present lecture notes are an attempt to present briefly the essential features of gauge-theory as used in the theory of the fundamental interactions. The lectures are divided into two Parts, each of six lectures. Part I deals with the theory, and Part II with the application to the weak, electromagnetic and strong interactions. In both Parts the emphasis is on the general principles and structure of the theory, although, of course, in the application, specific models (notably the standard SU(2) x U(1) model of the weak and electromagnetic interactions) are used.

In the theoretical Part, the gauge theory and the spontaneous symmetry breakdown are first treated as se-

ISSN:0094-243X/81/710135-42$1.50

parate and independent phenomena. Then they are amalgamated into a single scheme, and associated phenomena such as the Higgs-Kibble mechanism, discussed. The discussion is general so that it is applicable to any gauge-group and any representation of the matter-fields. An important omission in the theoretical Part is a discussion of renormalizability. This omission is due partly to lack of time and partly to the fact that excellent coverage is available elsewhere(4).

In the Part on applications, the particular models used are shown to follow uniquely from a small number of reasonable, qualitative, assumptions. In particular, it is shown that some of the most immediate experimental consequences, notably the existence of neutral currents and charmed particles, follow directly from quite general principles. In the case of the strong interactions, emphasis is placed on the dramatic change in our point of view which has followed the discovery(5) that non-abelian gauge-theories are "asymptotically free".

A detailed description of the topics considered in each Part is given in the list of Contents. Throughout the lectures, references have been restricted mainly to books and review articles as much more comprehensive references than could be given here can be found in these.

In conclusion the author would like to take this opportunity to thank the organizers of the summer school for their kind invitation to give these lectures; and to thank the participants for their attention and for many stimulating questions and discussions.

CONTENTS

Part I (Theory)

Part II (Applications)

Part I. Theory

1.- Rigid Symmetry.

Let G be any compact converted Lie group, with elements

$$g = e^{\lambda_a t_a} \in G \qquad a = 1, \dots, n \tag{1.1}$$

where λ_a are the parameters and t_a the (anti-hermetian) generators. Actually, since G is compact, it can only be(6) a product of one-dimensional abelian and simple groups (module some discrete central elements). For example G = U(1), U(1) x U(1), SU(2), SU(2) x U(1), SU(2) x SU(3) x U(1), SU(2) x SU(2)/$\mathbb{Z}_2$, U(1) x SU(3)/$\mathbb{Z}_4$, etc. If the elements g $\in$ G, or equivalently, the parameters λ_a are independent of space-time the group G is said to be rigid.

Suppose now that ϕ and ψ are a set of so-called matter-fields (i.e. spin 0 and spin $\frac{1}{2}$ fields respectively) and that

$$-\mathcal{L}(\phi, \psi, \partial_\mu \phi, \partial_\mu \psi) = \tfrac{1}{2}\partial_\mu \phi_k \partial_\mu \phi_k + \bar{\psi}_\alpha(\gamma^\mu \partial_\mu + m)\psi_\alpha + G^k \bar{\psi}_\alpha \psi_\beta \phi_k + V(\phi) , \tag{1.2}$$

where the potential $V(\phi)$ is a polynomial of most fourth degree, is a renormalizable Lagrangian for these fields. If the fields ϕ, ψ belong to unitary (but in general not irreducible) representations of the group G, and $\mathcal{L}$ is invariant with respect to the action of G then G is said to be a rigid symmetry group of $\mathcal{L}$. For example, if ψ belongs to the 2-dimensional representation of SU(2) and ϕ to the 3-dimensional vector representation then

$$-\mathcal{L}(\phi, \psi, \partial\phi, \partial\psi) = \tfrac{1}{2}\partial\underset{\sim}{\phi}\cdot\partial\underset{\sim}{\phi} + \bar{\psi}(\not\partial + m)\psi + G\bar{\psi}\underset{\sim}{\sigma}\psi\cdot\underset{\sim}{\phi} + V(\underset{\sim}{\phi}\cdot\underset{\sim}{\phi}) , \tag{1.3}$$

where $\underset{\sim}{\sigma}$ denotes the Pauli matrices, and inner product in SU(2) space for the spinors is understood, will have SU(2) as a rigid symmetry group. In general we will abbreviate (1.2) by writing

$$-\mathcal{L}(\phi,\psi,\partial\phi,\partial\psi)=\tfrac{1}{2}(\partial\phi)^2+\bar{\psi}(\not\partial+m)\psi+G\bar{\psi}\phi\psi+V(\phi)\,, \tag{1.4}$$

since the meaning of the notation is fairly obvious. We shall also denote the global and infinitesimal group transformations of ϕ by

$$\phi(x)\to\mu(g)\phi(x)\quad,\qquad \delta_a\phi(x)=\mu(t_a)\phi\,, \tag{1.5}$$

respectively, where $\mu(t_a)$ is the representative of t_a in the representation of G to which ϕ belongs, and similary for ψ.

The most immediate physical consequence of rigid symmetry is the existence of conserved charges, and conserved currents, as demonstrated by Noether's theorem[7]. Indeed in the rigid case Noether's theorem reduces to the simple observation that if we define the currents

$$j^a_\mu(x)=\frac{\partial\mathcal{L}}{\partial\phi_{,\mu}}\delta_a\phi+\frac{\partial\mathcal{L}}{\partial\psi_{,\mu}}\delta_a\psi\qquad, \tag{1.6}$$

where a summation over all components of ϕ and ψ is understood, then, as a consequence of the field equations we will have

$$\partial_\mu j^a_\mu=0 \quad\text{and hence}\quad \partial_t Q^a=0 \quad\text{where}\quad Q^a=\int d^3x\, j^a_0(x) \tag{1.7}$$

Proof: For $\Theta=\phi$ or ψ we have

$$\partial_\mu\left(\frac{\partial\mathcal{L}}{\partial\theta_{,\mu}}\delta^a\theta\right)=\left(\partial_\mu\frac{\partial\mathcal{L}}{\partial\theta_{,\mu}}\right)\delta^a\theta+\frac{\partial\mathcal{L}}{\partial\theta_{,\mu}}(\delta_a\theta_{,\mu})=\left(\frac{\partial\mathcal{L}}{\partial\theta}\right)\delta_a\theta+\frac{\partial\mathcal{L}}{\partial\theta_{,\mu}}(\delta_a\theta_{,\mu}) \tag{1.8}$$
$$=\delta_a\mathcal{L}=0$$

the last step coming from the assumption that $\mathcal{L}$ is invariant. Notice that there are exactly as many conserved currents as group generators.

An important corollary of Noether's theorem is that the charges Q^a are not only conserved, but actually generate the infinitesimal transformations (1.5) by commutation, because of the canonical commutation relations (or Poisson brackets)

$$[\pi(y),\phi(x)]=\delta^3(x-y)\quad,\ \text{where}\quad \pi=\frac{\partial\mathcal{L}}{\partial\phi_{,0}} \tag{1.9}$$

To see this we note from (1.6) and (1.7) that

$$Q^a = \int j_0^a(y)\, d^3y = \int \pi(y)\, \delta_a \phi(y)\, d^3y \quad , \tag{1.10}$$

and hence from (1.8) we have

$$\left[Q^a , \phi(x) \right] = \delta_a \phi(x) , \tag{1.11}$$

which is the required result. An analogous result holds for ψ with commutators replaced by anti-commutators.

The consistency of equation (1.11) with successive infinitesimal group transformations implies that the charges Q^a must also satisfy the commutation laws of the Lie algebra,

$$\left[Q^a , Q^b \right] = f_{abc}\, Q^c \quad , \tag{1.12}$$

where f_{abc} are the structure constants of G, and from the definition (1.7) and the canonical commutation relations it is easy to verify that this is true. Thus the Noether charges form a representation of the Lie algebra of G in the function space of the ϕ and ψ .

From the three properties of the Noether charges just discussed (the conservation law (1.7)), the generation of infinitesimal transformations (1.11), and the representation of the Lie algebra (1.12) one should expect them to play a fundamental role in any theory with rigid symmetry. Such is in fact the case, and we end this lecture with an example of their importance that comes from the Lie-algebraic property (1.12). From this property it follows that if two known Noether charges Q^a and Q^b are assumed to form part of a rigid structure then their commutator must again be a Noether charge and be the space-integral of the time component of a conserved Noether current. We shall see later that this apparently trivial statement becomes of great importance in gauge-theories, where it leads directly to the prediction of neutral weak currents and charmed particles.

2.- Gauge-Symmetry

Consider the Lagrangian density

$$-\mathcal{L}(\phi,\psi) = \tfrac{1}{2}(\partial\phi)^2 + \bar{\psi}(\not{\partial} + m)\psi + V(\phi) + G\bar{\psi}\phi\psi \quad , \tag{2.1}$$

of the first lecture, which is invariant with respect to the rigid symmetry group G with elements

$$g = e^{\lambda_a t_a} \in G \quad , \tag{2.2}$$

as suppose that we wish to make G local in the sense that there should be an independent transformation g at each point x of space-time. Then g (or equivalently the parameters λ) become x-dependent

$$g \to g(x) = e^{\lambda_a(x) t_a} \quad . \tag{2.3}$$

The transformation (2.3) is called a local group or gauge-transformation and under such a transformation the Lagrangian (2.1) is not invariant because the derivates $\partial\phi$, $\partial\psi$ no longer transform in the same homogeneous (invariant) manner as before,

$$\partial\left(\mathcal{U}(g(x))\,\phi(x)\right) \neq \mathcal{U}(g(x))\,\partial\phi(x) \quad . \tag{2.4}$$

The question is: can the Lagrangian be modified so as to make it invariant with respect to the gauge-transformation (2.3) ? The answer is that it can, and the resulting theory is called gauge-theory. In this lecture we wish to describe the basic principle of the modification which results in gauge theory.

For simplicity we first consider the case where the group G is irreducible. (We recall that G must then be either simple or 1-dimensional abelian). The first step is to introduce a set of n spin 1 (gauge) potentials $A^a_\mu(x)$, $a = 1 \dots n$, where n is the order of the Lie algebra (thus 3 for SU(2), eight for SU(3) and so on) and to define the covariant derivatives

$$D_\mu(A)\phi(x) = \partial_\mu\phi(x) + e\,\mathcal{U}(A_\mu)\,\phi(x) \equiv \partial_\mu\phi(x) + e A^a_\mu(x)\,\mathcal{U}(t_a)\,\phi(x) \tag{2.5}$$

(and similary for the ψ), where e is a (coupling) constant and $\mathcal{U}$ (t_a) is the representative of t_a in the representation $\mathcal{U}(g)$ of G to which ϕ belongs. The next step is to demand that the covariant derivates transform homogeneously, that is to demand that the A_μ (x) transform with respect to G so that

$$\mathcal{U}(g(x))\left(D_\mu(A)\phi\right) = D_\mu(A^g)\left(\mathcal{U}(g(x))\phi\right) \quad , \tag{2.6}$$

where A_μ^g denotes the transform of A_μ (and similary for ψ). Once (2.6) is satisfied it is clear that once the ordinary derivatives are replaced by the covariant ones everything will proceed exactly as in the rigid case, in particular if we replace the Lagrangian (2.1) by the quantity

$$-Q(\phi,\psi,\partial\phi,\partial\psi) = \frac{1}{2}(D\phi)^2 + \bar{\psi}(\not{D}+m)\psi + V(\phi) + G\bar{\psi}\phi\psi \quad , \tag{2.7}$$

this quantity will be automatically invariant with respect to the local gauge transformations (2.3).

There are two snags to this replacement, however. First we have to check that there exists a transformation of A_μ which permits (2.6). Second, the quantity (2.7) is no longer a Lagrangian because it contains no kinetic (derivative) terms for the A-potentials.

The first snag is easily disposed of, since a little algebra shows that (2.6) will be satisfied provided that the A_μ have the transformation law

$$A_\mu^g(x) = g(x)A_\mu(x)g^{-1}(x) + \frac{1}{e}g(x)\partial_\mu g^{-1}(x) \tag{2.8}$$

Indeed (2.8) is necessary and sufficient to (2.6) to hold for all representations of G. Note that (2.8) respects the group transformation, but is not homogeneous (covariant) in A_μ . However, the inhomogeneous term is independent of A_μ .

To dispose of the second snag we need to introduce a kinetic term for the A-potentials which is itself invariant with respect to the local gauge-transformations (2.3). The introduction of such a term constitutes the final step of the program and is made possible by the existence of the so-called curvature of field- strength

$$F_{\mu\nu} = \partial_\mu A_\nu - \partial_\nu A_\mu + e[A_\mu, A_\nu] \quad , \tag{2.9}$$

with the homogeneous, or covariant, transformation law,

$$F_{\mu\nu}(A^g) = g(x)F_{\mu\nu}g^{-1}(x) \quad . \tag{2.10}$$

(This law can be verified directly from (2.8) or seen by inspection from the fact that $\mathcal{U}(F_{\mu\nu}) = e^{-1}[D_\mu, D_\nu]$). From the analogy with Maxwell theory one sees at once that

$$\frac{1}{4}\,\mathrm{tr}\,(F_{\mu\nu}F^{\mu\nu}) , \tag{2.11}$$

where tr denotes the normalized trace of $F_{\mu\nu}F^{\mu\nu}$, is a natural kinetic term for the A_μ. Indeed, if we demand that the kinetic term be Lorentz and gauge-invariant, and quadratic in the first order derivatives, then (2.11) is unique up to a constant. Note, however, that in contrast to Maxwell theory, (2.11) contains cubic and quartic interactions as well as kinetic terms when G is non-abelian.

Thus finally we have the result. If the Lagrangian (2.1) is invariant with respect to the rigid symmetry group G, then the modified Lagrangian

$$-\mathcal{L}(\phi, \psi, D\phi, D\psi) = \frac{1}{4}\,\mathrm{tr}\,F^2 + \frac{1}{2}(D\phi)^2 + \bar{\psi}(\not{D} + m)\psi + V(\phi) , \tag{2.12}$$

obtained by replacing all ordinary derivatives by covariant ones, and adding $(\mathrm{tr}\,F^2)/4$, will be invariant with respect to the local gauge transformations (2.3). The Lagrangians (2.12) are the Lagrangians of gauge theory and in analogy with electromagnetism the induced spin-1 gauge fields A_μ are often called the radiation fields

So far we have dealt only with the case of irreducible G. When G is reducible the situation is slightly more complicated but has to be considered because of its physical relevance. In the reducible case the Lie algebra g of G is split into irreducible sets $t_a^{(r)}$, where r is fixed for each irreducible set and a is an index within each set.

Then the gauge fields and the covariant derivates (2.5) are generalized to be

$$F_{\mu\nu}(A) \quad \text{and} \quad D_\mu\phi = \partial_\mu\phi + \mathcal{U}(A_\mu)\phi \quad \text{where} \quad A_\mu = \sum_r e_{(r)} A_\mu^{(r)a} t_a^{(r)}$$

respectively, where there is a separate coupling constant $e_{(r)}$ for each irreducible component. Thus the important new feature of the reducible case is that there is an independent coupling constant for each irreducible component of G. It is important to note however that the independent coupling constants correspond only to the reducibility of the group G and have nothing to do with the irreducibility of the representation $\mathcal{U}(g)$ of G to which the matter-fields ϕ and ψ belong. In fact, if we now regard the gauge-invariant Lagrangian (2.12) from the point of view of parameters we see that

(i) For the matter-matter interaction

$$G\bar{\psi}\phi\psi + V(\phi), \tag{2.14}$$

there are as many parameters as were allowed by the original rigid symmetry and the usual renormalizability of spin-zero fields.

(ii) For the radiation-radiation interaction the only free parameters are the coupling constants $e_{(\alpha)}$ in (2.13),

(iii) For the radiation-matter interaction the only freedom lies in the choice of the representations $\mathcal{U}(g)$ for the matter fields ϕ and ψ.

From (ii) and (iii) we see that the interaction of the radiation field is universal. In particular two completely different kinds of matter fields (e.g. leptons and quarks) will interact with the gauge-fields in exactly the same way provided only that they belong to the same representation $\mathcal{U}(g)$ of G. Much of the strength of gauge theory lies in the small number of parameters which are permitted by (i), (ii), (iii) above.

3.- Spontaneous Symmetry Breaking

The gauge-invariant Lagrangian of the previous lecture with the scalar fields omitted

$$-\mathcal{L}(\psi, D\psi, F) = \frac{1}{4}\operatorname{tr} F^2 + \bar{\psi}(\not{D}+m)\psi \ , \tag{3.1}$$

is already adequate to describe classical and quantum electrodynamics (with G = U(1)) and is thought also to describe the strong-interactions at a fundamental, quark level (with G = SU(3)). But even though the weak interactions are assumed traditionally to be mediated by intermediate vector mesons $W_\mu^\pm$, and the gauge-fields would seem to be natural candidates for the $W_\mu^\pm$, the Lagrangian (3.1) cannot describe these mesons. The reason is that (experimentally) the intermediate vector mesons must be massive, and the gauge fields in the Lagrangian (3.1) are massless. Furthermore, the introduction of mass-terms m^2A^2 in an ad hoc manner would destroy the gauge-invariance. It took some time historically before it was realized that there existed a mechanism for introducing mass-terms without destroying the gauge-invariance. This mechanism consists in retaining the scalar fields ϕ in $\mathcal{L}(\phi,\psi,D\phi,D\psi,F)$ and allowing the scalar-field potential $V(\phi)$ to have what is called a <u>spontaneous symmetry breakdown</u>.

The resultant theory, with massive gauge fields, can then be used to describe the weak interactions.

Since spontaneous symmetry breaking is independent of gauge theory as such, we shall consider it as a separate phenomenon in this and the next lecture, and only afterwards return to its role in gauge theory. Thus in these two lectures we shall consider only rigid symmetries and scalar field Lagrangians of the form

$$-\mathcal{L}(\phi) = \frac{1}{2}(D\phi)^2 + V(\phi) \quad . \tag{3.2}$$

The essence of a spontaneous breakdown is that the potential V (ϕ) in (3.1) is invariant with respect to a symmetry-group G, but its minimum point $\overset{\circ}{\phi}$ is not. That is,

$$V(u(g)\phi) = V(\phi), \text{ but } \quad u(g)\overset{\circ}{\phi} \neq \overset{\circ}{\phi} \quad , \tag{3.3}$$

for all $g \in G$. The classic example is the single-field potential

$$V(\phi) = \lambda(\phi^2 - c^2)^2 \quad , \tag{3.4}$$

which is invariant under the reflexion symmetry $\phi \to -\phi$ but whose minimum points $\overset{\circ}{\phi} = \pm c$ are not separately invariant under the reflexion (though, of course, they transform into one another).

Before considering more complicated examples it is convenient to introduce two important concepts which arise out of the definition (3.3) namely, the concepts of <u>little groups</u> and <u>orbits</u>.

The little group concept arises from the fact that although $\overset{\circ}{\phi}$ is not invariant with respect to every element g of G, there may be a non-trivial subgroup H of elements for which $\overset{\circ}{\phi}$ is invariant,

$$u(h)\overset{\circ}{\phi} = \overset{\circ}{\phi} \quad \text{for} \quad h \in H \subset G \quad . \tag{3.5}$$

This group H is called variously the little group of $\overset{\circ}{\phi}$, the stability group of $\overset{\circ}{\phi}$, and the residual symmetry group of $V(\phi)$. The extreme cases are H = 1 (the trivial little group) and H = G (no spontaneous breakdown).

The concept of orbit arises by considering the set of all vectors of the form

$$\mathcal{U}(g)\overset{\circ}{\phi} \qquad \text{for} \qquad g \in G \; . \tag{3.6}$$

The set of all such vectors is called the orbit of $\overset{\circ}{\phi}$. In the example (3.3) the orbit consists of the two points $\pm c$. Comparing (3.7) and (3.8) we see that in general the points on the orbit are in one-one correspondence with the points in the coset space $G|H$. Clearly the orbit will be trivial (consists of a single point) if, and only if, the little group H is G itself, and thus a spontaneous breakdown may also be characterized by saying that the orbit of $\overset{\circ}{\phi}$ is non trivial. Since, because of the symmetry, $V(\phi)$ takes the same value for all points on an orbit, a spontaneous breakdown may also be characterized by saying that the potential minimum is degenerate. In general, even an irreducible group representation will contain more than one orbit.

Let us now consider some other examples. First, let ϕ belong to the n-dimensional representation of SO(n) and the potential be formally as in (3.3) but now with $\phi^2 = \phi_k\phi_k$, $k = 1,\ldots,n$. The minimum is then $\overset{\circ}{\phi}_k = cu_k$ where u_k is any unit vector, the little group is SO(n-1), and the (unique) orbit is the surface of the (n-1) sphere.

Next, let ϕ be a traceless symmetric tensor

$$\phi_{ij} = u_i v_j + v_i u_j - \frac{2}{3}\delta_{ij}(u\cdot v) \; , \tag{3.7}$$

in the 5-dimensional representation of SO(3). The most general renormalizable potential for $V(\phi)$ is then

$$V(\phi) = \lambda\left(\mathrm{tr}\,\phi^2 - c^2\right)^2 + \varepsilon\,\mathrm{tr}\,\phi^3 \; , \qquad \mathrm{tr}\,\phi = 0 \; , \tag{3.8}$$

and the extremal equation for this potential is easily seen to be

$$2\lambda\left(2\,\mathrm{tr}\,\overset{\circ}{\phi}{}^2 - c^2\right)\overset{\circ}{\phi} + \varepsilon\left(3\overset{\circ}{\phi}{}^2 - 3\,\mathrm{tr}\,\overset{\circ}{\phi}{}^2\right) = 0 \tag{3.9}$$

There are two kinds of orbits for ϕ, those for which u and v are parallel (H = 1) and not parallel (H=SO(2)).

If $\varepsilon = 0$ then (3.1) determines only the magnitude $tr\,\phi^2 = c^2/2$ of $\overset{\circ}{\phi}$ and allows $\overset{\circ}{\phi}$ to lie on either kind of orbit. But if $\varepsilon \neq 0$, (3.11) forces $\overset{\circ}{\phi}$ to satisfy a quadratic equation & hence to lie on the orbit with H = SO(2). Thus for $\varepsilon \neq 0$ the extremal equation determines not only the magnitude of $\overset{\circ}{\phi}$ but also the kind of orbit on which it must lie.

As a final example, let ϕ belong to the adjoint representation of SU(n). Writing ϕ as a traceless hermitian n x n matrix, and noting that every such matrix can be diagonalized by an SU(n) transformation, we see that every orbit has a point on it corresponding to a diagonal ϕ. Hence it suffices to consider only the diagonal ϕ's. Furthermore, we see that the little groups in this case consist exactly of those SU(n) matrices which commute with ϕ. It is then easy to see that the possible little groups are of the form

$$H = \mathcal{U}(m_1) \times \mathcal{U}(m_2) \times \ldots \times \mathcal{U}(m_s)/\mathcal{U}(1) \quad , \quad m_1 + m_2 + \ldots + m_s = n \ , \tag{3.10}$$

where the m_r, $r = 1, \ldots, s$ are the dimensions of the blocks of equal eigenvalues.

The most general renormalizable potential for this case is of the form

$$V(\phi) = \lambda \left(tr\,\phi^2 - c^2 \right)^2 + \varepsilon\, tr\,\phi^3 + \delta\, tr\,\phi^4 \quad , \tag{3.11}$$

and since the extremal equations for this potential are at most cubic, we see that (unless $\varepsilon = \delta = 0$) all the little groups except those corresponding to two or three blocks of equal eigenvalue are excluded . Thus, once again, the potential severly restricts the class of orbits open to $\overset{\circ}{\phi}$

With these examples in mind, it is not difficult to accept that for general compact Lie groups and their finite-dimensional representations there is the following kind of orbit structure: The little groups can be (partially) arranged in order of size (defined by inclusion up to conjugation). For example for SU(4) there are four non-trivial little groups with the ordering

$$\mathcal{U}(1) \times \mathcal{U}(1) \times \mathcal{U}(1) \subset \mathcal{U}(2) \times \mathcal{U}(1) \subset \begin{matrix} \mathcal{U}(2) \times \mathcal{U}(2)/\mathcal{U}(1) \\ \mathcal{U}(3) \end{matrix} \tag{3.12}$$

The minimal little group is unique, but the maximal one in general is not. On orbits belonging to the minimal little group no invariants except the norm ϕ^2 are extremal, whereas on orbits belonging to the maximal little group at least one other invariant is extremal The result is that if the potential contains only the norm-invariant, the minimum usually falls on a minimal orbit, whereas if V contains any other invariant, it usually falls on a maximal orbit. The orbits in between usually only occur for special ranges of the parameters in V .

4.- The Goldstone Theorem.

Before proceeding to use the spontaneous symmetry breaking mechanism in the gauge-invariant Lagrangian (2.13) there is an important property of the breakdown that must be considered, namely the appearance of zero mass fields. More precisely, there is a theorem due to Goldstone, which states that if a symmetry group is continuous its spontaneous breakdown must be accompanied by the appearance of mass zero scalar fields. For some time this property was believed to exclude spontaneous symmetry breaking for particle physics because there are no observed mass-zero scalar fields, but, as we shall see, the difficulty disappears in the context of gauge fields.

The simplest example of the Goldstone phenomena occurs in the SO(n) example of the previous lecture. If we expand the field in the form $\phi(x) = \overset{\circ}{\phi} + \theta(x)$ where $\overset{\circ}{\phi}$ is the minimum point of the potential, then $\theta(x)$ is the 'true' field (zero at the potential minimum) and the potential takes the form

$$V(\phi(x)) = W(\theta(x)) = \lambda\left\{4c^2(u\cdot\theta)^2 + 4c(u\cdot\theta)\theta^2 + \theta^4\right\} . \tag{4.1}$$

Ones sees by inspection of (4.1) that the only field with a mass-term is ($u \cdot \theta(x)$). Thus the other (n-1) components of $\theta(x)$ are massless, which illustrates the Goldstone theorem in this case.

The general proof of the Goldstone theorem is as follows. From the symmetry of the potential under the group G we have

$$V(u(g)\phi) = V(\phi) , \tag{4.2}$$

and if the group is continuous ,we can expand $U(g)\phi$ in the form

$$u(g)\phi = \phi + \lambda_a t_a \phi + \frac{1}{2!}\lambda_a\lambda_b t_a t_b \phi + \ldots \tag{4.3}$$

Inserting (4.3) in (4.2) we obtain

$$\lambda_a \frac{\partial V}{\partial \phi_k}(t_a\phi)_k + \frac{1}{2!}\lambda_a\lambda_b\left\{ \frac{\partial^2 V}{\partial\phi_k\partial\phi_j}(t_a\phi)_k(t_b\phi)_j + \frac{\partial V}{\partial \phi_k}(t_a t_b \phi)_k\right\} + \ldots = 0 , \tag{4.4}$$

for all λ_a . Since $\partial V/\partial\phi$ is zero for $\phi = \overset{\circ}{\phi}$ we then have in particular

$$\left(\frac{\partial^2 V}{\partial\overset{\circ}{\phi}_k\partial\overset{\circ}{\phi}_j}\right)(t_a\overset{\circ}{\phi})_k(t_b\overset{\circ}{\phi})_j = 0 . \tag{4.5}$$

Now because $\overset{\circ}{\phi}$ is not merely an extremum of $V(\phi)$ but a minimum, the matrix $\partial^2 V/\partial\phi^2$ is positive. Hence (4.5) implies that

$$\left(\frac{\partial^2 V}{\partial\overset{\circ}{\phi}_k\partial\overset{\circ}{\phi}_j}\right)\left(t_a\overset{\circ}{\phi}\right)_j = 0 \tag{4.6}$$

But $\partial^2 V/\partial\phi^2$ will also be recognized as the mass-matrix for the scalar fields since

$$V(\phi) = V(\overset{\circ}{\phi}) + \frac{1}{2}\left(\frac{\partial^2 V}{\partial\overset{\circ}{\phi}_k\partial\overset{\circ}{\phi}_j}\right)\theta_k\theta_j + O(\theta^3) , \tag{4.7}$$

where $\theta(x) = \phi(x) - \overset{\circ}{\phi}$. Hence (4.6) shows that the vectors $(t_a\overset{\circ}{\phi})_k$ are eigenvectors of the mass-matrix with eigenvalue zero. Thus there are (at least) as many zero-mass scalars as there are non-trivial independent directions $t_a\overset{\circ}{\phi}$ in the representation space of $\overset{\circ}{\phi}$.

To see how many such directions there are, we recall that, by definition of the little group H, $t_a\overset{\circ}{\phi}$ is zero if and only if, t_a is a generator of H. Hence $t_a\overset{\circ}{\phi} \neq 0$ if and only if, t_a is a generator of that part of the Lie algebra of G which is complementary to the Lie algebra of H. In other words $t_a\overset{\circ}{\phi} \neq 0$ if, and only if, t_a is a generator of the coset space G/H. Thus the number of independent directions for which $t_a\overset{\circ}{\phi} \neq 0$, and hence the number if Goldstone fields, is dim (G/H). In particular, if the spontaneous breakdown is non-trivial, H is strictly less than G and there exists at least one Goldstone field. This is the Goldstone theorem.

Note that the number of Goldstone fields depends only on the groups G & H and not directly on the representation of the Higgs field. The number of remaining fields will be dim R - dim G/H where R is the representation of G to which the scalar fields belong. If R happens to be the adjoint representation (or has the same dimensions as the adjoint representation) then the number of remaining fields is just dim H. In general the remaining scalar fields will be massive, but some may be fortuitously massless. Note that because the spontaneously broken potential is still invariant with respect to the little group H, the remaining fields will occur in H-invariant multiplets.

The above considerations are all classical, and it is important to realize that the Goldstone theorem holds also in the quantized case. Otherwise the Goldstone fields could acquire masses by renormalization, and become physically indistinguishable from ordinary massive particles. Proofs for the quantized case have been given using both the LSZ formulation of quantum field theory(8) and the functional integral formalism(9).

5.- Spontaneously Broken Gauge Symmetry

We are now in a position to return to our gauge-invariant Lagrangian

$$-\mathcal{L}(\phi,\psi,D\phi,D\psi,F) = \frac{1}{2}(D\phi)^2 + \bar{\psi}(\not{D}+m)\psi + \frac{1}{4}\mathrm{tr}\,F^2 + g\bar{\psi}\phi\psi + V(\phi) , \tag{5.1}$$

and see what happens when it undergoes a spontaneous breakdown, that is when we make the change of field $\phi(x) \rightarrow \theta(x)$ $\theta(x) = \phi(x) - \mathring{\phi}$. We already know what happens to $V(\phi)$ so we concentrate on the other terms in $\mathcal{L}(\phi, D\phi, \psi, D\psi, F)$. Making the change $\phi \rightarrow \theta$, we see from (5.1) that $\mathcal{L}(\phi, D\phi, \ldots)$ becomes

$$-\mathcal{L}(\phi,D\phi,\psi,D\psi,F) = -\mathcal{L}(\theta,D\theta,\psi,D\psi,F) + g\bar{\psi}\mathring{\phi}\psi + \frac{1}{2}(D\mathring{\phi})^2 + (D\mathring{\phi}, D\phi) . \tag{5.2}$$

Thus the net effect of the spontaneous symmetry breakdown is to induce three new terms as shown. Note that only the one new parameter, namely $\mathring{\phi}$, introduced, so that the new terms are strongly correlated to one another and to $\mathcal{L}(\theta, D\theta, \ldots)$. This observation, together with our previous discussion of the restricted nature of the parameters in the unbroken theory, shows that the Lagrangian (5.2) describes a theory with extremely few parameters and therefore with an entremely high predictive power.

Let us now discuss in the order shown in (5.2) the three terms which have been induced by the spontaneous symmetry breakdown.

(i) $g\bar{\psi}\mathring{\phi}\psi$. Since $\mathring{\phi}$ is a constant, this term represents a mass-term for the fermions. The mass-term breaks the G-invariance of the masses, but it preserves the H-invariance, where H is the little group. Comparing this mass-term with the Yukawa coupling in (5.1) we see that the fermion mass-differences and the Yukawa couplings are directly related. Either set can be chosen as the input parameters. At present, since we observe the fermion masses but not the Yukawa couplings, the fermion masses are usually used as input. But if and when the Yukawa couplings are observed, they will furnish a strong test of the theory.

(ii) $\frac{1}{2}(D\mathring{\phi})^2$. Since $\mathring{\phi}$ is constant we have

$$D_\mu \mathring{\phi} = \sum_r e_{(r)} A_\mu^{(r)a} \mu(t_a^{(r)}) \mathring{\phi} \quad , \tag{5.3}$$

and hence

$$\frac{1}{2}(D_\mu \mathring{\phi}, D_\mu \mathring{\phi}) = \frac{1}{2} M_{(r)a,(s)b} A_\mu^{(r)a} A_\mu^{(s)b}, \text{ where } \quad M_{(r)a,(s)b} = e_{(r)} e_{(s)} (\mathring{\phi}, \mu(t_a^{(r)} t_b^{(s)}) \mathring{\phi}) \tag{5.4}$$

Since $M_{(r)a,(s)b}$ is a constant matrix, we see that $(D\mathring{\phi})^2/2$ is just a mass-term for the gauge-fields, with $M_{(r)a,(s)b}$ as mass-matrix. In particular, if G is irreducible, the mass-matrix reduces to

$$M_{ab} = e^2 (\mu(t_a)\mathring{\phi} \ , \ \mu(t_b)\mathring{\phi}). \tag{5.5}$$

If we now recall that

$$\mu(t_a^{(r)})\mathring{\phi} = 0 \quad , \tag{5.6}$$

if, and only if, $t_a^{(r)}$ is a generator of the little group H, we see that the mass-matrix leaves the gauge-fields corresponding to the generators of H massless (as we would expect since H is the residual symmetry group) but it makes the gauge-fields corresponding to the generators of the coset space G/H massive. Furthermore, none of the coset-space masses is fortuitously zero, since otherwise the little group would be larger than H. As in the case of scalar fields, all the gauge-field masses fall into irreducible multiplets of the residual symmetry group.

Before going on the third induced term in (5.2) it may be worthwhile to consider some examples of the gauge-field mass-matrix

(a) G = SU(2). If H = 1 there are 3 massive gauge-fields. If H = U(1) there are two-massive gauge-fields and they form a U(1)-doublet.

(b) G simple, $\phi \in$ adjoint representation of G. Then from (5.4) we have

$$M_{ab} = e^2 tr \{[t_a, \overset{\circ}{\phi}][t_b, \overset{\circ}{\phi}]\} = e^2 tr\, t_a [\overset{\circ}{\phi} [\overset{\circ}{\phi}, t_b]] = e^2 q_b^2 \delta_{ab} \quad , \tag{5.7}$$

where q is the charge corresponding to $\overset{\circ}{\phi}$. Thus the mass-matrix is automatically diagonal and the masses are proportional to the squares of the q-charges.

(c) The standard model of the weak interactions, G = U(2), H = U(1). Then one gauge field (which is identified with the electromagnetic field) is massless, and three are massive. Two of the massive ones form a U(1) (electromagnetic) doublet and are identified with the traditional charged vector mesons $W^{\pm}$ of the weak interactions. The remaining field is a new neutral massive vector meson. If ϕ is a doublet, and $\overset{\circ}{\phi}$ is suitably normalized, then the mass-matrix takes the form

$$M_{ab} A_\mu^a A_\mu^b = (\overset{\circ}{\phi}, (g\sigma_a A_\mu + f A_\mu)^2 \overset{\circ}{\phi}) \quad , \tag{5.8}$$

which is not automatically diagonal.

Finally it may be worth comparing the masses obtained from the spontaneous symmetry breakdown for the gauge-fields and scalar fields respectively. The comparison is shown in the following box

	gauge-fields	scalar fields in adjoint	scalar fields in general
massive	dim G/H	dim H	dim R - dim G/H
massless	dim H	dim G/H	dim G/H

One sees that the results for the two sets of fields are complementary.

iii) $(D\phi, D\overset{\circ}{\phi})$. From the above discussions of the first two induced terms in the Lagrangian (5.7) we see that the main effect of the spontaneous symmetry breakdown is to provide mass-terms for the fermions and gauge-fields. However, the role of the third induced term $(D\phi, D\overset{\circ}{\phi})$ is a little different.

From (5.3) we see that the third term can be written in the form

$$\sum_r e_r A_\mu^{(r)a} \partial_\mu (\theta, u(t_a^{(r)})\overset{\circ}{\phi}) + \sum_{r,s} e_r e_s A_\mu^{(r)a} A_\mu^{(s)b} (u(t_a^{(r)})\theta, u(t_b^{(s)})\overset{\circ}{\phi}) \quad . \tag{5.9}$$

The first expression in (5.9) is not important, and, in fact, disappears in the Landau gauge (and also in the physical gauge which we shall define later). The second term represents an interaction between two gauge fields and a scalar-field. Such an interaction term does not occur in the original symmetric Lagrangian. Its role in the spontaneously broken Lagrangian is to help make the theory renormalizable by cancelling graphs which would otherwise increase their divergence with increasing order in the coupling constant .

We shall not show the cancellations here, but they are part of the general proof of renormalizability and simple examples of such cancellations have been given elsewhere(10). Indeed if one asks the question†: 'why is the present theory renormalizable while a theory with the fermion and gauge-field masses inserted by hand is not?' The answer is, partly because of the correlations between the parameters already mentioned, and partly because of the presence of the term (5.9).

It should perhaps be mentioned that in theories with axial currents $j_\mu^5 = \bar{\psi}\gamma_\mu\gamma_5\psi$ the cancellation is not quite complete because the classical Noether conservation law is replaced in the quantum case by the relation

$$\partial_\mu j_\mu^5 = \frac{e^2}{16\pi^2} \varepsilon_{\mu\nu\lambda\sigma} F_{\mu\nu} F_{\lambda\sigma} \quad , \tag{5.10}$$

where the $F_{\mu\nu}$ are the gauge-fields. (11) The contributions on the right-hand-side of (5.10) are called <u>anomalies</u> and in general they destroy the renormalizability. However, the sum of all such contributions may vanish for particular groups and representations and the conditions for their vanishing (which are purely algebraic) are given in reference (12). For the weak gauge-group SU(2)xU(1) to be discussed in the Part II the condition is simply that the electric charges of all the fermions must sum to zero.

† It is perhaps interesting to note that this question was asked independently by two participants (Pais and Sudarshan) after Salam's presentation of the theory at the Göteborg Nobel Symposium in 1968.

6. Renormalizability and Higgs-Kibble Mechanism

In the last lecture we have seen that the spontaneous symmetry breakdown supplies masses for the gauge-fields (and incidentally the fermion fields). But one has, of course, still to check whether the resulting theory is physically acceptable. There are a number of points to be checked. First is the crucial question as to whether the theory is renormalizable. Next is the question of what is to be done about the experimentally undesirable Goldstone fields. Third, there is the question of the negative metric introduced by the covariant canonical commutation relations

$$[A^a_\mu(x), A^b_\nu(y)] = i g_{\mu\nu}\, \delta_{ab}\, \delta^3(x-y) \quad , \tag{6.1}$$

for the gauge fields, since g_{00} is negative. For abelian massless gauge fields this problem disappears through the cancellation of the time-like and longitudinal contributions (Gupta-Bleuler mechanism[13]). But for massive gauge-fields such cancellations cannot take place because massive vector fields must have three physical components (three spin states).

To answer the renormalization question first, let me state at once that the theory is indeed renormalizable. Indeed that is why we are interested in it. For many years the renormalizability was in doubt because massive non-abelian vector meson theories are not in general renormalizable. But finally it was shown that the correlations discussed in the last section are so arranged that they eliminate the usual non-renormalizable divergences. We shall not consider the proof or renormalization here because it is a detailed technical subject which would require about six extra lectures, and because many excellent expositions are available in the literature [4]. However, we should perhaps stress the basic reason that the renormalization works (as is shown by proofs). The reason is that a spontaneous symmetry breakdown is such a 'soft' operation that it does not disturb the ultra-violet renormalizability of a theory. Hence if a theory is ultra-violet renormalizable before the spontaneous symmetry-breakdown (as an unbroken gauge theory is) it remains ultra-violet renormalizable after the breakdown. At the same time the infra-red problems of unbroken gauge-theory are removed by the introduction of masses by the spontaneous symmetry breakdown (provided that the residual symmetry group H is abelian). Thus spontaneous symmetry breaking has the magical property of preserving ultra-violet renormalizability while removing infra-red divergences.

But do we have to pay a price for this magic in the form of Goldstone scalars and indefinite metrics?

These are the other two problems which were to be considered, and perhaps the most beautiful part of the theory is that these two problems exactly eliminate one another! The mechanism by which they cancel is known as the Higgs-Kibble mechanism and it is analogous to the Gupta-Bleuler mechanism. What happens is that just as in the GB case there exists a gauge (transverse gauge) in which the timelike and longitudinal components of the massless gauge-fields cancel each other, in the HK case there exists a gauge in which the time-like components of the massive gauge fields and the Goldstone fields cancel each other. From the box of the previous lecture we recall that the number of Goldstone fields and massive gauge fields is exactly equal (dim G/H) so that a full cancellation can take place for any G and any H. Let us now consider the mechanism in more detail.

First let us consider the case of spontaneously broken scalar QED (super-conductivity). Since the scalar field ϕ is simply a complex number and the gauge-group just the group of phase transformations $\phi \to \phi \exp i\Lambda$, it is easy to see that if there is a spontaneous breakdown, with $|\overset{\circ}{\phi}| = c$, we can always choose $\overset{\circ}{\phi}$ real i.e. $\overset{\circ}{\phi} = c$ and then the Goldstone field will be just $\phi_2 = (\phi - \phi^*)/i\sqrt{2}$. Furthermore there will exist a gauge in which the Goldstone is zero. Thus for scalar QED the Goldstone field is not physical. But what about the time-like component of the gauge-field? For this field we consider the electromagnetic current, which is well-known to be

$$\overset{\circ}{j}_\mu(x) = ie\left(\phi\partial_\mu\phi^* - \phi^*\partial_\mu\phi\right) + e^2 A_\mu \phi^*\phi \quad . \tag{6.2}$$

Since $\overset{\circ}{j}_\mu$ is conserved as a consequence of the field equations, we have

$$\partial_\mu \overset{\circ}{j}_\mu = ie\left(\phi\Box\phi^* - \phi^*\Box\phi\right) + e^2\partial_\mu\left(A_\mu\phi^*\phi\right) = 0 \quad . \tag{6.3}$$

Expanding ϕ around $\overset{\circ}{\phi} = c$ we then have

$$e^2c^2(\partial_\mu A_\mu) = \sqrt{2}\, e\, \Box\phi_2 \quad + \quad \text{non-linear terms.} \tag{6.4}$$

The non-linear terms in (6.4) correspond to interactions and hence may be neglected for the asymptotic, or free, fields. Thus for the free fields we have

$$\partial_\mu A_\mu^f = \frac{\sqrt{2}}{ec^2} \Box \phi_2^f \quad . \tag{6.5}$$

It follows that in the gauge $\phi_2 = 0$ the divergence of A_μ vanishes. In particular, in the rest-frame of A_μ (recall that A_μ is massive after the spontaneous breakdown) we have $A_0 = 0$. We thus have the result: There exists a gauge in which the Goldstone field and the negative-metric component of the A-field vanish simultaneously. In this gauge there are only four fields, 3 space-like A-fields and 1 real ϕ-field, and all four fields are physical.

The problem now is to show that the corresponding result holds for the non-abelian case, and so the first thing we have to establish is that there exists a gauge in which the Goldstone fields vanish. For simplicity we shall assume that the group is irreducible and that the Higgs fields are real. Let $\mathring{\phi}$ be the potential minimum, and consider for each x the function

$$f(g) = (u(g)\mathring{\phi}, \theta) \quad , \tag{6.6}$$

over the group G. Since G is compact f(g) must take extremal values in the group and hence there exist elements, $\mathring{g}$ say, such that

$$\left(\frac{\partial f(g)}{\partial g}\right)_{g=\mathring{g}} = \left(W(\mathring{g})u(t_a)\mathring{\phi}, \theta\right) = \left(u(t_a)\mathring{\phi}, W^{+}(\mathring{g})\theta\right) \text{ where } \frac{\partial u(g)}{\partial \lambda_a} = W(g)u(t_a) \tag{6.7}$$

But since $u(t_a)\mathring{\phi}$ are the Goldstone directions in the group, we see that in the gauge where $\theta'(x) = (W^{+}(\mathring{g}(x)))\theta(x)$ the Goldstone fields vanish as required.

We next have to show that in the gauge $(u(t_a)\mathring{\phi}, \theta) = 0$ the negative-metric components of the free gauge-fields vanish. The argument is analogous to that for the abelian case. Consider the Noether currents

$$j_\mu^a(x) = e(\phi, t_a D_\mu \phi) = e^2(\phi, t_a t_b \phi)A_\mu^b + e(u(t_a)\phi, \partial_\mu \phi) \quad . \tag{6.8}$$

Since these are conserved as a consequence of the field equations, we have

$$\partial_\mu j^a_\mu(x) = \partial_\mu \left\{ (\phi, u(t_a t_b)\phi) A^b_\mu \right\} + \partial_\mu \left(u(t_a)\phi, \partial_\mu \phi \right) = 0 \quad , \tag{6.9}$$

and hence, if we expand ϕ around $\mathring{\phi}$, we obtain

$$e^2 M_{ab} \partial_\mu A^b_\mu = e \Box \left(u(t_a)\mathring{\phi}, \theta \right) \quad + \text{non-linear terms} \tag{6.10}$$

where M_{ab} is the mass-matrix (5.5). From (6.10) we see at once that the massive free-fields satisfy $\partial_\mu A_\mu = 0$ in the gauge where $(u(t_a)\mathring{\phi}, \theta) = 0$, and hence their negative metric component vanishes in that gauge. Note that the same is not true of the massless gauge fields, which drop out of (6.10). The gauge in which the Goldstone and negative-metric gauge-fields vanish simultaneously is called the physical or unitary gauge. If we use the transverse gauge for the massless gauge-fields all the fields appearing in this gauge are physical, and have a particle interpretation.

7. General Considerations.

As mentioned in the Introduction the consensus of opinion at present is that all the fundamental interactions are described by gauge-theory. The role of the gauge-group is different in each case and in many ways gravitation is the most exceptional case. In the gravitational case the gauge-group (the Lorentz group) is not even compact, and the A_μ and $F_{\mu\nu}$ are not primary but secondary objects, being derived from the metric in the form of Christoffel symbols and components of the Riemann tensor respectively. The Lagrangian is then linear, not quadratic, in the $F_{\mu\nu}$. Finally the force laws are universal and satisfy Einstein's equivalence principle. We shall not consider the case of gravitation in detail here as it has been discussed at length in the standard literature(2) on the subject. It might be worth mentioning, however, that the formulation of gravitational theory which comes closest to the spirit of gauge theory is that of Cartan(14). It should also be mentioned that there is, at present, no hint as to how gravitation fits in with the other three interactions.

In Part II, therefore, we shall consider the application of gauge theory to the other three interactions. Although the strong interaction gauge-group is the unbroken one, we shall follow the historical development and consider the weak and electromagnetic interactions first. The weak and electromagnetic interactions are considered together because it is now thought that these two interactions are generated by a single spontaneously broken gauge theory, the electromagnetic part corresponding to a residual, unbroken, U(1)-theory.

In considering the weak and electromagnetic interactions a central role will be played by the Noether matter currents, which, we recall enter into the field equations in the form

$$(D_\mu F_{\mu\nu})^a = J^a_\nu(\phi) + J^a_\nu(\psi) = (\phi, t_a D_\nu \phi) + \bar{\psi} t_a \gamma_\nu \psi \tag{7.1}$$

As remarked in lecture 1, Part I, the corresponding charges

$$Q^a = \int d^3x \; J^a_0(x) \quad , \tag{7.2}$$

are conserved and form a representation of the gauge algebra

$$[Q^a, Q^b] = f_{abc} Q^c \quad , \tag{7.3}$$

where f_{abc} are the structure constants. It was also remarked that if Q^a and Q^b are Noether charges, then, from (7.3), their commutator must be a Noether charge i.e. must be the space-integral of a Noether current as in (7.2). In adition, since commutators belong to the same irreducible part of the Lie algebra as the charges, it follows that Q^c must couple with the same strength as Q^a and Q^b (provided, of course, that the charges Q^a and Q^b lie completely in the irreducible part). This seemingly trivial remark is very useful to remember for model-building, and we shall refer to it as the Noether rule. Some of the most important experimental consequences of gauge theory, such as the existence of neutral currents and of charmed particles, are direct consequences of the Noether rule.

Another point worth stressing in connection with model-building is that since at present the only observable fields are the fermions (and the photon), models must be constructed along pre-Higgs lines. The Higgs field plays only the secondary role of providing gauge-field masses once the model is constructed. In fact, if one regards the basic spontaneously broken Lagrangian

$$-\mathcal{L} = \frac{1}{2}(D\phi)^2 + \bar{\psi}(\not{D}+m)\psi + \frac{1}{4}\mathrm{tr}\,F^2 + G\bar{\psi}\phi\psi + V(\phi) \, ,$$

it is important (and a little disturbing) to realize the the only non-QED part of it which is actually observed at present is the fermion part $\bar{\psi}(\not{\partial}+m)\psi$

Even this part is observed only indirectly by means of the four-fermi interaction mediated by the gauge fields. As in the case of general relativity, the acceptance of the theory is due as much to inner harmony as to experimental support.

8. Leptonic Weak and Electromagnetic Interactions.

The gauge-theoretical description of the weak and electromagnetic interactions originates in the discovery of the V-A structure of the weak interactions[15]. The analogy of the V-A structure with the vector structure of electromagnetism immediately suggests that the traditional local four-fermi interaction for weak decays is actually a non-local interaction mediated by a pair of charged vector mesons $W^{\pm}_{\mu}$. It looks local at low energies because the $W^{\pm}_{\mu}$ are massive. The local weak and electromagnetic interaction Lagrangian would then take the form

$$-\mathcal{L}_{int} = e_0\bar{\psi}\gamma_\mu\psi A^{el}_\mu + \frac{g_0}{\sqrt{2}}\bar{\psi}_L\gamma_\mu\psi_L W^+_\mu + h.c., \qquad \psi_L = \tfrac{1}{2}(1-\gamma_5)\psi \tag{8.1}$$

and A^{el}_μ and e_0 are the electromagnetic field and coupling constant respectively. In this picture the observed four-fermi interaction would come from a second-order Born graph[16] with $W^{\pm}_{\mu}$ intercharged between the fermions, and hence with an effective coupling constant

$$G = \frac{g_0^2}{2M_W^2}, \tag{8.2}$$

at low energies (the factor M_W^{-2} coming from the propagator $(k^2+M_W^2)^{-1}$ for $k^2 \ll M^2$.

The interaction (8.1) suggests, of course, some sort of universal coupling, but in order that the two (dimensionless) coupling constants e_0 and g_0 be of the same order of magnitude, the known experimental values of e_0 and G would require a very large value for M_W, something of the order of a hundred nuclear masses. While this number would certainly explain the apparent locality of the four-fermi interaction at low energies, it was thought at first to be too large to fit naturally into a nuclear theory.

Another problem with (8.1) was how to embed it in a Lagrangian with kinetic terms for the charged, massive gauge fields $W^{\pm}_{\mu}$. As we now know, gauge theory can provide the kinetic terms and spontaneous symmetry breaking can be used to make the $W^{\pm}_{\mu}$ massive. Hence the embedding reduces to a choice of gauge-group, and if representations for the matter fields. As mentioned in the previous section, the Higgs field plays only a secondary role in these choices and so the problem reduces to that of choosing the gauge group and fermion - representations for a standard gauge Lagrangian of the form

$$-\mathcal{L}_f = \frac{1}{4}\mathrm{tr}\, F^2 + \bar{\psi}(\not\partial + m)\psi + e_{(r)}\bar{\psi}\gamma^\mu t_a^{(r)}\psi A_\mu^{(r)a} \tag{8.3}$$

Of course, (8.3) should reduce to (8.1) on restriction to A_μ^{el} and $W_\mu^\pm$.

Optimally one should like to choose the gauge group and representations so that no fields other than A_μ^{el}, $W_\mu^\pm$ and the observed leptons (e, ν_e) or (μ, ν_μ) would appear in (8.3). But we can see at once from the Noether rule that this is not possible - the gauge structure is so tight that it necessarily implies the existence of new fields! To see this, we note that if there are only three gauge-fields A_μ^{el}, $W_\mu^\pm$, then there are only three Noether charges Q^{el}, $Q^\pm$. Furthermore, since $Q^\pm$ are charged they do not commute with the electric charge Q^{el} But the only compact three-dimensional non-commutative Lie algebra is SU(2). Thus the electric charge Q^{el} must be a generator of SU(2). But then in any representation its trace is zero and so the sum of the fermion charges must be zero also. Since this is manifestly not the case for the lepton pairs above, we see that we must introduce either new gauge-fields, or new leptons (or both).

Models with all three possibilities have been tried. In particular, when there are no new leptons, the charge

$$Q^3 = [Q^+, Q^-] \quad , \tag{8.4}$$

cannot be identified with the electric charge, for the reasons outlined in the previous paragraph. Then Q^3 is a new neutral charge and so (by the Noether rule) the theory must have at least one new neutral gauge field and one new neutral current. The discovery(17) of such a new neutral current in 1974 was the first experimental indication that gauge theories were on the right track.

The existence of the new neutral current favoured also the models without new leptons and recent experimental results indicate that, in fact, the simplest such model, namely one with a single new neutral charge Q^3, is the correct one. This model, with no new leptons and one new gauge-field, has become the standard model of the weak and electric interactions, and we now wish to describe it for the (e, ν_e)-sector. Since there are, by hypotesis, only four Noether charges Q^{el}, $Q^\pm$, Q^3, with two of them electrically charged (not commuting with Q^{el})

The only possible Lie algebra for the charges is SU(2) x U(1). Furthermore the $Q^{\pm}$ must belong to the SU(2) part, and Q^{el} must be some linear combination of Q^3 and the generator Q of U(1), with non-zero projection on Q because $Q^{el} \neq Q^3$. Thus the choice of G for the standard model is fixed and the only freedom left is the choice of representation for (e, ν_e). We now show that if we require (8.3) to contain (8.1) with $\psi = (e, \nu_e)$ then the choice of representation is also fixed, and so the model is completely determined.

First, since $W_\mu^{\pm}$ is an SU(2) vector and (e_L, ν_e) couples to $W_\mu^{\pm}$ while e_R does not, we see that (e_L, ν_e) must form an SU(2) doublet ψ_L, and e_R an SU(2) single ψ_R. Next since trace $Q^3 = 0$ and $Q^{el} \neq Q^3$ we have

$$\mathrm{tr}\, Q^{el} = k\, \mathrm{tr}\, Q \quad , \tag{8.5}$$

where k is a non-zero constant. But since Q takes a fixed value q for each irreducible representation of SU(2) we can apply (8.5) to ψ_L and ψ_R and then we obtain

$$2 q_L = q_R \tag{8.6}$$

This equation fixes the relative U(1) characters of ψ_L and ψ_R. (It is not necessary to fix the absolute character, but by suitably normalizing Q we can obtain the conventional normalization $q_L = -1$). We then have the following SU(2) x U(1) assignment for the fermions.

	SU(2)	U(1)
$\psi_L = (e_L, \nu_e)$	doublet	character q_L
$\psi_R = e_R$	singlet	character $2q_L$

Thus G and the fermion assigment are determined, as stated. According to the general principles of gauge theory the Lagrangian (8.3) then takes the unique form

$$-\mathcal{L} = \frac{1}{4} F_{\mu\nu} F_{\mu\nu} + \frac{1}{4} F^a_{\mu\nu} F^a_{\mu\nu} + \bar{\psi} \not{\partial} \psi + \frac{g}{2} \bar{\psi}_L \sigma_a \gamma_\mu \psi_L W^a_\mu - \frac{f}{2} (\bar{\psi}_L \gamma_\mu \psi_L + 2 \bar{\psi}_R \gamma_\mu \psi_R) W_\mu \tag{8.7}$$

where W^a_μ and $F_{\mu\nu}(W^a_\nu)$ belong to SU(2), W_μ and $F_{\mu\nu}(W_\nu)$ to U(1), σ_a are the Pauli-matrices, and the factor ($-\frac{1}{2}$) has been placed in front of the coupling constant f for later convenience. Note that there is no

SU(2) invariant mass-term for the fermions. The Lagrangian (8.7) is the Lagrangian of the standard model with the Higgs sector omitted, and as we have seen, we are led uniquely to it by the requirements that G be minimal, that there be no new leptons, and that (8.7) contain (8.1).

It should be mentioned that when confined to the lepton sector alone the standard model has the anomalies discussed in connection with renormalization. But this is not necessarily a drawback, because, as shown in the next section, the anomalies may disappear when the non-leptonic sector is added.

If we now replace the fields ψ_L and ψ_R in (8.7) by the more physical fields e and ν, we obtain for the interaction terms the expression

$$\frac{gf}{\sqrt{g^2+f^2}}\,\bar{e}\gamma_\mu e A^{el}_\mu + \frac{g}{2}(\bar{\nu}\gamma_\mu e_L W^+_\mu + h.c.) + \sqrt{g^2+f^2}\left\{\bar{\nu}\gamma_\mu\nu + \bar{e}\gamma_\mu\gamma_5 e + \frac{3f^2-g^2}{f^2+g^2}\,\bar{e}\gamma_\mu e\right\} Z^\circ_\mu$$

where

$$A^{el}_\mu = \frac{gW^3_\mu + fW_\mu}{\sqrt{g^2+f^2}} \quad \text{and} \quad Z^\circ_\mu = \frac{gW_\mu - fW^3_\mu}{\sqrt{g^2+f^2}} \qquad (8.8)$$

Comparing (8.8) with (8.1) we identify A^{el}_μ as the electromagnetic field and g and $gf/\sqrt{g^2+f^2}$ as the dimensionless weak and electromagnetic coupling constants g_0 and e_0 respectively. But we see that the gauge-invariance has induced an extra term in (8.8) with a new neutral gauge field Z°_μ coupled to a new neutral current J°_μ. It is the experimental discovery of J°_μ and the verification of its structure (polarization and parity violation), that has led to the acceptance of this model as the standard model.

9. Hadronic Weak and Electromagnetic Interactions.

If the leptonic theory of the previous section is to be universal, it must be extended to the strongly interacting fields also. But it turns out that the theory can not be extended to the strong-interacting fields without making a modification of the traditional (Gell-Mann SU(3)) theory of the strong internal symmetries[18]. We wish to describe here the necessary modification, which is known as the GIM mechanism. As we shall see, the modification leads immediately to a more universal picture of all the fundamental fermions. In this universal picture the lepton pairs (e,ν_e) (μ,ν_μ) ... are supplemented by quark pairs (u,d) (s,c) ... with each pair playing exactly the same role vis-a-vis the weak gauge group SU(2) x U(1). Why nature arranges the fermions in a series of "xerox" copies of (e , ν_e) remains, however, a mystery.

To consider the strongly-interacting field let us return to the fermion part of the traditional left-handed V -A weak current, and expand it in the leptonic and baryonic parts

$$\overset{\circ}{j}{}^{\pm}_{\mu} = \overset{\circ}{j}{}^{\pm}_{\mu}(\ell) + \overset{\circ}{j}{}^{\pm}_{\mu}(b) \quad . \tag{9.1}$$

If we let $Q^{\pm}$ (b) be the charges corresponding to $\overset{\circ}{j}{}^{\pm}_{\mu}(b)$ and Q^{el}_L and S_L be the left-handed projections of the electromagnetic and strange charges respectively, then, according to traditional SU(3) theory the normalizations and relative positions of all the charges in the fundamental representation of $SU(3)_L$ are given by

$$Q^{el}_L = \frac{1}{3}\begin{pmatrix} 1 & & \\ & 1 & \\ & & -2 \end{pmatrix} \qquad S_L = \begin{pmatrix} 1 & & \\ & 0 & \\ & & 0 \end{pmatrix} \qquad Q^{+}_L = \begin{pmatrix} 0 & 0 & \sin\theta_c \\ 0 & 0 & \cos\theta_c \\ 0 & 0 & 0 \end{pmatrix} \tag{9.2}$$

where θ_c is the Cabbibo angle ($\simeq$ 0.25). But if the SU(2) x U(1) leptonic gauge theory is to be extended to the strongly interacting fields, the Noether rule tells us that the charge

$$Q^{3}(b) = \left[Q^{+}_L(b) \, , \, Q^{-}_L(b) \right] \qquad Q^{-} = \left(Q^{+}\right)^{\dagger} \; ,$$

must be associated with a current $\overset{\circ}{j}{}^{3}_{\mu}$ (b) which couples to a gauge field (in fact to the gauge-field W_μ of the previous section) with strength comparable to the coupling of $\overset{\circ}{j}{}^{\pm}_{\mu}$ (b) to $W^{\pm}_{\mu}$. From (9.2) we see that the place of Q^{3}_L in the fundamental representation is

$$Q^{3}_L = \begin{bmatrix} \sin^2\theta_c & \cos\theta_c \sin\theta_c & 0 \\ \cos\theta_c \sin\theta_c & \cos^2\theta_c & 0 \\ 0 & 0 & -1 \end{bmatrix} \tag{9.3}$$

Since Q^{3}_L then commutes with Q^{el}_L , but not with S_L , we see that will be a strangeness-changing neutral current. But it has been known for a long time experimentally that contributions from such currents, if they exist, are suppressed by a factor of at least 10^{-5} compared to contributions from the charged currents [15]. Since such a tiny factor cannot be explained by the smallness of the Cabbibo angle, we are left with a dilemma: either the leptonic gauge theory SU(2) x U(1) cannot be extended to the baryons or the traditional

Gell-Mann-Cabbibo theory must be modified!

It turns out that there is a simple modification of the GMB theory, namely, the enlargement of SU(3) to U(2n), that allows the gauge theory SU(2) x U(1) to be extended to the baryons, and to be extended in a smooth and elegant manner (GIM mechanism). Furthermore, the new fields that are necessitated by the enlargement (the so-called charmed fields) have been found to exist experimentally. Let us now consider the modification of the GMB theory in detail.

The basic assumption is that the enlarged group is unitary and that there are no exotically charged fields ($Q^{el} \neq 0, \pm 1$) in the adjoint representation (in agreement with experiment). Since charges are additive, there can then be only two values of charge in the fundamental representation and these values must differ by an integer. Thus Q_L^{el} must take the form

$$Q_L^{el} = \begin{bmatrix} eI_1 & 0 \\ 0 & (e-1)I_2 \end{bmatrix} , \qquad (9.4)$$

in the fundamental representation where I_1 and I_2 are unit matrices (not necessarily of the same dimension). Since strangeness S commutes with electric charge, and $Q_L^{\pm}$ have the usual commutation relations with Q_L^{el}, we see that S_L and $Q_L^{\pm}$ must then have the form

$$S_L = \begin{bmatrix} S_1 & 0 \\ 0 & S_2 \end{bmatrix} , \quad Q_L^{+} = \begin{bmatrix} 0 & C \\ 0 & 0 \end{bmatrix} , \quad Q_L^{-} = \left(Q_L^{+}\right)^{\dagger} , \qquad (9.5)$$

in the fundamental representation, where the blocks are of the same dimension as in (9.4), and S_1, S_2 and C are arbitrary (except that S must be hermitian). Then Q_L^3 takes the form

$$Q_L^3 = \left[Q_L^{+} , Q_L^{-}\right] = \begin{bmatrix} C^{\dagger}C & 0 \\ 0 & -CC^{\dagger} \end{bmatrix} . \qquad (9.6)$$

and the neutral current commutes with strangeness provided that

$$\left[S_1 , C^{\dagger}C\right] = \left[S_2 , CC^{\dagger}\right] = 0 . \qquad (9.7)$$

Suppose now that the $Q_L^{el}, Q_L^{\pm}, Q_L^3$ belong to the standard gauge-group SU(2) x U(1). Then the commutator of Q_L^3

with Q_L^+, Q_L^- must close to form an SU(2) algebra, and we have

$$[Q_L^3, Q_L^+] = Q_L^+ \quad \text{or} \quad CC^+C = C \tag{9.8}$$

But then C^+C and CC^+ are projection operators. If we diagonalize them the zero projection has only the trivial algebra $Q_L^+ = Q_R^+ = 0$. Hence if we discard the zero projection, and use $\mathrm{tr}\, C^+C = \mathrm{tr}\, CC^+$ we see that the blocks in (9.4) are of equal dimension, that $CC^+ = C^+C = 1$ on each block, and that (9.7) is automatically satisfied. Thus, assuming only the (experimental) result that there are no exotic charges, and that the neutral current commutes with strangeness, we find that the standard gauge-group SU(2) x U(1) can be embedded in the strong symmetry group according to the scheme

$$Q_L^{el} = \begin{pmatrix} e & 0 \\ 0 & e-1 \end{pmatrix}, \quad Q_L^3 = \begin{pmatrix} 1 & 0 \\ 0 & -1 \end{pmatrix}, \quad Q_L^+ = \begin{pmatrix} 0 & C \\ 0 & 0 \end{pmatrix} \quad \text{where } C^+C = CC^+ = 1 \tag{9.9}$$

and that this is the only possible embedding.

Note that the charges in (9.9) actually lie within an even dimensional subgroup U(2n) of the unitary group. The traditional SU(3) group does not work because for SU(3) the U(2n) would have to be U(2), and that would force $Q_L^{el} = Q_L^3$, or zero Cabbibo angle. The next simplest case is U(2n) = U(4), which contains both traditional SU(3) and (9.9) with

$$C = \begin{pmatrix} \sin\theta_c & -\cos\theta_c \\ \cos\theta_c & \sin\theta_c \end{pmatrix} \tag{9.10}$$

This case works for all the weak interactions, except the one involving CP violation[14]. It does not work for CP-violation because C in (9.10) is a real matrix, and CP-violation requires complex charges. However, all the higher cases SU(2n), $n \geq 3$, allow CP-violation because if C is n x n for $n \geq 3$ it contains phases which cannot be removed by a choice of basis. For example, for U(6) there is one such phase and the conventional form of C in (9.9) is

$$C = \begin{bmatrix} c_1 & -s_1 s_2 & s_1 c_2 \\ -s_1 s_3 & -s_2 s_3 c_1 + c_2 c_3 e^{i\beta} & s_3 c_1 c_2 + s_2 c_3 e^{i\beta} \\ s_1 c_3 & s_2 c_1 c_3 + s_3 c_2 e^{i\beta} & -c_1 c_2 c_3 + s_2 s_3 e^{i\beta} \end{bmatrix}, \quad c_a = \cos\theta_a, \quad s_a = \sin\theta_a \tag{9.11}$$

Note that in general the embedding of the algebra (9.9) in U(2n) does not commute with strangeness (hence the off-diagonal form of C in a strangeness-basis). If we ignore strangeness (and other SU(2n) quantum numbers) then (9.9) can be decomposed within U(2n) into a direct sum of n copies of the 2-dimensional representation of SU(2) x U(1). Then the fundamental representation space of U(2n) decomposes into n SU(2) x U(1) doublets and if we identify these doublets as quarks we obtain the Xerox picture of the fundamental fermions mentioned earlier, that is the fundamental fermions consisting of lepton pairs (e, ν_e), (μ, ν_μ) ,,, and quark pairs (u, d), (s, c). Note, however, that the quark pairs are massive and hence have two right-handed components, both of which are SU(2) x U(1) singlets. Furthermore, since strangeness is not diagonal in the SU(2) x U(1) basis, the pairs (u,d) etc. do not have their masses diagonal. To obtain the 'physical' quarks with diagonal masses we must rotate so that the matrix C takes the form (9.10) (9.11) etc.

The upshot of the discussion, however, is that the Lagrangian

$$-\mathcal{L} = \frac{1}{4}F^2 + \frac{1}{4}F^aF^a + \bar{\psi}\not{\partial}\psi + \frac{g}{2}\bar{\psi}_L\sigma_a\psi_L W^a_\mu - \frac{f}{2}(\bar{\psi}_L\gamma_\mu\psi_L + 2\bar{\psi}_R\gamma_\mu\psi_R)W_\mu \tag{9.12}$$

which is simply the extension of (8.7) to all fundamental fermion fields, forms a universal Lagrangian for the weak and electromagnetic interactions (apart from the Higgs sector, to be discussed in the next lecture). An added bonus is that the anomalies mentioned earlier cancel if the number of lepton and quark pairs are the same[18].

The group SU(2n) x SU(2n) obtained by adding to the SU(2n) just discussed, its right-handed counter part, is often called the chiral flavor group, and the (parity-preserving) diagonal SU(2n) the flavor group. The latter group is the generalization of Gell-Mann's SU(3) and is used to classify particles. The fact that the gauge-group SU(2) x U(1) is a subgroup of the chiral flavor group suggests, of course, that the latter group should be gauged, but at present there is no experimental evidence either for or against this suggestion.

10. Higgs Sector and Experimental Tests of the Neutral Current.

So far we have not considered the Higgs field because, as mentioned earlier, the model building proceeds along pre-Higgs lines, and the representation of the Higgs field is afterwards chosen to provide suitable masses. For the standard model the Higgs representation is largely determined by the fact that since there are no group invariant fermion masses, the observed masses must be generated by the spontaneous symmetry breakdown.

For that to happen there must exist a non-trivial Yukawa interaction, and since all fermions have the ψ_L -doublet- ψ_R -singlet assignment, this is possible only if the Higgs field has an irreducible component which is an SU(2) doublet. In particular, if the representation of the Higgs field is irreducible, it must be the doublet, and this is what is assumed in the standard model. It is then easy to see that, if the Yukawa interaction is to conserve electric charge, the Higgs field must have the U(1) character - q_L .

With these assignments for the Higgs field the Higgs sector of the Lagrangian (9.12) must take the form

$$-\mathcal{L}_\phi = \left[\left(\partial_\mu + \tfrac{1}{2} g W_\mu^a \sigma_a - \tfrac{1}{2} f w_\mu\right)\phi\right]^2 + G(\bar{\psi}_L, \phi)\psi_R^{(1)} + (\bar{\psi}_L, c\phi^*)\psi_R^{(2)} + V(|\phi|^2), \tag{10.1}$$

where V is quadratic in $|\phi|^2$ and $c_{ab} = \varepsilon_{ab}$. Since any constant SU(2) 2-spinor can be transformed into a given 2-spinor $\mathring{\phi}$ by an SU(2) x U(1) transformation, and $\mathring{\phi}$ will be U(1)-invariant, there is only one orbit possible for the spontaneous breakdown. After the breakdown the gauge-field and fermion mass-terms will take forms

$$\left[\left(\tfrac{1}{2} g W_\mu^a \sigma_a - \tfrac{1}{2} f w_\mu\right)\mathring{\phi}\right]^2 \qquad \text{and} \qquad \bar{\psi}\begin{pmatrix} G & 0 \\ 0 & F \end{pmatrix}\psi , \tag{10.2}$$

respectively.

Although in principle the choice of $\mathring{\phi}$ is free, we have already fixed some conventions by the identification of the fields e , ν_e and A_μ^{el} in (8.8). If we require that A_μ^{el} be massless so that it can be identified with the electromagnetic field, then $\mathring{\phi}$ should not have an interaction with A_μ^{el} in (10.1), i.e. should be neutral. From (10.1) we see that the necessary and sufficient condition the neutrality of $\mathring{\phi}$ is

$$\sigma_3 \mathring{\phi} = \mathring{\phi} . \tag{10.3}$$

Then (10.2) reduces to

$$g^2 c^2 (W_\mu^\pm)^2 + (g^2 + f^2) c^2 (Z_\mu^0)^2 + Gc(\bar{e}e) + (G_1 c\bar{u}u + F_1 c\bar{d}d) + \ldots \tag{10.4}$$

where $W_\mu^\pm$ and Z_μ° are the fields defined in (8.8). Notice that the neutrino does not acquire a mass-term because ϕ° is neutral. One of the most important consequences of (10.4) is that the ratio of the gauge-field masses

$$\frac{M_W^2}{M_Z^2} = \frac{g^2}{g^2 + f^2} , \tag{10.5}$$

is independent of c . However, it should be emphasized that the relation (10.5) is a consequence of the assumption that the representation of the Higgs field is irreducible. If the representation is only a reducible one, which contains a doublet, then (10.5) no longer holds.

The mass of the Higgs field itself is not fixed by the symmetry breaking since

$$V = \lambda\left(|\phi|^2 - c^2\right)^2 \longrightarrow 4\lambda c^2 (R\ell\theta)^2 + 4\lambda c (R\ell\theta)|\theta|^2 + \lambda|\theta|^4 , \tag{10.6}$$

where λ is arbitrary, but it is worth noting that once λ and c are fixed, then so are the interactions in (10.6). In particular if $\lambda \ll 1$ so that perturbation theory is valid for V, then $m_\theta \ll m_W/g$.

The discussion of the Higgs sector completes the description of the standard model and we turn now to a brief discussion of the experimental verification. For this purpose we first consider how many parameters are involved. By inspection of (8.7) (10.1) (10.2) and (10.6) we see that in fact there are only four fundamental parameters, namely, f, g, c and λ. If we disregard λ which occurs only in the (presently unobserved) Higgs potential, there are only three fundamental parameters, namely f, g and c. Furthermore, since $e_0 = gf/\sqrt{g^2+f^2}$ and $G = g^2/2m_W^2 = c^2/2$. two of these parameters are equivalent to the traditional electromagnetic and fermi coupling constants. Thus, apart from λ, the theory contains only one new parameter. (Note, however, that without (10.5), which follows from the irreducibility of the Higgs field, there would be two new parameters). There are various choices of the independent parameters, for example (g, f, M_W) or (e_0, G, M_W), but the most popular choice seems to be e_0, G and $\tan\theta = f/g$.

On account of the small number of parameters, the predictions of the standard model are very strong. Unfortunately, only a small sector of these predictions is available to present-day experiments. This sector is the gauge-field fermion interaction (8.8), in which the new part is the neutral current term

$$j^{\circ}_{\mu} Z^{\circ}_{\mu} = \sqrt{g^2 + f^2} \left\{ \bar{\nu} \gamma_{\mu} \nu + \bar{e} \gamma_{\mu} \gamma_5 e + \left(\frac{3f^2 - g^2}{f^2 + g^2} \right) \bar{e} \gamma_{\mu} e \right\} Z^{\circ}_{\mu} \tag{10.7}$$

Thus all the experiments focus on this term, or rather on the four-fermi interaction

$$j^{\circ}_{\mu} j^{\circ}_{\mu} / M_Z^2 \quad , \tag{10.8}$$

which is produced by the neutral current in second order of perturbation. Thus what the present experiments really test is the structure of the new neutral current. They do not directly test the gauge-structure of the theory, much less the Higgs sector and the spontaneous symmetry breakdown. However, the motivation for the structure of j°_{μ} comes from the general principles of gauge theory, and the correlation (10.5) comes from the spontaneous symmetry breakdown, so there is an indirect test of the gauge-structure and the spontaneous symmetry breakdown.

Considered as a test of the neutral-current structure, the experimental results have turned out to be quite spectacular(19). There are a total of about six independent experiments that can be carried out, and all agree (10.7) and (10.8). They give a common value of 30° for the angle θ . One sees by inspection of (10.7) (10.8) that the possibilities for testing are

(i) neutrino scattering on leptons

(ii) neutrino scattering on nucleons (considered as composites of quarks)

(iii) corrections to ordinary electromagnetic effects due to Z°_{μ} exchange.

The Z°_{μ} correlations to electromagnetism become detectable in spite of the heavy mass of the Z°_{μ} because they are parity violating. The neutrino-lepton scattering experiments confirm e-μ universality, and determine the value of tan θ , leaving no free parameters for experiments (ii) and (iii). There are four independent neutrino-nucleon scattering experiments (2 polarizations x two nucleons), and the experimental results for all four are is excellent agreement with the theory. Finally the contributions to the electromagnetic current have been measured at both low and high energies and again the agreement is excellent. The high energy experiment (and SLAC) is particularly impressive, because it is clean both theoretically (no nuclear corrections) and experimentally, and this clean experiment gives the best agreement with the theory.

11. The Strong Interactions and their Rigid Symmetry Groups.

The traditional view of the strong interaction is that they are a simple extension of nuclear physics, and are therefore due to Yukawa interactions between baryon and pseudo-scalar mesons e.g. interactions of the form $\bar{N}\gamma_5 N\pi$, $N\gamma_5\Sigma K$. Although such interactions would seem to have little to do with gauge theories, the view that the strong interactions are fundamentally gauge-interactions (with a non-abelian gauge group) has been steadily gaining ground. According to this view the observed strong interactions between baryons are phenomenological in much the same way that molecular forces are phenomenological, the basic interactions in that case being electromagnetic. This is certainly a revolutionary point of view and it originated in a dramatic discovery made in the early seventies concerning the scaling behaviour of non-abelian gauge-theories. This behaviour is known as asymptotic freedom.

In order to discuss asymptotic freedom we need to digress a little and consider two other concepts, namely the symmetry groups of the strong interactions and the renormalization group of quantum field theory. In this lecture we shall consider these two concepts.

First, with regard to the strong symmetry groups the best known are the isospin group SU(2) and Gell-Mann's group SU(3). But as we have seen in the section 9, these groups have had to be enlarged to the (broken) flavor group $G_f = \mathcal{U}(2n)$, $n \geq 2$. The flavor group is used to classify the observed hadrons (measons and baryons such as π, K and N, Σ). However, the hadrons do not belong to the fundamental representations of G_f , and it turns out that their spectroscopy is enormously simplified if they are assumed to be composite fields made up of quarks[20], which are spin $\frac{1}{2}$ fields belonging to the fundamental representation of G_f . The baryons are assumed to transform like qqq and the measons like $\bar{q}q$.

Although the quark hypothesis works very well phenomenologically, it runs into some immediate theoretical difficulties. First, the hadron spectroscopy requires that the quarks have the 'wrong' statistics (be spin $\frac{1}{2}$ bosons). Second, quarks have never been observed experimentally. To explain their non-observance it is usually assumed that the quarks are 'confined' i.e. bound together by forces which increase with distance. There is, as yet, no proper dynamical justification for this assumption, but there are a number of qualitative indications. The strongest indications come, as we shall see, precisely from the assumption that the strong interactions are gauge-interactions.

The statistical problem for the quarks is resolved by assuming that they possess a further internal quantum number, called colour. A further internal symmetry group, the colour group $\mathcal{G}_c$, is introduced[21] to accommodate the new quantum number, and, unlike $\mathcal{G}_f$, the group is assumed to be unbroken. Since quarks must come in at least three colours in order to correct the statistics (and at present) need not come in more than three colours, the group $\mathcal{G}_c$ is usually taken to be an SU(3) group (called $SU(3)_c$ to distinguish it from Gell-Mann's SU(3)). Apart from the statistics there are two pieces of experimental evidence for the existence of colour (indeed for just three colours). These concern $\pi_0 \to 2\gamma$ decay, and inelastic e^+e^--scattering[22]. Without colour the experiments and theory would differ by a factor of about three.

The upshot of all this discussion is that at present the strong-interaction symmetry group is assumed to be of the direct-product form

$$\mathcal{G} = \mathcal{G}_c \times \mathcal{G}_f \quad ,$$

where $\mathcal{G}_f$, the flavour group, is probably U(n) for $n \geq 6$ and $\mathcal{G}_c$, the colour group, is probably SU(3). The flavour group is broken, contains Gell-Mann's SU(3) and isospin, and contains the weak-gauge-group SU(2) x U(1) in the lefthanded part of its chiral extension $\mathcal{G}_f \times \mathcal{G}_f$. The colour group, in contrast, is unbroken. The observed hadrons are $\mathcal{G}_c$ singlets and belong to non-fundamental representations of $\mathcal{G}_f$. The quarks belong to the fundamental representations of both $\mathcal{G}_c$ and $\mathcal{G}_f$.

The second concept that we need to consider before going on to asymptotic freedom is that of the renormalization group[9] [23]. This concept originates in the behaviour of renormalized quantities such as coupling constants and Green's functions with respect to changes of scale in momentum space. In classical physics there is no problem because dimensionless coupling constants remain unchanged and if $\mathcal{G}^{el}(p,g)$ is a Green's function with generic coupling constants g and generic momenta (including masses) p, then the scaling relation is simply

$$\mathcal{G}^{el}(sp,g) = s^d \mathcal{G}^{el}(p,g) \qquad (11.1)$$

where d is the dimension of G. But in quantum theory the situation is more subtle because of renormalization, and the analogue[24] of (11.1) is

$$G^{q\mu}(sp,g) = s^{d-\delta(s)} G^{q\mu}(p, g(s)) \quad , \tag{11.2}$$

where the functions δ (s) (the anomalous dimension and g(s) satisfy the the differential equations

$$s\frac{d}{ds}\delta(s) = \gamma(g(s)) \quad , \quad s\frac{d}{ds}g(s) = \beta(g(s)) \quad . \tag{11.3}$$

In these equations the input functions γ and β are obtained by log-differentiating the renormalization constants with respect to the cut-off Λ, expressing the result in terms of the renormalized quantities and then taking the limit $\Lambda \to \infty$. Thus

$$\gamma(g) = \mathcal{L}t_{\Lambda\to\infty} \Lambda\frac{d}{d\Lambda} z(g_0,\Lambda) \quad , \quad \beta(g) = \mathcal{L}t_{\Lambda\to\infty} \Lambda\frac{d}{d\Lambda} g(g_0,\Lambda) \quad . \tag{11.4}$$

Of course, it is not obvious that the indicated limits will exist, but the point is that they will exist if the theory is renormalizable. Eqs. (11.2) (11.3) (11.4) are known as the renormalization group equations.

Now once the coupling constant g becomes s-dependant one can enquire whether $g(s) \to 0$ as s approaches any limit, s_0 say. If so the theory becomes a free theory at $s = s_0$, and near s_0 perturbation theory becomes valid. The nice feature is that one can check for consistency because, if perturbation theory is valid, then β (g) in (11.4) can be computed perturbatively and the result inserted in (11.3) to see of dg/ds is negative.

In particular, if $s_0 = \infty$ (g (s)$\to 0$ as $s \to \infty$) then the coupling constant decreases with increasing scale of the momentum, and the theory is said to be asymptotically free. Now for QED, Yukawa and $\lambda\phi^4$ interactions the perturbative (one-loop) result for $\beta(g)$ has been known for some time[25] and is

$$\beta(e) = \frac{1}{12\pi^2}e^3 \quad , \quad \beta(G) = \frac{1}{16\pi^2}G^3 \quad , \quad \beta(\lambda) = \frac{1}{(16\pi)^2}\lambda^2 \quad , \tag{11.5}$$

respectively. In each of these cases $\beta(g)$ is positive, so from (11.3) we see that QED, Yukawa and $\lambda\phi^4$ interactions are not asymptotically free.

The dramatic discovery referred to earlier is that, in contrast to QED, Yukawa and $\lambda\phi^4$, non-abelian gauge theories are asymtotically free. More precisely, if we compute $\beta_5(g)$ to lowest-order for a non-abelian theory we obtain(5)

$$\beta(g) = \frac{-g^3}{16\pi^2} C_2 \left\{ \frac{11}{3} - \frac{2}{3} n_f W_f - \frac{1}{6} n_s W_s \right\} \quad \text{where} \quad , W = \frac{\mathrm{tr}\, C}{\mathrm{tr}\, C_2} \tag{11.6}$$

C_2 the Casimir invariant for the adjoint representation, C the Casimirs for the fermion and scalar representations, and n_f and n_s the numbers of such representations. Hence if the numbers of fermions and scalars is not too large the coupling constant g(s) is asymtotically free. In particular a pure (non-abelian) gauge-theory is asymptotically free. From (11.3) we then have

$$g(s) \sim (\ln s)^{-1} \qquad s \to \infty \quad , \tag{11.7}$$

The reason that the fermions and scalars contribute a "wrong" sign in (11.6) is that their interaction with the gauge fields is the same as in QED. The contribution with the "correct" sign comes from the interaction of the gauge-field with itself. With hindsight, one can produce more or less plausible physical arguments as to why the self-interaction of non-abelian fields should have this property(26).

Finally it should be mentioned that if scalar fields are present the coupling constant λ for the self-interaction of these fields must also be considered. In general, on account of (11.5), λ will not be asymptotically free (though for small numbers of scalars in special representations the gauge-field contribution may reverse the sign of $\beta(\lambda)$). Thus in general spontaneously broken gauge theories will not be asymptotically free.

12. Gauge Theory of the Strong Interactions.

The two discoveries discussed in the last lecture, namely, the existence of a non-abelian colour group $\mathcal{G}_c$ and the asymptotic freedom of such groups, have led to a radical change in our concept of the strong interactions. The belief now is that the strong-interactions are nothing but the interactions obtained by gauging the (unbroken) colour group $\mathcal{G}_c$, and thus are nothing but a non-abelian (probably SU(3)) generalization of the electromagnetic interactions. (The theory is called QCD or chromodynamics in analogy to QED). Although there is no firm quantitative support for the theory at present, the qualitative evidence is so persuasive that it has become rapidly accepted. We shall now sketch the qualitative advantages of such theory, and some experimental evidence in its favour.

The first great advantage, of course, is that the strong interactions come into line with the other three fundamental interactions which are already gauge theories. Furthermore, the scheme gives a dynamical meaning to the colour group, which hitherto was introduced mainly for statistical reasons. Of course, the gauging of the colour group introduces a new set of vector fields (gluons) and since these are not colour singlets they remain unobserved (confined) just like the quarks.

At first sight, the introduction of yet another set of confined fields would appear to be a disadvantage, but the second great feature of the strong-interaction gauge theory is that it provides a qualitative explanation of confinement. The argument is that the limit $g(s)\to 0$ as $s\to\infty$ (asymptotic freedom) should, by (11.3), imply also the converse limit $g(s)\to\infty$ as $s\to 0$. This limit means that the strength of the coupling constant increases with increasing momentum, and hence with increasing distance. In that case, the further the quarks or gluons are pulled apart the stronger is the force trying to pull them together. Hence (unless the fields are colour singlets and do not experience these forces) they stand a good chance of being confined.

A third qualitative advantage of the strong gauge-theory is that it resolves an old paradox which arises from the fact that quarks (which are confined and therefore tightly bound) behave as if they were free particles in high energy collisions. The free behaviour manifests itself in the classical (see (11.1)) scaling of the Green's functions[27]. Now if we insert $g(s)\to(\ln s)^{-1}$ into (11.3) with $\gamma(g)\sim g^2$ we find that the anomalous dimension is $\delta(s)\to\ln\ln s$ and hence $d+\delta(s)$ is practically constant as $s\to\infty$, and we have essentially classical scaling. Thus the strong interaction gauge-theory not only allows confinement and the free quark (scaling) behaviour to coexist, but actually predicts their coexistence, since it implies $g(s)\to\infty$ as $s\to 0$, and simultaneously $\delta(s)\to\ln\ln s$ as $s\to\infty$.

What about quantitative evidence? Naturally, the search for such evidence has been a major field of activity in recent years. The problem is difficult, because the evidence is necessarily indirect-like the evidence for atomic QED in molecular experiments. Two main lines of research may be mentioned. First are the QCD corrections to 'naive' quark model computations. These corrections are calculated in exactly the same way as QED corrections[26], since the gauge-field-fermion interactions does not exhibit any non-abelian characteristics. The non-abelian character of the theory is put in phenomenologically in the form of quark-confinement and the value of the coupling constant g(s). The second line of research concerns the so-called 'jets' of particles emerging from inelastic collisions. If the gluons really exist, then the particles emerging from strong-interactions collisions should not emerge isotropically but in narrow jets. The dominant effect would be two jets in opposite directions, but secondary three-jet effects should also be detectable. The experimental evidence at present would appear to favour the existence of such jets[28].

Finally there is a curious piece of numerology that might be mentioned. It turns out that the value of the scale parameter s which gives the best fit for g(s) is approximately half a nucleon mass. At the same time, if one thinks of the hadrons as composites of quark, one finds that the hadron masses are not equal to the sum of the constituent quark masses but exceed it by a universal mass

$$\text{Hadron mass} = M_0 + \text{quark-masses}, \tag{12.1}$$

and the mass M_0 also turns out to be about half a nucleon mass. The mass M_0 is assumed to come from a renormalization effect, and so there seems to be some universal mass of order M_0 associated with renormalization.

Once the SU(3)-colour and SU(2) x U(1) groups are gauged, the idea of unifying the weak, electromagnetic and strong interactions into a single gauge scheme immediately suggests itself. Many models have been proposed, the prototype[29] using the group SU(5), which is the minimal group containing SU(2) x U(1) x SU(3). Since the quarks and leptons couple through the gauge-fields, such schemes necessarily imply the possibility of quark -lepton and hence proton - lepton decay[30]. The apparent stability of the proton would then be due only to the large mass of the mediating gauge-fields.

REFERENCES

(1) H. Weyl, Z. Phys. 56, 330 (1929)

(2) C. Misner, K. Thorne, J. Wheeler, Gravitation (Freeman, New York 1973)

(3) J.C. Taylor, Gauge Theories of Weak Interactions (Cambridge Univ. Press 1976)

(4) G. 't Hooft, Nucl. Phys. B33, 173, B35, 167 (1971)
B. Lee, J. Zinn-Justin, Phys. Rev. D5, 3121 (1972)
D7, 1049 (1973)
E. Abers, B. Lee, Phys. Reports 9C, 2 (1973)
C. Becchi, A. Rouet, R. Stora, Phys. Lett. 52B, 344 (1974)
Comm. Math. Phys. 42, 127 (1975)

(5) D. Politzer, Phys. Reports 14C, 129 (1974)

(6) R. Gilmore, Lie Algebras and Applications (Wiley, New York 1974)

(7) E. Noether, Nachr. Kgl. Wiss. Göttingen, 235 (1918)

(8) G. Goldstone, A. Salam, S. Weinberg, Phys. Rev. 127, 965 (1962)

(9) D. Amit, Field Theory, Renormalization Group and Critical Phenomena (McGraw-Hill, New York 1978)

(10) M. Beg, A. Sirlin, Ann. Rev. Nucl. Sci. 24, 379 (1974)
L. O'Raifeartaigh, Rep. on Prog. in Phys. 42, 159 (1979)

(11) S. Adler, Phys. 140B, 736 (1965) 177B 2426 (1969)
J. Bell, R. Jackiw, Nuovo Cim. 60A 47 (1969)

(12) D. Gross, R. Jackiw, Phys. Rev. D6 477 (1972)

(13) S. Gupta, Proc. Phys. Soc. A63 681 (1950)
K. Bleuler, Helv. Phys. Acta, 23 567 (1950)

(14) E. Cartan, Leçons sur la Theorie des Spineurs (Hermann, Paris 1938)

(15) R. Marshak, C. Ryan and Riazzudin, Theory of Weak Interactions (Wiley, New York 1969)
E. Commins, Weak Interactions (McGraw-Hill, New York, 1973)

(16) S. Schweber, Introduction to Rel. Quantum Field Theory (Row-Peterson, New York 1961)

(17) F. Hasert et al. Phys. Lett. 46B, 138 (1973)
T. Donnelly, R. Peccei, Phys. Reports 50, 1 (1979)

(18) M. Gourdin, Unitary Symmetry (North-Holland, Amsterdam 1967)
P. Carruthers, Introduction to Unitary Symmetry (Interscience, New York 1966)

(19) C. Baltay (Experiment), S. Weinberg (Theory) Proc. 19th (Tokyo) International Conference on High Energy Physics (Phys. Soc. Japan, Tokyo 1979)

(20) J. Kokkedee, The Quark Model (Benjamin, New York 1969)
F. Close, Introduction to Quarks and Partons (Academic Press, New York 1979)

(21) O. Greenberg, C. Nelson, Phys. Reports 32C, 69 (1977)
H. Fritsch, M. Gell-Mann, H.Leutwyler Phys. Lett. 47B, 365 (1973)
M. Gell-Mann, P. Ramond, R. Slansky, Rev. Mod. Phys. 50, 721 (1978)

(22) S. Glashow, J. Iliopolous, L. Maiani, Phys. Rev. D2, 1285 (1970)
M. Gaillard et al. Rev. Mod. Phys. 47, 277 (1975)

(23) E. Stueckelberg, A. Peterman, Helv. Phys. Acta 26, 499 (1953)
M. Gell-Mann, F. Low, Phys. Rev. 95, 1300 (1954)
C. Callan, Phys. Rev. D2, 1541 (1970)
K. Symanzik, Comm. Math. Phys. 18, 227 (1970)
K. Wilson, Phys. Rev. 179, 1499 (1969)

(24) S. Weinberg, Phys. Rev. D8, 3497 (1973)

(25) D. Gross, F. Wilczek, Phys. Rev. D8, 3633 (1973)
S. Coleman, D. Gross, Phys. Rev. Lett. 31, 851 (1973)

(26) R. Field in Quantum Chromodynamics (La Jolla Workshop 1978, eds. Frazer et al. Amer. Inst. Phys. Conf. Proceedings no. 55, New York 1979)

(27) A. Buras, Rev. Mod. Phys. 52, 199 (1980)
A. Peterman, Phys. Reports 53C, 158 (1979)
G. Fox, Nucl. Phys. B131, 107 (1977)
W. Bardeen et al. Scale and Conformal Symmetry in Hadron Physics (Wiley, New York 1973)

(28) J. Ellis, M. Gaillard, G. Ross, Nucl. Phys. B111, 253 (1976)
C. Llewellyn-Smith, Phys. Lett. 79B, 83 (1978)
H. Fritsch, P. Minkowski, Phys. Lett. 69B, 316 (1977)
K.Koller et al. Zeit. f. Physik C6, 131 (1980)

(29) H. Georgi, S. Glashow, Phys. Rev. Lett. 32, 438(1974)

(30) S. Glashow, Sci. American 233, 38 (1978)
J. Pati, A. Salam, Comm. Nucl. Particle Phys. 6, 183, 7, 1, (1976)

THE MICROSCOPIC REALIZATION OF NUCLEAR COLLECTIVE MODELS

D.J. Rowe
Department of Physics, University of Toronto,
Toronto, Ontario M5S 1A7, Canada

1. INTRODUCTION

A model by its very nature can characterize only some aspects of the system it represents. If one looks closely at the system, there is always more going on than can be described by any simple model. A highly desirable feature of a model is therefore that it can be embedded in a more detailed model. Our objective in these lectures is to formulate phenomenological collective models in such a way that they are microscopically realizable; i.e., we want models that are expressible in shell model terms.

We begin by formulating a collective model in algebraic terms. To do this we must identify a Lie algebra of collective observables. The model is then expressible as an irreducible unitary representation of its Lie algebra. Furthermore, if the collective observables can be expressed in microscopic terms one obtains a realization of the collective Lie algebra as a subalgebra of the microscopic shell model algebra of observables. In this way one obtains the collective model as a submodel of the shell model. One also obtains a microscopic expression of collective states by finding the collective unirreps. in microscopic shell model space.

Consider what this achieves. First of all, by employing only microscopically realizable models one ensures that any predictions are at least compatible with one's deeper understanding of the system. This enables one to avoid unphysical models and to profit from the possibility of probing the dynamical structure of collective states in a way that is not possible within the confines of a phenomenological model. For example, one can never learn about the dynamical current flows in a rotational nucleus simply by interpreting the data with the phenomenological rotational model. This is because the current operator is not one of the observables of the model. However, by embedding the rotational model in the shell model and by realizing rotational states as shell model states one acquires the full arsenal of shell model observables, including the current operator, with which to investigate the dynamical content of rotational states.

In the first part of this course, we express the mass quadrupole collective model[1,2] as a microscopically realizable algebraic model. The associated algebraic model is the cm(3) model[3,4]. This model is then extended to the more general, but computationally more tractable, $sp(3,\mathbb{R})$ model[5]. In the second part of the course we give a brief review of some elements of differential geometry and the geometrical aspects of Lie groups in preparation for part three. The latter concerns the complete decomposition of many-fermion Hilbert space into irreducible collective subspaces. In this way we learn what representations of the collective models exist in

ISSN:0094-243X/81/710177-43$1.50

many-fermion space. At the same time we expose the complementary intrinsic aspects of the space and learn how to do shell model calculations in a basis which explicitly reveals the collective content of nuclear states.

The approach reviewed here has been developed in collaboration with G. Rosensteel but draws heavily from the works of others as the references will hopefully indicate.

2. THE cm(3) MODEL

2.1 The algebra of observables

The Bohr-Mottelson model starts with the assumption that, before quantization, the nucleus has the character of a liquid drop with a well-defined surface whose radius at angle (θ, ψ) can be expressed by means of a multipole expansion

$$R(\theta, \psi) = R_o \, [1 + \sum_{\lambda\mu} \alpha^*_{\lambda\mu} Y_{\lambda\mu}(\theta, \psi)] \, . \tag{2.1}$$

Thus the coefficients $(\alpha_{\lambda\mu})$ serve as coordinates for a model of surface vibrational dynamics. Corresponding to these coordinates one introduces conjugate momenta $(\pi_{\lambda\mu})$ and a Heisenberg-Weyl algebraic structure

$$[\alpha_{\lambda\mu}, \pi_{\lambda'\mu'}] = i \, \delta_{\lambda\lambda'} \, \delta_{\mu\mu'} \tag{2.2}$$

where we put $\hbar = 1$. For quadrupole dynamics one restricts consideration to $\lambda = 2$.

Our objective is to rephrase and if necessary modify this model so that its observables become expressible in microscopic terms. The first step is straightforward. In the Bohr-Mottelson model the nuclear quadrupole moments $(Q_{2\mu})$ are in simple one-to-one correspondence with the $(\alpha_{2\mu})$ deformation parameters. $Q_{2\mu}$ is simply expressible microscopically

$$Q_{2\mu} = \sum_{n=1}^{N} r_n^2 \, Y_{2\mu}(\theta_n, \psi_n) \tag{2.3}$$

where $n = 1, \ldots , N$ labels nucleons. Thus we replace $(\alpha_{2\mu})$ by $(Q_{2\mu})$. We next enquire as to whether or not one can discover microscopically realizable canonical momenta $(P_{2\mu})$ having the property

$$[Q_{2\mu}, P_{2\nu}] = i \, \delta_{\mu\nu} \quad . \tag{2.4}$$

It appears that one can but that they have a complicated structure and are not very useful. One therefore questions whether the Heisenberg-Weyl algebraic structure of eq. (2.4) is really appropriate for the collective observables of finite nuclei. The construction of Weaver and Biedenharn[6] suggests an alternative.

Weaver and Biedenharn considered the 'velocity' operators defined (in a Cartesian basis) by

$$\dot{Q}_{ij} = i \, [H, Q_{ij}] \tag{2.5}$$

where

$$Q_{ij} = \sum_{n=1}^{N} x_{ni}\, x_{nj} \tag{2.6}$$

and where (x_{ni}; n = 1,...,N, i = 1,2,3) are Cartesian coordinates for $\mathbb{R}^{3N}$, the configuration space of N particles in three dimensions. Now if the microscopic Hamiltonian is of the form

$$H = \sum_{n=1}^{N} \frac{1}{2m} p_n^2 + V , \tag{2.7}$$

where the nucleon-nucleon interaction V is assumed to commute with (Q_{ij}), then the momentum observables

$$\tau_{ij} \equiv m\, \dot{Q}_{ij} = im[H, Q_{ij}] \tag{2.8}$$

have the microscopic expression

$$\tau_{ij} = \tau_{ji} = \sum_{n=1}^{N} (x_{ni}p_{nj} + p_{ni}x_{nj}) . \tag{2.9}$$

However these momenta do not close under commutation. For example,

$$[\tau_{12}, \tau_{23}] = iL_{31} = iL_2 \tag{2.10}$$

where

$$L_i = L_{jk} = \sum_{n=1}^{N} (x_{nj}p_{nk} - x_{nk}p_{nj}) \tag{2.11}$$

(i, j, k cyclic) is an angular momentum operator. Together with the angular momenta, the (τ_{ij}) do close and span the Lie algebra gl(3,$\mathbb{R}$) of GL(3,$\mathbb{R}$), the general linear group.

If now we adjoin the Cartesian quadrupole moments (Q_{ij}) to (τ_{ij}) and (L_k), we obtain a basis for the semi-direct sum Lie algebra called cm(3), where

$$\mathrm{cm}(3) \equiv [\mathbb{R}^6]\ \mathrm{gl}(3,\mathbb{R}) . \tag{2.12}$$

Note that the six quadrupole moments (Q_{ij}) commute among themselves and span the Abelian Lie algebra $\mathbb{R}^6$. On the other hand, the (Q_{ij}) transform under gl(3,$\mathbb{R}$) into themselves. For example, from the above realizations one readily obtains

$$[Q_{12}, \tau_{23}] = i\, Q_{13} . \tag{2.13}$$

It is emphasized that although the simple Hamiltonian (2.7) was introduced to discover the gl(3,$\mathbb{R}$) momentum observables, we now discard this Hamiltonian together with the associated identification (2.8) of τ_{ij} with the velocity m $\dot{Q}_{ij}$. Thus we shall be concerned only with the exact algebraic properties of the gl(3,$\mathbb{R}$) and cm(3) momentum operators and think of them as infinitesimal generators of quadrupole deformations and rotations rather than velocities.

2.2 Actions of the cm(3) observables

Consider first the action of $GL(1,\mathbb{R})$ on a single particle in one dimension. The Lie algebra $gl(1,\mathbb{R})$ has a single basis vector given by

$$\tau = xp + px \,. \tag{2.14}$$

Thus, if p acts on single-particle wave functions $\psi(x)$ as the usual differential operator $p \sim -id/dx$, τ acts by

$$\tau \sim -i(x \frac{d}{dx} + \frac{d}{dx} x) = -i\,(1 + 2x \frac{d}{dx})\,. \tag{2.15}$$

Exponentiation generates the group action. We find

$$\psi_\alpha(x) \equiv e^{i\alpha\tau}\,\psi_o(x) = e^{\alpha}\psi_o(e^{2\alpha}x) \tag{2.16}$$

where α is a real parameter. Thus $GL(1,\mathbb{R})$ acts unitarily on wave functions and by scale transformations on $x \in \mathbb{R}$. The action on a wave function is illustrated in Fig. 2.1

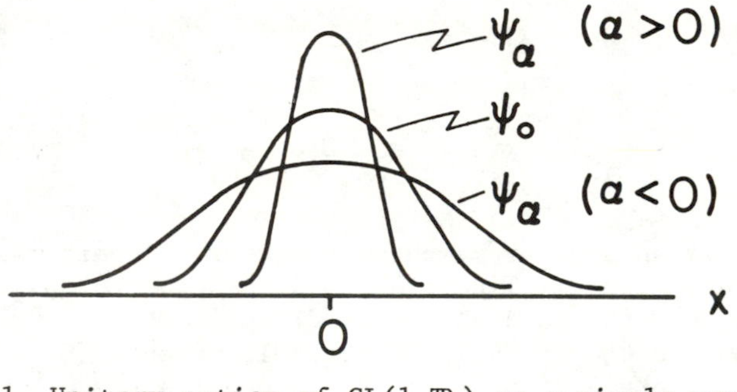

Figure 2.1 Unitary action of $GL(1,\mathbb{R})$ on a simple wave function.

The unitary action of $GL(3,\mathbb{R})$ on a many-body wave function is similarly discovered to be

$$U(g)\Psi(x) = N(g)\Psi(g^{-1}\cdot x), \quad g \in GL(3,\mathbb{R}) \tag{2.17}$$

where N(g) is a normalization factor and

$$(g^{-1}\cdot x)_{ni} = x_{nj}\, G_{ji}(g)\,. \tag{2.18}$$

G is the fundamental 3 x 3 matrix representation of $GL(3,\mathbb{R})$; i.e., G(g) is a 3 x 3 real matrix of non-zero determinant sometimes written simply

$$G_{ji}(g) = g_{ji},\ g \in GL(3,\mathbb{R})\ . \tag{2.19}$$

Let (t_{ij}; i, j = 1,2,3) be the basis

$$t_{ij} = \sum_{n=1}^{N} x_{ni} p_{nj} \tag{2.20}$$

for gl(3,$\mathbb{R}$). From eq. (2.18) one then obtains immediately

$$G_{\alpha\beta}(t_{ij}) = -\, i\, \delta_{\alpha i}\, \delta_{\beta j}\ . \tag{2.21}$$

One also obtains the corresponding Hermitian operator (generator of a unitary transformation)

$$\begin{aligned} U(t_{ij}) &= \frac{1}{2} \sum_{n=1}^{N} (x_{ni} p_{nj} + p_{nj} x_{ni})\ . \\ &= t_{ij} - \frac{1}{2}\, iN\ . \end{aligned} \tag{2.22}$$

Some of the possible shape changes of a nucleus induced by GL(3,$\mathbb{R}$) transformations are illustrated in Fig. 2.2. They include both monopole and quadrupole deformations and rotations. A shape rotation can be generated either by a rigid rotation or by continuously shrinking along one axis while at the same time expanding along another. In this way an irrotational surface wave is generated corresponding to a shape rotation with no real rotation of the matter density. These shape changes illustrate the major collective degrees of freedom embraced by the cm(3) model.

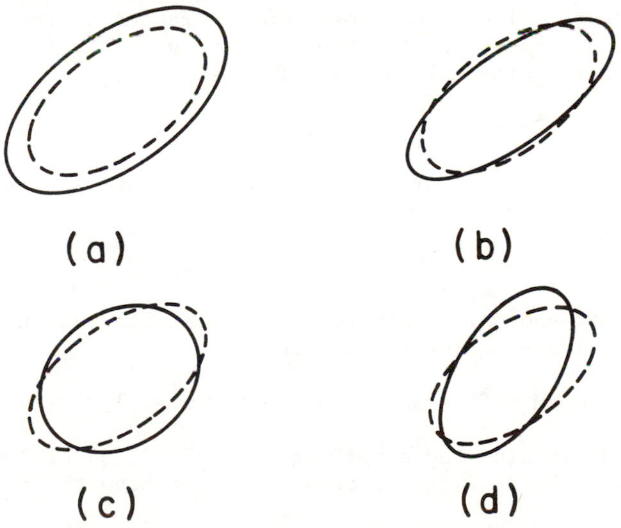

Figure 2.2 Possible shape changes induced by GL(3,$\mathbb{R}$) transformations.

The quadrupole moments (Q_{ij}) act on many-body wave functions simply as multiplicative operators with the expansion (2.6). They themselves transform under GL(3,ℝ) in a manner that is obtained directly from eq. (2.18); i.e.,

$$Q_{ij} \to Q_{ij}(g) = Q_{lk}\, G_{li}(g) G_{kj}(g),\ g \in GL(3,\mathbb{R}) \quad . \tag{2.23}$$

2.3 The phenomenological cm(3) model

Recall that a basis of rotational model wave functions is given by the Wigner rotation matrices $\mathcal{D}^L_{KM}(\alpha,\beta,\gamma)$. If we think of the Euler angles as corresponding to a factorization of an arbitrary SO(3) rotation

$$r = e^{i\gamma L_z}\, e^{i\beta L_y}\, e^{i\alpha L_z} \in SO(3) \tag{2.24}$$

and think of SO(3) as a configuration space rather than a group (cf section 5), we may regard the Wigner matrices as a basis of square integrable functions on SO(3). Thus the rotational model Hilbert space is $L^2(SO(3))$, the space of square integrable functions on SO(3). The Euler angles are simply coordinates for the SO(3) configuration space.

Similarly, the cm(3) Hilbert space is defined as $L^2(GL(3,\mathbb{R}))$, the square integrable functions on GL(3,ℝ). (The appropriate volume element on GL(3,ℝ) is given in refs. 7, 8). It can be shown that a factorization parallel to (2.24) also exists for GL(3,ℝ). Any GL(3,ℝ) matrix g can be expressed as a product

$$g = r_2^{-1} \cdot \lambda \cdot r_1 \tag{2.25}$$

where $r_1 \in SO(3)$, $\lambda = \mathrm{diag}\,(\lambda_1, \lambda_2, \lambda_3)$ is a diagonal GL(3,ℝ) matrix and $r_2 \in O(3)$. (The inverse r_2^{-1} rather than r_2 is purely conventional). However, the decomposition (2.25) is not unique unless some restrictions are imposed on the ranges of the group elements. First observe that there exist rotations $r \in SO(3)$ that simply reorder the diagonal entries of λ; e.g.,

$$r^{-1} \cdot \lambda \cdot r = \mathrm{diag}(\lambda_2, \lambda_1, \lambda_3) \tag{2.26}$$

giving an alternative factorization. To avoid this ambiguity one simply requires the ordering

$$\lambda_1 \geq \lambda_2 \geq \lambda_3 > 0 \quad . \tag{2.27}$$

Even this is not enough. There remains a discrete subgroup $M \subset SO(3)$ that leaves λ invariant; i.e.,

$$M = \{m \in SO(3);\ m^{-1} \cdot \lambda \cdot m = \lambda\} \tag{2.28}$$

For example, M contains the set ($R_A(\pi)$, A = 1, 2, 3) of 180° rotations about the 1, 2, and 3 axes. Thus we have the ambiguity

$$g = r_2^{-1} \cdot \lambda \cdot r_1 = (mr_2)^{-1} \cdot \lambda \cdot mr_1, \; m \in M \, . \tag{2.29}$$

In constructing functions of g one must therefore select those functions of r_1, λ and r_2 that are invariant under a change to a different but equivalent set mr_1, λ, mr_2.

Consider the functions

$$\mathcal{D}^{L}_{KM}(r_1)\; f^{L\mathcal{L}}_{K\mathcal{K}[\nu]}(\lambda)\mathcal{D}^{\mathcal{L}}_{\mathcal{K}\mathcal{M}}(r_2) \; . \tag{2.30}$$

where $f^{L\mathcal{L}}_{K\mathcal{K}[\nu]}$ are square integrable functions of $(\lambda_1, \lambda_2, \lambda_3)$ indexed by quantum numbers $[\nu]$ for each $L\mathcal{L}K\mathcal{K}$. We have, from the standard properties of Wigner matrices,

$$\begin{aligned} R_3(\pi)\mathcal{D}^{L}_{KM}(r_1) &= (-1)^{K}\mathcal{D}^{L}_{KM}(r_1) \\ R_2(\pi)\mathcal{D}^{L}_{KM}(r_1) &= (-1)^{L-K}\mathcal{D}^{L}_{-KM}(r_1) \\ R_1(\pi)\mathcal{D}^{L}_{KM}(r_1) &= R_2(\pi)R_3(\pi)\mathcal{D}^{L}_{KM}(r_1) \; . \end{aligned} \tag{2.31}$$

Thus a basis of square integrable functions invariant under the transformation $r_1 \to mr_1$, $r_2 \to mr_2$, $m \in M$ is given by the symmetrized functions

$$\begin{aligned} \Psi^{\mathcal{L}L}_{\mathcal{K}\mathcal{M}[\nu]KM}(g) &= [1 + (-1)^{K+\mathcal{K}}]\; f^{L\mathcal{L}}_{K\mathcal{K}[\nu]}(\lambda) \\ &\times[\mathcal{D}^{L}_{KM}(r_1)\mathcal{D}^{\mathcal{L}}_{\mathcal{K}\mathcal{M}}(r_2)+(-1)^{L-K+\mathcal{L}-\mathcal{K}}\mathcal{D}^{L}_{-KM}(r_1)\mathcal{D}^{\mathcal{L}}_{-\mathcal{K}\mathcal{M}}(r_2)] \; . \end{aligned} \tag{2.32}$$

To obtain a physical interpretation of the variables (r_1, λ, r_2), consider the action of the cm(3) observables on $\mathcal{L}^2(GL(3,\mathbb{R}))$. One readily ascertains that the unitary GL(3,ℝ) action corresponding to eq. (2.17) is

$$U(g)\Psi(\alpha) = N(g)\Psi(\alpha g), \quad \Psi \in \mathcal{L}^2(GL(3,\mathbb{R})) \, . \tag{2.33}$$

Now, in consideration of the action (2.23) of GL(3,ℝ) on Q_{ij} and recognizing that any Q_{ij} can be expressed

$$Q_{ij} = \sum_{\ell} G_{\ell i}(g)G_{\ell j}(g) \tag{2.34}$$

for some $g \in GL(3,\mathbb{R})$, we use the factorization (2.25) to write eq. (2.34)

$$Q_{ij} = \sum_{A} \mathcal{D}^{1}_{Ai}(r_1)\lambda_A^2\; \mathcal{D}^{1}_{Aj}(r_1) \; , \tag{2.35}$$

where $\mathcal{D}^1$ is the fundamental 3 x 3 matrix representation of SO(3).

We may therefore define the action of Q_{ij} on $L^2(GL(3,\mathbb{R}))$ as multiplication by the expression (2.35). Thus we obtain the interpretation of (λ_A^2) as the principal values of the quadrupole tensor and r_1 as the rotation from the space-fixed to the principal axes.

The cm(3) momentum operators are the elements of the gl(3,$\mathbb{R}$) Lie algebra. A Cartesian basis (t_{ij}) for gl(3,$\mathbb{R}$) is defined in the fundamental matrix representation by eq. (2.21) and the unitary actions of this basis on $L^2(GL(3,\mathbb{R}))$ are obtained by differentiation of eq. (2.33), i.e., by regarding them as generators of infinitesimal GL(3,$\mathbb{R}$) transformations. However, since we have constructed the cm(3) wave functions as functions of the subgroup elements r_1, λ, r_2 of eq. (2.25), it is convenient to use the infinitesimal generators of transformations in these subgroups. Thus, following ref. 7, we introduce the intrinsic angular momenta

$$L_A = \sum_{BC} \varepsilon_{ABC}\, r_1^{-1}\, t_{BC}\, r_1 \tag{2.36}$$

the vibrational momenta

$$t_A = r_1^{-1}\, t_{AA}\, r_1 \tag{2.37}$$

and the vortex spin momenta

$$\mathcal{L}_A = \sum_{BC} \varepsilon_{ABC}\, r_1^{-1}\, \lambda^{-1} t_{BC}\, \lambda\, r_1 \tag{2.38}$$

where ε_{ABC} is the totally anti-symmetric Levi-Civita tensor. These momentum operators act on function of r_1, λ and r_2 respectively. For example

$$L_3\, \mathcal{D}^L_{KM}(r_1) = K\, \mathcal{D}^L_{KM}(r_1) \tag{2.39}$$

$$t_A\, f(\lambda) = -i\, \lambda_A\, \partial f(\lambda)/\partial\lambda_A \tag{2.40}$$

$$\mathcal{L}_3\, \mathcal{D}^{\mathcal{L}}_{\mathcal{K}\mathcal{M}}(r_2) = -\mathcal{K}\, \mathcal{D}^{\mathcal{L}}_{\mathcal{K}\mathcal{M}}(r_2)\ . \tag{2.41}$$

One observes that none of the cm(3) observables is able to change the quantum number $\mathcal{L}$.[3] Furthermore, each unirrep $\mathcal{L}$ occurs in $L^2(GL(3,\mathbb{R}))$ a number $(2\mathcal{L}+1)$ times, the corresponding subspaces being labelled by the index $\mathcal{M} = -\mathcal{L}, \ldots, \mathcal{L}$.

One can show that the Bohr-Mottelson irrotational flow model corresponds to the cm(3) representation with $\mathcal{L} = 0$.[7] One can also show that $\mathcal{L}$ is a measure of the departure of the collective flows from irrotational flow[7,9]. Thus we refer to $\mathcal{L}$ as the vortex spin momentum. Which vorticities are most appropriate in the description of nuclear collective states or indeed whether or not vortex spin is a good quantum number, as implicitly assumed in the extreme cm(3) model, are questions that we are currently actively investigating.

Finally, we consider the cm(3) Hamiltonian. The kinetic energy is readily constructed from microscopic considerations by evaluating the cm(3) component of the many-particle kinetic energy $\Sigma p_n^2/2m$. The result was derived in ref. 7 and again more simply using the standard techniques of differential geometry in ref. 8. It has also been discussed in ref. 9. The general technique and application to cm(3) is presented in sections 4 and 5 respectively. The result is

$$T_{cm(3)} = \frac{1}{2m}\sum_A \frac{\lambda_B^2+\lambda_C^2}{(\lambda_B^2-\lambda_C^2)^2}\,(L_A^2 + \mathcal{L}_A^2) - \frac{4\lambda_B\lambda_C}{(\lambda_B^2-\lambda_C^2)^2}\,L_A\mathcal{L}_A$$

$$+ \frac{1}{2m}\sum_A \Big[(t_A - i(N-2))\,\frac{1}{\lambda_A^2} - 2i\sum_{B\neq A}\frac{1}{\lambda_A^2-\lambda_B^2}\Big]t_A \qquad (2.42)$$

(A, B, C cyclic). The potential energy should be an SO(3) scalar function of the (Q_{ij}). We have at our disposal the three scalars

$$a_1 = \mathrm{tr}\, Q, \quad a_2 = \frac{1}{2}\,\mathrm{tr}\, Q_2^2, \quad a_3 = \frac{1}{3}\,\mathrm{tr}\, Q_2^3\ . \qquad (2.43)$$

where Q_2 is the traceless part of (Q_{ij}), i.e. the second rank SO(3) tensor component. The first scalar is just the monopole moment whose expectation is the mean nuclear radius $\langle a_1\rangle = \langle\sum_n r_n^2\rangle$, related to the liquid drop radius R_o of eq. (2.1) by

$$a_1 = \frac{3}{5}\,N\,R_o^2\ . \qquad (2.44)$$

The second and third define the quadrupole shape. They are related to the more familiar β, γ deformation parameters by[10,5]

$$a_2 = \frac{3}{20\pi}\,N^2R_o^4\beta^2 \qquad (2.45)$$

$$a_3 = \frac{1}{20\pi\sqrt{5\pi}}\,N^3R_o^6\beta^3\cos 3\gamma\ . \qquad (2.46)$$

Thus a simple possibility is to take the shell model harmonic oscillator potential

$$V_o = \frac{1}{2}\,m\omega^2 a_1 \qquad (2.47)$$

for the monopole part and the potential

$$V(\beta,\gamma) = b_3a_3 + b_4a_2^2 = c_3\beta^3\cos 3\gamma + c_4\beta^4 \qquad (2.48)$$

for the quadrupole. This particular potential illustrated in fig. 2.3, has an axially symmetric minimum at some non-zero deformation, depending on the values of the parameters.

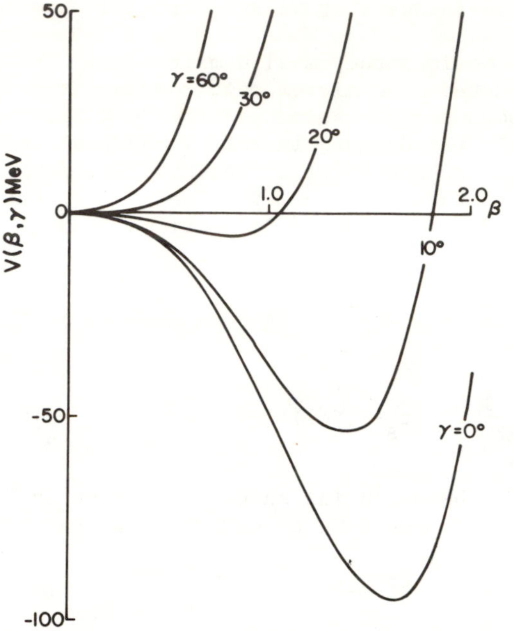

Figure 2.3 A possible cm(3) potential V(β, γ).

To date, no cm(3) model calculations have been completed but calculations and extensions to include vortex spin mixing in order to investigate the transition from irrotational to rigid flow rotations are being pursued by Rosensteel and myself. The major outstanding problem is the construction of a suitable basis of vibrational wave functions $\{f^{LL}_{KK[\nu]}\}$ and the subsequent evaluation of matrix elements. Attention to the solution of these problems has been delayed because of the advent of the computationally simpler sp(3,ℝ) model.

Since the cm(3) algebra was explicitly constructed to be microscopically realizable it follows that the phenomenological cm(3) model becomes a microscopic model simply by interpreting the observables as microscopic operators acting on shell model space. A cm(3) Hamiltonian can therefore, in principle, be diagonalized in shell model space. However to reduce the dimensions of such a diagonalization it is necessary to first reduce the shell model space into invariant cm(3) subspaces. A first reduction is obtained in the following section by reducing the shell-model space into irreducible subspaces of the sp(3,ℝ) algebra. Since sp(3,ℝ) contains cm(3) as a subalgebra, these subspaces are evidently cm(3) invariant. However, they are, in general, reducible. A complete reduction by construction of a shell model basis which fully reduces the cm(3) algebra is given in section 6. We shall see that this basis also reduces the space permutation group and so naturally enables one to study both the collective and the complementary intrinsic properties of nuclear wave functions. Having decomposed shell model space into irreducible cm(3) invariant subspaces a microscopic cm(3) Hamiltonian can be defined as the restriction of a fully microscopic shell model Hamiltonian to the relevant irreducible cm(3) subspace. However the techniques for doing this remain to be developed.

3. THE sp(3,ℝ) MODEL

3.1 The sp(3,ℝ) algebra

The cm(3) algebra is microscopically realized as the span of the self-adjoint observables

$$Q_{ij} = \sum_{n=1}^{N} x_{ni}\, x_{nj} \tag{3.1}$$

$$\tau_{ij} = \sum_{n} (x_{ni}\, p_{nj} + p_{ni}\, x_{nj}) \tag{3.2}$$

$$L_k = \sum_{n} (x_{ni}\, p_{nj} - x_{nj}\, p_{ni}) \quad (i, j, k \text{ cyclic}) . \tag{3.3}$$

To this set the symplectic sp(3,ℝ) algebra simply adds the quadratics in momenta

$$K_{ij} = \sum_{n} p_{ni}\, p_{nj} . \tag{3.4}$$

It is clear that sp(3, ℝ) contains the collective subalgebras

$$sp(3,\mathbb{R}) \supset cm(3) \supset [\mathbb{R}^5]so(3) \supset so(3) . \tag{3.5}$$

(Note that sp(3,ℝ) is frequently called sp(6,ℝ)).

There are several major advantages to expanding cm(3) to sp(3,ℝ). From a physical point of view, sp(3,ℝ) already contains the full many-particle kinetic energy

$$T = \frac{1}{2m} \sum_{n} p_n^2 = \frac{1}{2m} \sum_{i} K_{ii} \tag{3.6}$$

and thus obviates the need for the more complicated cm(3) expression (2.42). However there are even more significant advantages. First of all, it is semi-simple. Its complexification is denoted by C_3 in Cartan's classification. Thus it is a well-studied Lie algebra. Furthermore its unitary representations are lowest weight representations and hence easily constructed. Secondly, it contains the subalgebras

$$sp(3,\mathbb{R}) \supset u(3) \supset su(3) \supset so(3) \tag{3.7}$$

where u(3) is the Lie algebra of the standard degeneracy group U(3) of the three dimensional harmonic oscillator. Thus its representations are readily constructed in a harmonic oscillator shell model basis. This is an extremely significant technical advantage because it means that it is much more straightforward to restrict a microscopic many-particle Hamiltonian to an $sp(3,\mathbb{R})$ unirrep than to a cm(3) unirrep. By the same token, the use of shell model observables to probe the microscopic structure of $sp(3,\mathbb{R})$ states becomes a much more attainable objective.

3.2 $sp(3,\mathbb{R})$ unirreps

In order to construct a basis of states for a unitary representation labelled according to the subgroup chain (3.7), it is useful to introduce the usual step operators

$$x_{ni} = \frac{1}{\sqrt{2m\omega}} (b^{\dagger}_{ni} + b_{ni})$$

$$p_{ni} = i \sqrt{\frac{2\omega}{2}} (b^{\dagger}_{ni} - b_{ni}) \ . \qquad (3.8)$$

A basis for C_3, the complexified $sp(3,\mathbb{R})$ algebra, is then given by the 21 quadratics

$$A_{ij} = \sum_n b^{\dagger}_{ni} b^{\dagger}_{nj}$$

$$B_{ij} = \sum_n b_{ni} b_{nj} \qquad (3.9)$$

$$C_{ij} = \frac{1}{2} \sum_n (b^{\dagger}_{ni} b_{nj} + b_{nj} b^{\dagger}_{ni}) \ .$$

In this basis one immediately identifies the generators of U(3); namely the elements (C_{ij}; i, j = 1, 2, 3).

A Cartan subalgebra for C_3 is spanned by the three vectors

$$C_{ii} = \frac{1}{2} \sum_n (b^{\dagger}_{ni} b_{ni} + b_{ni} b^{\dagger}_{ni}) \ . \qquad (3.10)$$

which one recognizes as three simple harmonic oscillator Hamiltonians (boson number operators). We define the raising operators

$$(A_{ij};\ i, j = 1, 2, 3), \quad (C_{ij};\ i > j) \qquad (3.11)$$

and the lowering operators

$$(B_{ij};\ i, j = 1, 2, 3), \quad (C_{ij};\ i < j). \qquad (3.12)$$

A lowest weight state $|\phi\rangle$ is a state having the property that it is annihilated by all lowering operators; i.e.,

$$\begin{aligned} B_{ij}\ |\phi\rangle &= 0, \quad i, j = 1, 2, 3 \\ C_{ij}\ |\phi\rangle &= 0, \quad i < j\ . \end{aligned} \tag{3.13}$$

An irreducible representation is characterized by the eigenvalues of the three vectors (3.10) of the Cartan subalgebra on the lowest weight state; i.e., by the numbers $(\omega_1, \omega_2, \omega_3)$ where

$$C_{ii}\ |\phi\rangle = \omega_i\ |\phi\rangle,\ i = 1, 2, 3\ . \tag{3.14}$$

A basis for this representation is generated by acting on the lowest weight state repeatedly with the step up operators.

Observe that the Cartan subalgebra is the same for both sp(3,$\mathbb{R}$) and its u(3) subalgebra. The Cartan subalgebra for su(3)$\subset$u(3) is spanned by the vectors $C_\lambda = C_{11} - C_{22}$, $C_\mu = C_{22} - C_{33}$. Thus by acting on the lowest weight state $|\phi\rangle$ with just the u(3) raising operators, one generates a u(3) and an su(3) unirrep. On the other hand acting with the (A_{ij}) raising operators generates states from other u(3) unirreps contained in the single sp(3,$\mathbb{R}$) unirrep.

For example, the ^{16}O closed-shell state $|\phi\rangle$ is a lowest weight state with

$$\begin{aligned} W &= \omega_1 + \omega_2 + \omega_3 = 36 \\ \lambda &= \omega_1 - \omega_2 = 0 \\ \mu &= \omega_2 - \omega_3 = 0 \end{aligned} \tag{3.15}$$

where $(\lambda, \mu) = (0, 0)$ are the SU(3) labels. This state carries a one-dimensional representation of U(3) and its subgroup SU(3).

The lowest sp(3,$\mathbb{R}$) states for ^{16}O belonging to the $W(\lambda, \mu) = 36\ (0, 0)$ representation are shown in Figure 3.1.

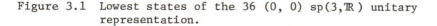

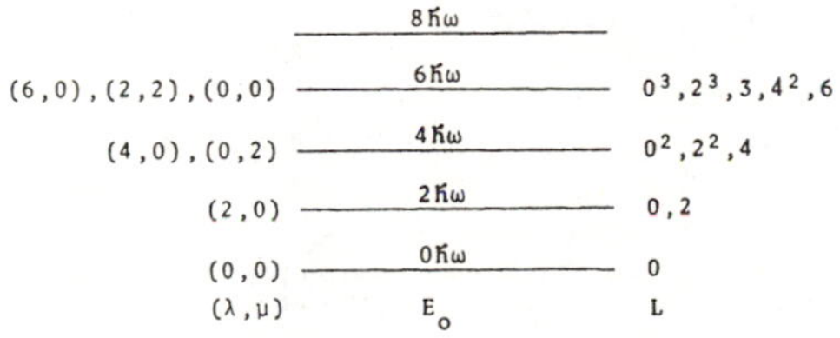

Figure 3.1 Lowest states of the 36 (0, 0) sp(3,$\mathbb{R}$) unitary representation.

For an arbitrary nucleus, the shell model valence space (the $0\hbar\omega$ space) is defined as the space of all shell model states having the lowest possible energy with respect to the three-dimensional harmonic oscillator Hamiltonian. Thus if $|\phi\rangle$ belongs to the $0\hbar\omega$ space it is annihilated by all the (B_{ij}) lowering operators. To construct sp(3,ℝ) unirreps we must therefore find the su(3) lowest weight states within the $0\hbar\omega$ space. It follows that these states are then sp(3,ℝ) lowest weight states. Pursuing this construction one obtains a complete decomposition of the shell model space into sp(3,ℝ) irreducible subspaces as illustrated schematically in Figure 3.2.

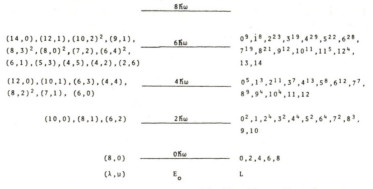

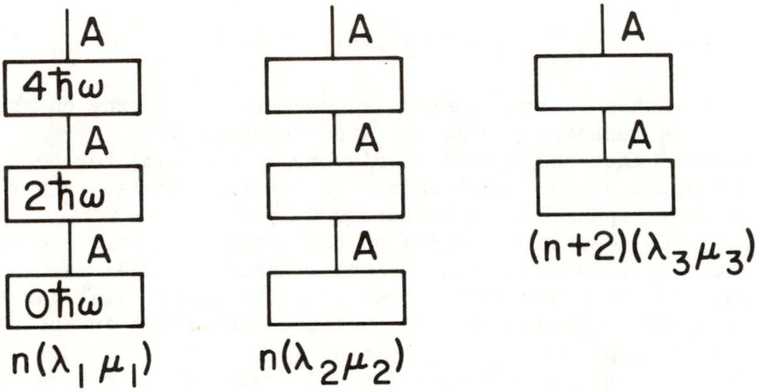

Figure 3.2 Decomposition of the positive parity shell model space into sp(3,ℝ) ⊃ U(3) ⊃ SU(3) subspaces.

For example, the most symmetric SU(3) representation in the valence space for ^{20}Ne is the (8, 0) representation. The states of the sp(3,ℝ) representation built on the lowest weight state for this representation are shown in Figure 3.3.

(λ,μ)	E_o	L
	$8\hbar\omega$	
$(14,0),(12,1),(10,2)^2,(9,1),(8,3)^2,(8,0)^2,(7,2),(6,4)^2,(6,1),(5,3),(4,5),(4,2),(2,6)$	$6\hbar\omega$	$0^9,1^8,2^{23},3^{19},4^{29},5^{22},6^{28},7^{19},8^{21},9^{12},10^{11},11^5,12^4,13,14$
$(12,0),(10,1),(6,3),(4,4),(8,2)^2,(7,1),(6,0)$	$4\hbar\omega$	$0^5,1^3,2^{11},3^7,4^{13},5^8,6^{12},7^7,8^9,9^4,10^4,11,12$
$(10,0),(8,1),(6,2)$	$2\hbar\omega$	$0^2,1,2^4,3^2,4^4,5^2,6^4,7^2,8^3,9,10$
$(8,0)$	$0\hbar\omega$	$0,2,4,6,8$

Figure 3.3 Lowest state of the 50 (8, 0) sp(3,ℝ) unirrep.

3.3 The sp(3,ℝ) Hamiltonian

A natural phenomenological sp(3,ℝ) Hamiltonian is one of the form

$$H = \frac{1}{2m} \sum_i K_{ii} + \frac{1}{2} m\omega^2 \sum_i Q_{ii} + V(\beta, \gamma) \qquad (3.16)$$

which includes the many-particle kinetic energy, cf. eq. (3.6), the harmonic oscillator shell model potential, cf. eq. (2.47) and a collective potential $V(\beta, \gamma)$ such as, for example, that given by eq. (2.48). The first two, as already noted, are in the sp(3,ℝ) algebra and the third is in the sp(3,ℝ) enveloping algebra. More general potentials can of course be constructed. But in order that their matrix elements be readily calculable, they should be low order polynomials in the sp(3,ℝ) algebra.

The spectrum obtained for ^{20}Ne by diagonalizing a Hamiltonian of the form (3.16) within the space of the 50 (8, 0) sp(3,ℝ) representation is shown in Fig. 3.4. The results are taken from ref. 5. Since the space is infinite dimensional an exact diagonalization is clearly impossible and we were obliged to truncate. The results show the effects of truncating at the 0, 4 and 8 $\hbar\omega$ shell model levels. In this way we could investigate the rate of convergence of the results with the dimensionality of the space. The proposition is that, if the harmonic oscillator shell model makes any sense, the lowest states of the harmonic oscillator should dominate. We indeed find this to be the case although significant contributions to collective states from harmonic oscillator shells of up to 10 $\hbar\omega$ are indicated. The results at the 8 $\hbar\omega$ level of truncation appear to have almost converged.

A couple of points are worthy of note. First of all, it would appear that the effective moment of inertia for what appears to be a rotational band of states is too large; i.e., the levels are too compressed. Expanding the space and modifying the parameters in $V(\beta, \gamma)$ can improve the situation somewhat. However, it is not obvious that the experimental results can be reached within a single sp(3,ℝ) irrep without introducing a potential $V(\beta, \gamma)$ with, for example, a minimum very different from the quadrupole deformation indicated for ^{20}Ne by the observed electric quadrupole transition data. In contrast the 0$\hbar\omega$ shell model can fit the energy levels almost perfectly by adjusting the matrix elements of the effective interaction. However, by fitting levels in that way one loses the possibility of learning very much about the real validity of the model or of the character of the states fitted. A major advantage of the phenomenological sp(3,ℝ) model is that one retains tight reins on the physical acceptability of a proposed potential $V(\beta, \gamma)$. Thus failure to achieve agreement with experiment can be as informative as success. The kinetic energy component of the energies predicted in the calculations are indicated in parenthesis above the levels. One notes that the kinetic energy contribution increases from zero in the 0$\hbar\omega$ limit to essentially 100% in the 8 $\hbar\omega$ limit, the latter being of course what one would intuitively

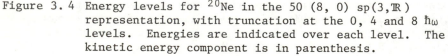

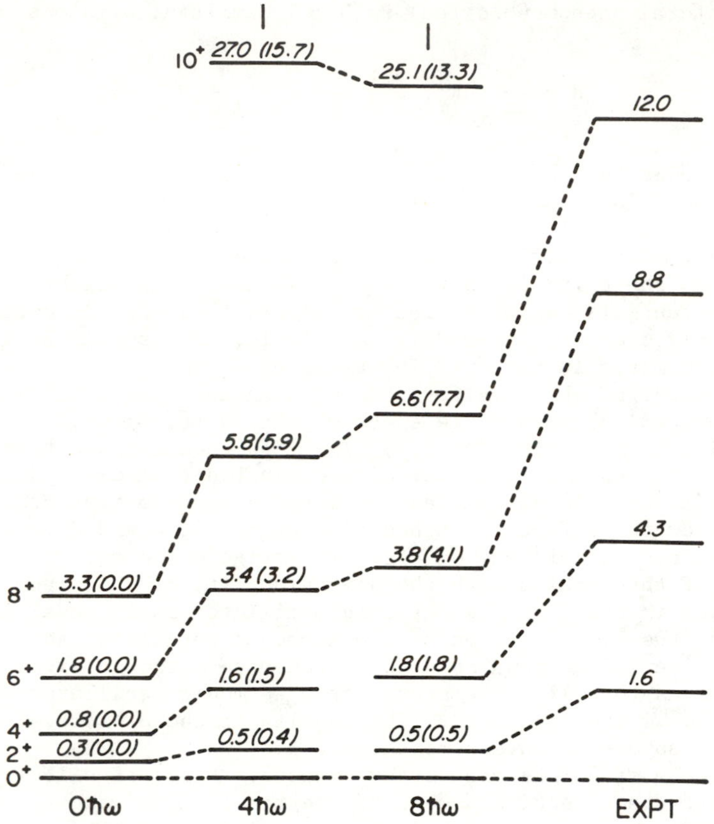

Figure 3. 4 Energy levels for ^{20}Ne in the 50 (8, 0) sp(3,ℝ) representation, with truncation at the 0, 4 and 8 ħω levels. Energies are indicated over each level. The kinetic energy component is in parenthesis.

expect for a rotor. In this context it is important to note that, in a low-dimensional truncated space, one tends to lose track of what parts of the Hamiltonian depend on the spatial coordinates and what on the momentum coordinates. For example, the potential $\frac{1}{2} m\omega^2 \Sigma_{ni} x_{ni}^2$ has exactly the same matrix elements as the kinetic energy in the 0ħω space. Another point worth noting is the substantial gap that appears in the calculated spectrum between the 8^+ and 10^+ levels. This is just one of some of the shell model effects that are in evidence in the ^{20}Ne spectrum. Others include the fall-off in the predicted reduced electric quadrupole (E2) transition rates between levels, shown in Figure 3.5.

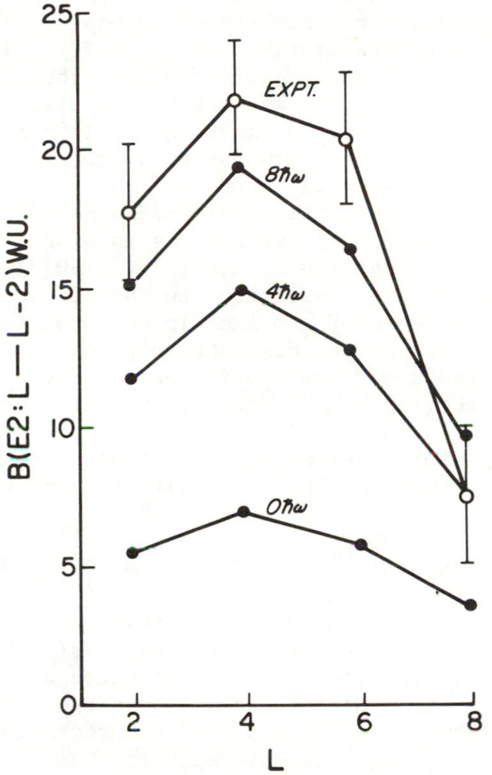

Figure 3.5 Reduced E2 transition rates in ^{20}Ne.

3.4 Permutation symmetry

In classifying su(3) states one exploits space permutation symmetry to label states. The point is that the SU(3) group acts on the spatial part of wave functions only; it is indifferent to particle index. Indeed the generators are symmetric in particle index and commute with space permutation operators. This is important because we need spatial wave functions of well-defined permutation symmetry in order that they can be combined with spin/isospin SU(4) wave functions of contragredient symmetry to make totally anti-symmetric states. We can do exactly the same for sp(3,$\mathbb{R}$). In fact for sp(3,$\mathbb{R}$) one can exploit a larger group.

Observe that the boson operators ($b^{\dagger}_{ni}$, b_{ni}; n = 1, ... , N, i = 1, 2, 3) are the building blocks for the Lie algebra sp(3N,$\mathbb{R}$) with basis

$$b^{\dagger}_{ni}\, b^{\dagger}_{mj},\ b_{ni}\, b_{mj},\ \frac{1}{2}\,(b^{\dagger}_{ni}\, b_{mj} + b_{mj}\, b^{\dagger}_{ni}) \ .$$

Now consider the space spanned by all boson states constructed by starting with the vacuum state and forming the states

$$|0\rangle,\ b^{\dagger}_{ni}\ |0\rangle,\ b^{\dagger}_{ni}\ b^{\dagger}_{mj}\ |0\rangle, \text{ etc.}$$

This space decomposes into just two unirreps of $sp(3N,\mathbb{R})$ spanned respectively by the states of even and odd numbers of bosons. They are the fundamental (metaplectic) representations of $sp(3N,\mathbb{R})$. It is important to note that this space includes more than the shell model space since it takes no account of the indistinguishability of nucleons (fermion statistics). Thus it is important to look for subspaces of states which have the correct permutation symmetry properties. It turns out that this is easily done.

First note that, in considering the actions of subgroups of $Sp(3N,\mathbb{R})$ on the building blocks $(b^{\dagger}_{ni}, b_{ni})$, we can distinguish groups which act only on the space index i = 1, 2, 3, e.g., SO(3), $GL(3,\mathbb{R})$, $Sp(3,\mathbb{R})$, etc., and which are indifferent to the particle index n = 1, ..., N; i.e., the elements of the Lie algebras sum over the particle index. Similarly, we can distinguish groups which act only on the particle index and are indifferent to the space index, e.g., SO(N), $GL(N,\mathbb{R})$, $Sp(N,\mathbb{R})$, etc. Of particular interest is the subgroup O(N).

The subgroup O(N) is of special interest because it has been shown that the decomposition of irreps of $Sp(3N,\mathbb{R})$ with respect to the direct product subgroup

$$Sp(3,\mathbb{R}) \times O(N) \subset Sp(3N,\mathbb{R}) \qquad (3.17)$$

is multiplicity free[11†] (i.e., there are no missing labels and we can label states according to the way they transform under $Sp(3,\mathbb{R})$ and O(N)). For the fundamental irreps there is an even stronger result; namely that the $Sp(3,\mathbb{R})$ and O(N) decompositions are complementary[12]. This means that all the states that transform as the α component of the $Sp(3,\mathbb{R})$ $(\omega_1, \omega_2, \omega_3)$ unirrep carry an irreducible unirrep of O(N). Thus we can use the same labels $(\omega_1, \omega_2, \omega_3)$ to label both $Sp(3,\mathbb{R})$ and O(N) unirreps.

Of still more significance is the fact that O(N) contains the subgroups

$$O(N) \supset O(N-1) \supset S_N \qquad (3.18)$$

where S_N is the symmetric (permutation) group. Since there are just four spin-isospin states of a nucleon, the shell model space is now identified as the subspace of the $Sp(3N,\mathbb{R})$ space spanned by states which transform under S_N with Young tableaux having not more than four columns; i.e., these are the states which can be combined with spin-isospin states to make totally anti-symmetric states.

4.5 Decompositions of the many-fermion Hilbert space

We have seen in this section how to decompose many-fermion

---

†We are grateful to S. Sternberg for pointing this out to us.

Hilbert space into Sp(3,$\mathbb{R}$) irreducible subspaces and label basis states with respect to the chain of subgroups

$$\begin{array}{ccccccc} Sp(3,\mathbb{R}) & \supset & U(3) & \supset & SO(3) & \supset & SO(2)\ . \\ \omega & & [n_1, n_2, n_3] & & L & & M \end{array} \tag{3.19}$$

We have also observed that, since the Sp(3,$\mathbb{R}$) and O(N) group actions commute, we can simultaneously label states with respect to the subgroup chain

$$\begin{array}{ccc} O(N) & \supset & S_N\ . \\ \omega & \alpha & \lambda \end{array} \tag{3.20}$$

Furthermore, since the Sp(3,$\mathbb{R}$) and O(N) actions on many-fermion space are complementary, the same label ω serves to label both the Sp(3,$\mathbb{R}$) and O(N) unirreps.

Note that separation of the centre-of-mass degrees of freedom is almost trivial since the many-body configuration space separates into a direct sum

$$\mathbb{R}^{3N} = \mathbb{R}^{3} \oplus \mathbb{R}^{3(N-1)} \tag{3.21}$$

of centre-of-mass configuration space and an orthogonal complementary space $\mathbb{R}^{3(N-1)}$ of relative configurations. We may then restrict the action of Sp(3,$\mathbb{R}$) and its subgroups to $\mathbb{R}^{3(N-1)}$ in the natural way and replace O(N) by the corresponding group O(N-1) acting on $\mathbb{R}^{3(N-1)}$. Instead of eq. (3.20), we then label states by the subgroup chain

$$\begin{array}{ccc} O(N-1) & \supset & S_N \\ \omega & \alpha & \lambda \end{array} \tag{3.22}$$

In section 6 we shall show how to simultaneously reduce the many-fermion Hilbert space with respect to $O(N) \supset S_N$ or $O(N-1) \supset S_N$, and the subgroup chain

$$\begin{array}{ccccccc} Sp(3,\mathbb{R}) & \supset & CM(3) & \supset & SO(3) & \supset & SO(2) \\ \omega & & \delta \quad \mathcal{L} & & L & & M\ . \end{array} \tag{3.23}$$

Ideally we would like to perform Sp(3,$\mathbb{R}$) model calculations in the cm(3) rather than the u(3) basis in order that we can interpret more readily the collective content of the states that emerge. In particular it would be nice to diagonalize a realistic microscopic Hamiltonian within a single sp(3, $\mathbb{R}$) unirrep and discover what vorticities occur in the low energy spectrum of a collective nucleus, how the admixtures effect moments of inertia, etc. However, the u(3) basis, being naturally adapted to the shell model, lends itself much more readily to a microscopic calculation. Clearly what one would really like is the transformation between the chains (3.19) and (3.23). But this is not at present available.

4. SOME ELEMENTS OF DIFFERENTIAL GEOMETRY

4.1 Smooth Manifold

An n-dimensional manifold M is a topological space that looks locally like a neighbourhood of an n-dimensional Euclidian space $\mathbb{R}^n$, c.f. Figure 4.1. It is equipped with a complete set of coordinate charts. A coordinate chart ϕ is a map from a neighbourhood U(M) of M into $\mathbb{R}^n$

$$\phi: \; U(M) \to \mathbb{R}^n; \; m \mapsto (x^1(m), \ldots, x^n(m)) \; . \qquad (4.1)$$

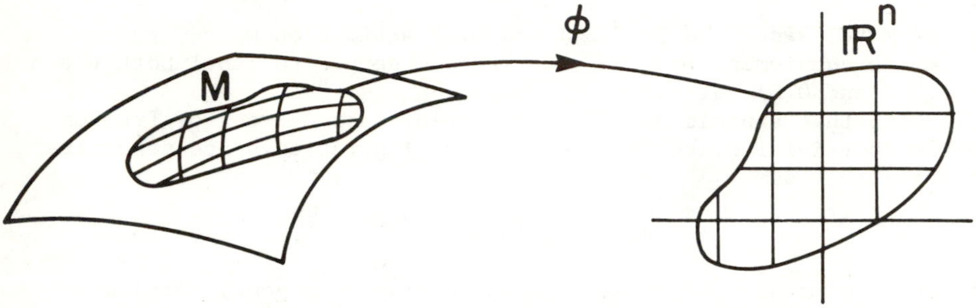

Fig. 4.1 A map ϕ from a neighbourhood of an n-dimensional manifold M into $\mathbb{R}^n$.

Thus it assigns coordinates $(x^1 \ldots x^n)$ to any point m in the domain of the chart. A requirement of a coordinate chart is that it be invertible i.e., the inverse map ϕ^{-1} must be well defined. There are evidently many ways to coordinatize a manifold and it is stressed that no physical result can possibly depend on what particular coordinates are used to describe it.

A function f on a manifold is a scalar field i.e., the assignment of a real number f(m) to each point m in the domain of the function. Given a coordinate chart ϕ we can express a function in terms of coordinates and so differentiate it with respect to the coordinates. If a function can be differentiated infinitely many times we say that it is smooth. Of course a function on a manifold could be smooth when expressed in terms of one set of coordinates but not smooth in terms of another set. If, however, a function being smooth in terms of one set implies it is smooth in any set, we say that the manifold is smooth. The set of all smooth functions on a smooth manifold M is denoted $C^\infty(M)$.

Analytic functions and manifolds are defined in a parallel way. A function f of coordinates (x^i) is analytic over a neighbourhood if for any x and x + y in the neighbourhood the Taylor expansion

$$f(x + y) = f(x) + y^i \, \partial f/\partial x^i + \ldots \qquad (4.2)$$

is absolutely convergent. If any function which is analytic on a manifold when expressed in terms of one set of coordinates is analytic in terms of any set, we say that the manifold is analytic.

In subsequent sections we shall implicitly assume that a manifold is smooth or analytic without always explicitly saying so, whenever we need to differentiate functions or make Taylor expansions.

4.2 Tangent Vectors and Vector Fields

Let m(t) be a path on M indexed by a real parameter t as illustrated in Figure 4.2 and let $(x^i(t))$ be coordinates for points

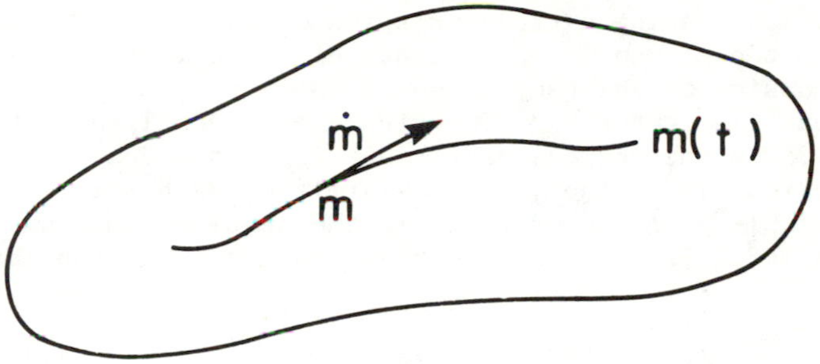

Fig. 4.2 A path m(t) in M with tangent vector $\dot{m}$.

on the path in some chart. We can then define the tangent vector $\dot{m}$ at each point on the path as a vector with components $(\dot{x}^i = dx^i/dt)$. We are justified in calling $\dot{m}$ a vector because under a change of coordinates $(x^i) \to (y^i)$ the components of $\dot{m}$ transform linearly

$$\dot{x}^i \to \dot{y}^i = \dot{x}^j \ \partial y^i/\partial x^j \ . \qquad (4.3)$$

Thus $\dot{m}$ at a point m on the path belongs to a linear space TM(m) called the tangent space of M at m ε M.

The tangent space TM(m) is defined as the real linear space of all possible tangent vectors at m ε M, i.e., X ε TM(m) is an object with components (X^i) defined with respect to the coordinate chart (x^i) in such a way that under a change of chart $(x^i) \to (y^i)$ the components of X transform

$$X^i \to Y^i = X^j \ \partial y^i/\partial x^j \ . \qquad (4.4)$$

Note that without more information about the manifold M we are not able to compare vectors at two different points m and m' in a coordinate independent way. To do this we need a connection.

A vector field X is the assignment of a vector X(m) at each m ε M.

4.3 Action of a Vector on a Function

Let X be a tangent vector in TM(m) and f a smooth function in $C^\infty(M)$. We define Xf to be the real number

$$Xf = X^i \, (\partial f/\partial x^i)_m \, . \tag{4.5}$$

Note that in conventional Euclidian geometry one would more familiarly call this quantity $X \cdot \nabla f$. However, it is more convenient to think of X itself as a differential operator and so express it

$$X = X^i \, (\partial/\partial x^i)_m \, . \tag{4.6}$$

The reason is that this expression is coordinate independent as one readily observes by making a transformation of coordinates. Thus we can think of the elementary coordinate differentials $(\partial/\partial x^i)_m$ as defining a basis for the tangent space TM(m).

The action of a vector on a function can evidently be extended to vector fields in the obvious way. Thus if X is a vector field, Xf is the function with value (Xf)(m) = X(m)f at each m in its domain. An example of a vector field is the angular momentum operator L_z on $\mathbb{R}^3$. If (x, y, z) is the standard Euclidian chart for $\mathbb{R}^3$ then

$$iL_z = y \, \partial/\partial x - x\partial/\partial y \tag{4.7}$$

is a vector field on $\mathbb{R}^3$. The vector fields $(\partial/\partial x^\nu)$ defined by a coordinate chart (x^ν) on a manifold are called coordinate fields.

4.4 The Metric

The metric g is a bilinear form on the tangent space at each point of a manifold. For example, consider the Euclidian space with orthonormal coordinates $(x^1 \ldots x^n)$ and let $ip_\nu = \partial/\partial x^\nu$ denote the associated coordinate fields. We can then define the metric by

$$g(p_\mu, p_\nu) = \delta_{\mu\nu} \, . \tag{4.8}$$

Hence for the angular momentum

$$L_i = \Sigma \, \varepsilon_{ijk} \, x_j p_k \tag{4.9}$$

on $\mathbb{R}^3$, for example, we obtain

$$g(L_i, L_j) = I_{ij} \tag{4.10}$$

where I is the rigid-body inertia tensor.

4.5 The Laplace-Beltrami Operator

The Laplace-Beltrami operator on an arbitrary Riemannian manifold is defined in terms of coordinate fields

$$\Delta = \frac{1}{\sqrt{g}} \frac{\partial}{\partial x^{\mu}} \sqrt{g}\, g^{\mu\nu} \frac{\partial}{\partial x^{\nu}} , \tag{4.11}$$

where $g = \det(g_{\mu\nu})$, $(g^{\mu\nu}) = (g_{\mu\nu})^{-1}$ and $g_{\mu\nu} = g(\partial/\partial x^{\mu}, \partial/\partial x^{\nu})$. This expression is known to be coordinate independent. For a Euclidian space with metric (4.8) it reduces to

$$\Delta = \sum_{\mu} p_{\mu}^{2} \tag{4.12}$$

which to within a factor (-1) is the standard Laplacian. Note that the factor (-1) could be incorporated into the metric but since we are interested in the kinetic energy it is convenient to define the sign of the metric as in eq. (4.8).

For practical purposes it is often more convenient to express Δ in terms of an arbitrary set of base fields rather than coordinate fields. By base fields we mean a set of vector fields (π_{μ}) which form a basis of tangent vectors at each point over some neighbourhood of the manifold. In terms of an arbitrary set of base fields one can show that Δ of eq. (4.12) becomes

$$\Delta = \frac{1}{\sqrt{g}} \pi_{\mu} \sqrt{g}\, g^{\mu\nu} \pi_{\nu} + C^{\lambda}_{\lambda\mu} g^{\mu\nu} \pi_{\nu} \tag{4.13}$$

where now

$$g_{\mu\nu} = g(\pi_{\mu}, \pi_{\nu})$$

$$g = \det(g_{\mu\nu}), \quad g^{\mu\nu} = (g_{\mu\nu})^{-1} \tag{4.14}$$

and $C^{\lambda}_{\mu\nu}$ is defined by

$$[\pi_{\mu}, \pi_{\nu}] = C^{\lambda}_{\mu\nu} \pi_{\nu} . \tag{4.15}$$

Note that if the base fields are coordinate fields they commute with one another hence the structure factors $C^{\lambda}_{\mu\nu}$ vanish and eq. (4.13) reduces to eq. (4.11).

In the special case that the manifold is a Euclidian space with metric (4.8) a still simpler expression results namely

$$\Delta = \pi_{\mu} g^{\mu\nu} \pi_{\nu} + [p_{\alpha}, \pi^{\alpha}_{\mu}] g^{\mu\nu} \pi_{\nu} \tag{4.16}$$

where π^{α}_{μ} is the coefficient of p_{α} in the expansion

$$\pi_\mu = \Sigma_\alpha \pi_\mu^\alpha (x) p_\alpha \tag{4.17}$$

and

$$[p_\alpha, \pi_\mu^\alpha] = -i \, \partial\pi_\mu^\alpha/\partial x^\alpha \ . \tag{4.18}$$

These expressions for the Laplace-Beltrami operator are derived in ref. 8.

5. GROUP ORBITS AND HOMOGENEOUS SPACES

5.1 Lie Groups and Group Actions

A Lie group is a group and an analytic manifold with an analytic composition function. This means, among other things, that we may make Taylor expansions of analytic functions on a Lie group.

Lie groups are often defined as groups of transformations of a space; eg., the classical matrix groups are groups of linear transformations of linear spaces. If a group G is realized as a group of transformations of a manifold M, it follows that there must be an action of G on M which we shall signify by

$$\begin{aligned} G &: M \to M \\ g &: m \mapsto g\cdot m \ . \end{aligned} \tag{5.1}$$

For example, the group of translations of the Euclidian plane has natural action

$$(\alpha, \beta)\cdot(x, y) \mapsto (x + \alpha, y + \beta) \ . \tag{5.2}$$

A group G is said to act transitively on a manifold M if for every p, q ε M there exists some g ε G such that $g\cdot p = q$. One observes that every group acts transitively on itself. A manifold which supports a transitive group action is called a homogeneous space.

The action of a group G on a smooth manifold M naturally induces an action of G on the smooth functions $C^\infty(M)$. If f ε $C^\infty(M)$, this action is defined by

$$(P_g f)(m) = f(g^{-1}\cdot m) \ . \tag{5.3}$$

In this way one constructs representations of the group carried by linear spaces of functions on the manifold. When M is the group G itself we observe that there are in fact two natural actions given respectively by left and right group multiplication. Thus if f ε $C^\infty(M)$ we have the left and right actions

$$(L_\alpha f)(g) = f(\alpha^{-1}g) \tag{5.4}$$

$$(R_\alpha f)(g) = f(g\alpha) \quad . \tag{5.5}$$

5.2 The Lie Algebra

The Lie algebra $\underline{g}$ of a Lie group G embodies the local but not all the global properties of G. If f is a function on G that is analytic about an element $g \in G$ and α is a group element close to the identity element in G, we can make a Taylor expansion

$$(R_\alpha f)(g) = f(g\alpha) = f(g) + X_\alpha(g)\ f + \ldots . \tag{5.6}$$

This defines a tangent vector $X_\alpha(g)$ at each $g \in G$. The set of vector fields (X_α) span the Lie algebra of G. These vector fields are called left invariant vector fields or right shift operators. An alternative realization of the Lie algebra $\underline{g}$ is given by the right invariant vector fields or left shift operators obtained by expanding

$$(L_\alpha f)(g) = f(\alpha^{-1}g) = f(g) + Y_\alpha(g)\ f + \ldots . \tag{5.7}$$

If the group G acts on a manifold M as discussed above, then a realization of the Lie algebra of G is given by the vector fields on M defined by

$$(P_\alpha f)(m) = f(\alpha^{-1}\cdot m) = f(m) + X_\alpha(m) f + \ldots . \tag{5.8}$$

In this way one obtains, for example, the realization of the angular momentum algebra so(3) as vector fields on $\mathbb{R}^3$; c.f. eq. (4.7).

Note that as physicists we are often negligent in using the same symbol to denote an element of a Lie algebra and its realization as a vector field on some manifold. For example, for so(3) we will write simultaneously

$$L_z\ \Phi_{LM}(x, y, z) = M\ \Phi_{LM}(x, y, z) \tag{5.9}$$

$$L_z\ \mathcal{D}^L_{KM}(\alpha, \beta, \gamma) = M\ \mathcal{D}^L_{KM}(\alpha, \beta, \gamma) \tag{5.10}$$

where Φ_{LM} is a function on $\mathbb{R}^3$ and $\mathcal{D}^L_{KM}$ is a function on SO(3). The distinction, however is often very important.

5.3 Group Orbits

Let m be some fixed point of a manifold M. The G orbit containing m is the set of points

$$O_m = \{g\cdot m;\ g \in G\} \ . \tag{5.11}$$

O_m is a hypersurface in M. For example, if G is SO(3) and M is $\mathbb{R}^3$, the orbit O_x through any point $x \neq 0$ is a sphere centered about the origin of $\mathbb{R}^3$. Three points are worthy of note: (i) if G acts transitively on M, then every orbit O_m in M is equal to M; i.e., the orbits of G cover M. This is the situation for example when M = G. (ii) G acts transitively on each orbit (iii) in general, M is stratified into classes of orbits. Orbits belonging to the same stratum are diffeomorphic. Before discussing how to characterize strata we first give an example.

Let $\mathbb{R}^5$ be the five dimensional space of real traceless 2 x 2 symmetric matrices and let SO(3) act on $\mathbb{R}^5$ by

$$g \cdot Q = gQ\tilde{g}, \quad g \in SO(3), \quad Q \in \mathbb{R}^5 \tag{5.12}$$

where $\tilde{g}$ is the transpose of g. Since every real symmetric matrix can be diagonalized by an orthogonal transformation of the type (5.12), it follows that every SO(3) orbit in $\mathbb{R}^5$ contains a diagonal matrix with entries $(\lambda_1, \lambda_2, \lambda_3)$ ordered such that $\lambda_1 \geq \lambda_2 \geq \lambda_3$. It follows then that we can distinguish four kinds of orbits characterized respectively by the diagonal matrices (a, a, a), (a, a, b), (a, b, b) and (a, b, c) where it is understood that $a > b > c$. If the elements of $\mathbb{R}^5$ are interpreted as the quadrupole moments of a nucleus, we see that these four kinds of orbit correspond to rotations of spherical, prolate, oblate and triaxial nuclei respectively. One observes that the spherical orbits each consist of a single point and are thus manifolds of zero dimensionality. The prolate and oblate orbits are two dimensional and the triaxial orbits are three-dimensional. From a geometrical point of view we shall see that the prolate and the oblate orbits are in fact, diffeomorphic and there are three strata in $\mathbb{R}^5$ consisting respectively of spherical, axially symmetric and generic triaxial orbits.

The concept of group orbits is very important in the theory of collective motion. For example, any path on the surface of an SO(3) orbit in N-particle configuration $\mathbb{R}^{3N}$ corresponds to a pure rotational motion of the system. More general rotational vibrational collective motions correspond to paths on SL(3, $\mathbb{R}$) or GL(3, $\mathbb{R}$) orbits.

5.4 Coset Spaces

Let H be a subgroup of a group G. The right coset gH of $g \in G$ is the set

$$gH = \{gh;\ h \in H\} . \tag{5.13}$$

It is a subset, but in general not a subgroup, of G. The coset space G/H is the set of all right cosets

$$G/H = \{gH;\ g \in G\} . \tag{5.14}$$

G/H like G is a manifold, as we shall show presently by constructing coordinate charts for G/H. The points of G/H clearly correspond to sets of points (cosets) in G and as one can easily ascertain these cosets have no point in common. A useful way to visualize G/H is to regard it as the image of a projection

$$G \rightarrow G/H; \quad g \mapsto gH \tag{5.15}$$

in which the elements of the coset gH project onto the single coset $gH \in G/H$. The projection is illustrated in fig. 5.1.

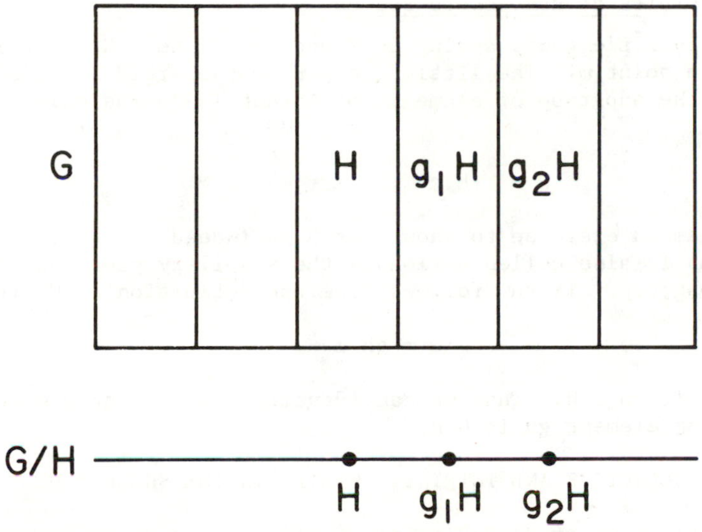

Fig. 5.1 A group G and its projection onto the coset space G/H.

The dimensionality of G/H is given by

$$\dim G/H = \dim G - \dim H \,. \tag{5.16}$$

As an example, suppose that G is SO(3) and H is SO(2). Any element r of SO(3) can be factored

$$r = r_z(\alpha)\, r_y(\beta)\, r_z(\gamma) \,, \tag{5.17}$$

where r_y and r_z are respectively rotations about the y and z axes. If H = SO(2) is regarded as the group of rotations about the z-axis then

$$rSO(2) = r_z(\alpha)\, r_y(\beta)\, SO(2) \tag{5.18}$$

and we see that all we need to characterize a coset rSO(2) in SO(3)/SO(2) are the polar angles (α, β) of a point on the surface of a sphere. Thus SO(3)/SO(2) is seen to be isomorphic to the sphere.

An important characteristic of a coset space G/H is that G acts transitively on it by

$$G : G/H \to G/H$$
$$\alpha : \quad gH \mapsto \alpha gH \tag{5.19}$$

thus G/H is a homogenous space.

A left coset space $H\backslash G$ is similarly defined as a space of left cosets $\{Hg;\ g \in G\}$. One notes that if a group element α belongs to the right coset gH, its inverse belongs to the left coset Hg^{-1}.

5.5 Group Orbits as Homogenous Spaces

Let G be a Lie group acting on M and let O_m be a G orbit in M containing a point m. The little group H of the orbit at m is defined as the subgroup of elements of G that leave the point g fixed, i.e.,

$$H = \{h \in G;\ h\cdot m = m\} . \tag{5.20}$$

It is left as an exercise to show that H is indeed a group. The little group is also called variously the stability group and the isotropy subgroup. It now follows from the definition of H, that

$$g\cdot m = gh\cdot m \tag{5.21}$$

if and only if $h \in H$. Thus we can identify every point $g\cdot m$ on O_m with a unique element gH in G/H.

6. COLLECTIVE AND INTRINSIC STATES IN THE SHELL MODEL

6.1 Stratification of Many-Particle Configuration Space

Many attempts have been made to obtain a transformation from the Cartesian coordinates of $\mathbb{R}^{3N}$ to collective and intrinsic coordinates. Although handicapped by conceptual difficulties such as the non-integrability of a set of coordinates defined infinitesimally, the attempts have met with considerable success and have revealed some surprisingly simple results. These results we now understand to have a very simple geometric origin.

Essentially the idea is to stratify the configuration space into orbits of a kinematical Lie group. Recall that the kinematical group for rotations is SO(3) and for the mass quadrupole collective model it is $GL(3, \mathbb{R})$. Recall also that any motion of the many-body system corresponding to a path in configuration space lying on an orbit of the kinematical group is by definition regarded as purely collective. We may therefore regard the generic orbit of the kinematical group as the configuration space for collective dynamics and define intrinsic motion as motion along some smooth transversal of the collective orbits, where a transversal is a hypersurface in $\mathbb{R}^{3N}$ that intercepts each collective orbit at just one point. Collective coordinates can then be introduced as coordinates for the generic orbit and intrinsic coordinates are a complimentary set for the transversal.

We illustrate by considering the simple rotation model. Let x^o be a generic point in $\mathbb{R}^{3N}$ and $O(x^o)$ the SO(3) orbit in $\mathbb{R}^{3N}$ containing x^o. Now let $\mathcal{D}^1(g)$ denote the fundamental 3 x 3 matrix representation of an SO(3) rotation g and let $(x_{ni};\ n = 1, \ldots, N,\ i = 1, 2, 3)$ be Cartesian coordinates for the point $x \in \mathbb{R}^{3N}$. The Cartesian coordinates for a point $g \cdot x^o$ on the orbit are then given by

$$(g \cdot x^o)_{ni} = x^o_{nj}\ \mathcal{D}^1_{ji}(g^{-1}) \tag{6.1}$$

c.f. eq. (2.18). According to section 5 the orbit is diffeomorphic to a homogeneous space SO(3)/H where H is the little group at x^o. Thus eq. (6.1) reduces to

$$(g \cdot x^o)_{ni} = (gH \cdot x^o)_{ni} = x^o_{nj}\ \mathcal{D}^1_{ji}(Hg^{-1}) \ . \tag{6.2}$$

We can now introduce collective coordinates (θ) for G/H and intrinsic coordinates (ξ) for a transversal $x^o(\xi)$ of the SO(3) orbits and write

$$x_{ni} = x^o_{nj}(\xi)\ \mathcal{D}^1_{ji}\ (Hg^{-1}(\theta)) \ . \tag{6.3}$$

Thus we obtain a transformation from Cartesian to collective and intrinsic coordinates.

For example, suppose N = 1. The SO(3) orbits are spheres centered about the origin in $\mathbb{R}^3$. We may take as transversal the positive 3-axis and use r as intrinsic coordinate by writing

$$x^o_A(r) = r\delta_{A3} \ . \tag{6.4}$$

The little group H is recognised as the group SO(2) of rotations about the 3-axis. If we use the standard factorisation of SO(3) to write an arbitrary element

$$g = e^{-i\alpha L_3}\ e^{-i\beta L_2}\ e^{-i\gamma L_3} \ , \tag{6.5}$$

we obtain

$$gSO(2) = e^{-i\alpha L_3}\ e^{-i\beta L_2}\ SO(2) \tag{6.6}$$

with the third rotation absorbed into SO(2). Thus eq. (6.3) gives

$$x_i = r\mathcal{D}^1_{3i}(SO(2)e^{i\beta L_2}\ e^{i\alpha L_3}) = r\mathcal{D}^1_{3i}(e^{i\beta L_2}\ e^{i\alpha L_3}) \tag{6.7}$$

and we obtain the familiar spherical polar coordinate decomposition

$$x_1 = r \sin\beta \cos\alpha$$

$$x_2 = r \sin\beta \sin\alpha \qquad (6.8)$$

$$x_3 = r \cos\beta \ .$$

For $N > 1$, the only SO(3) element that leaves a generic $\mathbb{R}^{3N}$ point invariant is the identity element. The little group H is then the identity subgroup I and we find that $O(x^o)$ is diffeomorphic to SO(3). We therefore need all three rotational Euler angles (α, β, γ) and a complimentary set (ξ) of 3N-3 intrinsic coordinates to write

$$x_{ni} = x^o_{nA}(\xi) \, \mathcal{D}^1_{Ai}(\alpha, \beta, \gamma) \ . \qquad (6.9)$$

This construction can clearly be applied whenever one is given a kinematical collective group of transformations of $\mathbb{R}^{3N}$.

The second problem we consider is the associated decomposition of the kinetic energy into collective and intrinsic components. The many-body kinetic energy $(1/2m) \Sigma p^2_{ni}$ is by eq. (4.12) simply proportional to the Laplace-Beltrami operator Δ on $\mathbb{R}^{3N}$. We therefore use the general expression (4.16) to express Δ in terms of infinitesimal generators of collective and intrinsic motions. Since we start with a kinematical group as a group of transformations of $\mathbb{R}^{3N}$ a natural set of infinitesimal generators of collective motions (collective momenta) is obtained directly from the corresponding realization of the Lie algebra. From among these we must select a set of base fields which, by definition of base fields, form a basis of generators of infinitesimal collective motion at each point of a neighbourhood of the orbit. Let (π_μ) be a set of base fields for the generic collective orbit and (X_α) a complimentary orthogonal set for the intrinsic motion. By orthogonal we mean that

$$g(\pi_\mu, X_\alpha) = 0 \qquad (6.10)$$

for all (π_μ) and (X_α), where g is the Euclidian metric for $\mathbb{R}^{3N}$ defined by

$$g(p_{mi}, p_{nj}) = \delta_{mn} \, \delta_{ij} \ . \qquad (6.11)$$

Since the tangent space to a manifold is a linear space, we can evidently always find a complimentary set of orthogonal intrinsic momenta at any point. Furthermore, since the collective momenta obtained from the Lie algebra are smooth we can always find an orthogonal intrinsic set which are likewise smooth over a neighbourhood of a collective orbit. It follows therefore that we can use eq. (4.16) to express the Laplace-Beltrami operator in terms of such a set of collective and intrinsic momenta and obtain

$$\Delta = \Delta_{coll} + \Delta_{intr} \qquad (6.12)$$

where

$$\Delta_{coll} = \pi_\mu \, g^{\mu\nu}_{coll} \, \pi_\nu + [p_{ni}, \pi^{ni}_\mu] \, g^{\mu\nu}_{coll} \, \pi_\nu \qquad (6.13)$$

with

$$(g_{coll})_{\mu\nu} = g(\pi_\mu, \pi_\nu) \quad . \qquad (6.14)$$

The intrinsic component of Δ is similarly defined in terms of the intrinsic momenta. The interesting observation is that Δ separates into a sum of collective and intrinsic components with no dynamical coupling terms, i.e., there are no cross terms in the collective and intrinsic momenta. This result follows directly from our ability to choose collective and intrinsic momenta which are orthogonal.

For example, a set of base fields for the SO(3) orbits in $\mathbb{R}^{3N}$ are given by the angular momenta

$$L_i = \Sigma \, \varepsilon_{ijk} \, x_{nj} \, p_{nk} \; . \qquad (6.15)$$

From eq. (6.11) we obtain immediately

$$g(L_i, L_j) = \mathcal{I}_{ij} \qquad (6.16)$$

where $(\mathcal{I}_{ij})$ is the rigid-body inertia tensor and from eq. (6.13)

$$\Delta_{coll} = L_i \, \mathcal{I}^{-1}_{ij} \, L_j = L \cdot \mathcal{I}^{-1} \cdot L \qquad (6.17)$$

giving the familiar rigid-body expression for the rotational kinetic energy. This expression was obtained with much more difficulty in ref. 13 using the method of infinitesimal canonical transformations (i.e., chain-rule) method.

We conclude this subsection with two observations. Firstly one should not be mislead into thinking that the intrinsic momenta are tangent to some intrinsic transversal. For if they were, it would mean that there existed a transversal orthogonal to every collective orbit and such is not in general the case. Thus we must introduce a non-orthogonal transversal if we wish to define intrinsic coordinates and accept the fact that our intrinsic momenta, although everywhere orthogonal to the collective orbit surfaces, are not tangent to the transversal. It follows therefore that there do not, in general, exist intrinsic coordinates and momenta which are canonical with respect to each other. The stratification of $\mathbb{R}^{3N}$ into collective group orbits in fact endows $\mathbb{R}^{3N}$ with an interesting fibre bundle structure which reveals what is going on but which we do not discuss here. Fortunately, and this is the second observation, if one is interested only in the collective sub-dynamics it is not necessary to construct either an intrinsic transversal or intrinsic momenta in order to examine the structure of the collective configuration space or to derive the collective kinetic energy.

6.2 The CM(3) Kinetic Energy

The kinematical group for mass quadrupole collective dynamics is the general linear group GL(3, $\mathbb{R}$). This group acts on the Cartesian coordinates of $\mathbb{R}^{3N}$ as indicated in eq. (2.18)

$$(g^{-1}\cdot x)_{ni} = x_{n\alpha}\, G_{\alpha i}(g), \qquad g \in GL(3, \mathbb{R}) \tag{6.18}$$

where G is the fundamental 3 x 3 matrix representation of GL(3, $\mathbb{R}$). Now the only element of GL(3, $\mathbb{R}$) that leaves a generic point x^o in $\mathbb{R}^{3N}$ fixed is the identity element. Therefore the little group is the identity subgroup and the generic orbits are diffeomorphic to GL(3, $\mathbb{R}$).

If we take as generators of infinitesimal general linear collective displacements the collective momenta t_{ij} which are realized as the vector fields on $\mathbb{R}^{3N}$

$$t_{ij} = \sum_{n=1}^{N} x_{ni}\, p_{nj} \tag{6.19}$$

then we immediately obtain from the $\mathbb{R}^{3N}$ metric (6.11)

$$g(t_{ij}, t_{\ell k}) = \delta_{jk}\, Q_{i\ell} \tag{6.20}$$

where

$$Q_{i\ell} = \sum_{n=1}^{N} x_{ni}\, x_{n\ell} \; . \tag{6.21}$$

In other words, the microscopic realization of gl(3, $\mathbb{R}$) naturally lets us induce a metric on GL(3, $\mathbb{R}$), from the Euclidian $\mathbb{R}^{3N}$ metric. We also find that

$$\sum_{ni} [p_{ni}, t^{ni}_{\ell k}] = -iN\delta_{\ell k} \tag{6.22}$$

and hence obtain the CM(3) kinetic energy

$$\begin{aligned} T_{CM(3)} &= \frac{1}{2m}\, \Delta_{coll} \\ &= \frac{1}{2m} \sum_{i\ell} [\sum_{j} t_{ij} Q^{-1}_{i\ell} t_{j\ell} - iNQ^{-1}_{i\ell} t_{i\ell}] \; . \end{aligned} \tag{6.23}$$

To obtain the CM(3) kinetic energy in the form (2.42) we must express it in terms of the nine momenta (L_A, t_A, $\mathcal{L}_A$; A = 1, 2, 3) of equations (2.36) - (2.38) rather than the (t_{ij}). Before doing this it is important to appreciate precisely what the somewhat schematic definitions (2.36) - (2.38) of these momenta really mean. First of all the (t_{ij}) are infinitesimal generators of the group action

$$\alpha\psi(g) = \psi(g\alpha), \quad \alpha,\ g \in GL(3,\ \mathbb{R}) \tag{6.24}$$

where ψ is a smooth function on GL(3, $\mathbb{R}$). The definition of L_A, for example, is then to be understood by using the expansion (2.29) of g and writing

$$L_A\ \psi(r_2^{-1}\cdot\lambda\cdot r_1) = \Sigma\ \varepsilon_{ABC}\ r_1^{-1}\ t_{BC}\ r_1\ \psi(r_2^{-1}\cdot\lambda\cdot r_1)\ . \tag{6.25}$$

We now wish to regard these momenta simply as vector fields on GL(3, $\mathbb{R}$). Recall, as discussed in section 5.2, that a basis for a Lie algebra is a set of base fields on the group and so we can expand the momenta (L_A, t_A, $\mathcal{L}_A$) on the (t_{ij}) basis at each point

$$g = r_2^{-1}\cdot\lambda\cdot r_1 \tag{6.26}$$

on the orbit. Recall also that $r_1^{-1}\cdot t_{BC}\cdot r_1$ for a given r_1 is a linear transformation of t_{BC} into a new element of gl(3, $\mathbb{R}$). Now for each g we have defined by eq. (2.21) the matrix

$$\begin{aligned} G_{ij}(L_A) &= \Sigma\ \varepsilon_{ABC}\ \mathcal{D}^1_{i\ell}(r_1^{-1})G_{\ell k}(t_{BC})\ \mathcal{D}^1_{kj}(r_1) \\ &= -i\ \Sigma\ \varepsilon_{ABC}\ \mathcal{D}^1_{Bi}(r_1)\ \mathcal{D}^1_{Cj}(r_1)\ . \end{aligned} \tag{6.27}$$

It follows therefore, using eq. (2.21) now in reverse, that L_A is realised as the vector field

$$L_A = \Sigma\ \varepsilon_{ABC}\ \mathcal{D}^1_{Bi}(r_1)\ \mathcal{D}^1_{Cj}(r_1)\ t_{ij}\ . \tag{6.28}$$

Similarly for t_A and $\mathcal{L}_A$ we obtain

$$t_A = \Sigma\ \mathcal{D}^1_{Ai}(r_1)\ \mathcal{D}^1_{Aj}(r_1)\ t_{ij} \tag{6.29}$$

$$\mathcal{L}_A = \Sigma\ \varepsilon_{ABC}\ \frac{\lambda_C}{\lambda_B}\ \mathcal{D}^1_{Bi}(r_1)\ \mathcal{D}^1_{Cj}(r_1)\ t_{ij}\ . \tag{6.30}$$

Using the induced GL(3, $\mathbb{R}$) metric (6.20) we now obtain the expression (2.42) for the CM(3) kinetic energy. Details of the derivation which is quite straighforward are given in ref. 8.

6.3 Collective and Intrinsic Configuration Spaces

The configuration space for N particles in three dimensions is $\mathbb{R}^{3N}$. Wave functions on this space are customarily expressed in terms of Cartesian coordinates (x_{ni}; $n = 1, \ldots, N$, $i = 1, 2, 3$). In order to construct a basis of wave functions that naturally reduces cm(3), we seek a more convenient coordinatization. The coordinates we use are those introduced by Zickendraht[14] and independently by Dzyublik et al.[15], which we present here using the method of group orbits.

The elements of $\mathbb{R}^{3N}$ can be regarded either as an N-tuple $(r_1, \ldots, r_N)$ of $\mathbb{R}^3$ vectors or as a triple (R_1, R_2, R_3) of $\mathbb{R}^N$ vectors. Thus there are natural actions of $GL(3, \mathbb{R})$ and $GL(N, \mathbb{R})$, and hence of their subgroups, on $\mathbb{R}^{3N}$. In particular, we are interested in $GL(3, \mathbb{R})$ and $O(N) \subset GL(N, \mathbb{R})$. The action of $GL(3, \mathbb{R})$ on $\mathbb{R}^{3N}$ was expressed in terms of Cartesian coordinates and the fundamental matrix representation G of $GL(3, \mathbb{R})$ by eq. (2.18), viz.

$$(g\cdot x)_{ni} = \sum_{j=1}^{3} x_{nj}\, G_{ji}(g^{-1}), \quad g \in GL(3, \mathbb{R}) \ . \qquad (6.31)$$

The action of O(N) is similarly defined

$$(R\cdot x)_{ni} = \sum_{m=1}^{N} x_{mi}\, D^1_{mn}(R^{-1})$$
$$= \sum_{m=1}^{N} D^1_{nm}(R)\, x_{mi}, \quad R \in O(N) \ , \qquad (6.32)$$

where D^1 is the fundamental $N \times N$ matrix representation of O(N). Thus the direct product group $GL(3, \mathbb{R}) \times O(N)$ acts by

$$(g\cdot R\cdot X)_{ni} = \sum_{mj} D^1_{nm}(R)\, x_{mj}\, G_{ji}(g^{-1}) \ . \qquad (6.33)$$

We now claim that almost all $x \in \mathbb{R}^{3N}$ can be expressed

$$x = r^{-1}\cdot\lambda^{-1}\cdot R^{-1}\cdot x^o \qquad (6.34)$$

for some fixed $x^o \in \mathbb{R}^{3N}$ and for some $r \in SO(3) \subset GL(3, \mathbb{R})$, $\lambda = \mathrm{diag}(\lambda_1, \lambda_2, \lambda_3) \in GL(3, \mathbb{R})$ and $R \in O(N) \subset GL(N, \mathbb{R})$ depending on the point x. (The use of inverses is for later convenience.) In terms of Cartesian coordinates, eq. (6.34) is expressed, using (6.33),

$$x_{ni} = \sum_{m}^{N} \sum_{A=1}^{3} D^1_{mn}(R)\, x^o_{mA}\, \lambda_A\, \mathcal{D}^1_{Ai}(r) \ . \qquad (6.35)$$

The proof is by construction. The fixed vector x^o is chosen to be the N particle configuration with particles 1, 2 and 3 respectively at unit distance along the 1, 2, and 3 axes and particles 4, ..., N at the origin; i.e.,

$$x^o_{mA} = \delta_{mA} \ . \qquad (6.36)$$

Equation (6.35) then becomes

$$x_{ni} = \sum_{A=1}^{3} \mathcal{D}^1_{Ai}(r)\ \lambda_A\ D^1_{An}(R)\ . \tag{6.37}$$

Now, given $x \in \mathbb{R}^{3N}$, define r and λ by diagonalization of the 3 x 3 matrix

$$Q_{ij} = \sum_{n=1}^{N} x_{ni}\ x_{nj}\ , \tag{6.38}$$

i.e.

$$\mathcal{D}^1_{Ai}(r)\ Q_{ij}\ \mathcal{D}^1_{Bj}(r) = \lambda^2_A\ \delta_{AB}\ . \tag{6.39}$$

There is considerable ambiguity in the choice of $r \in SO(3)$ that satisfies this equation. Some of the ambiguity is resolved by imposing the constraint

$$\lambda^2_1 \geq \lambda^2_2 \geq \lambda^2_3 \geq 0\ ; \tag{6.40}$$

cf. eq. (2.27). The rotation r is then defined generically to within a rotation m from the discrete subgroup M of rotations that leave the diagonal matrix λ invariant (cf. eq. (2.28)). We must therefore select a representative r from the coset $Mr \in M\backslash SO(3)$. Having made a choice of r, define

$$D^1_{An}(R) = \sum_{i=1}^{3} \lambda^{-1}_A\ \mathcal{D}^1_{Ai}(r)\ x_{ni}\ , \tag{6.41}$$

where λ_A is the positive square root of λ^2_A. One observes that $D^1_{An}(R) D^1_{Bn}(R) = \delta_{AB}$ implying that $D^1_{An}(R)$ may be regarded as the first three rows of an orthogonal N x N matrix. However, since only three rows of $D^1(R)$ are specified, it follows that $R \in O(N)$ is not uniquely defined. In fact R is defined modulo any R' having the property

$$D^1_{An}(R'\cdot R) = D^1_{An}(R) \tag{6.42}$$

which implies that

$$D^1_{Am}(R') = \delta_{Am} \tag{6.43}$$

and hence, since A = 1, 2, 3, that $R' \in O(N-3)$. Thus any R in the coset $O(N-3)R$ in fact satisfies eq. (6.41). This completes the proof.

To identify an element $O(N-3)R$ of the coset space $O(N-3)\backslash O(N)$ we need 3N-6 coordinates, since

$$\dim(H\backslash G) = \dim G - \dim H \tag{6.44}$$

and

$$\dim O(N) = \frac{1}{2} N(N-1)\ . \tag{6.45}$$

Three rotational coordinates, e.g., Euler angles, for $r \in SO(3)$ and the elements $(\lambda_1, \lambda_2, \lambda_3)$ then make up a set of 3N coordinate for $\mathbb{R}^{3N}$, related to the Cartesian coordinates by eq. (6.37).

It is clear that one can identify r and λ with the r_1 and λ featuring in the decomposition $g = r_2^{-1}\cdot\lambda\cdot r_1$ of $g \in GL(3, \mathbb{R})$. Regarding $x \in \mathbb{R}^{3N}$ in eq. (6.34) as a $GL(3, \mathbb{R}) \times O(N)$ transform of the fixed vector $x^o \in \mathbb{R}^{3N}$, one may wonder what has happened to the $O(3) \subset GL(3, \mathbb{R})$ transformation r_2. The answer is that it is absorbed into the O(N) rotation. To see how this comes about observe that, while eq. (6.37) can be written

$$x_{ni} = \sum_{A=1}^{3} G_{Ai}(\lambda\cdot r)\, D^1_{An}(R) \ , \tag{6.46}$$

it can also be written

$$x_{ni} = \sum_{A=1}^{3} G_{Ai}(r_2^{-1}\cdot\lambda\cdot r)\, D_{An}(r_2^{-1}\cdot R) \tag{6.47}$$

where r_2 is any O(3) element in $GL(3, \mathbb{R})$ and $r_2 \in O(3) \subset O(N)$ is defined by

$$D^1_{Am}(r_2^{-1}) = \sum_{B=1}^{3} \delta_{mB}\, \mathcal{D}^1_{AB}(r_2^{-1}) \ . \tag{6.48}$$

Thus it is possible to define $GL(3, \mathbb{R})$ coordinates, corresponding precisely to those of the phenomenological cm(3) model, and a complementary set of 3N-9 coordinates for an intrinsic hypersurface in $O(N-3)\backslash O(N)$ to make up a complete set for $\mathbb{R}^{3N}$. However, since the O(3) part of the $GL(3, \mathbb{R})$ decomposition missing in eq. (6.46) is seen to be simply embedded in O(N) in a way clearly exhibited by eq. (6.47) there is no need to complicate matters by pulling it out.

6.4 The Many-Particle Kinetic Energy

Having in mind that a many-particle wave function $\phi(x)$ is to be expressed as a function

$$\psi(r, \lambda, R) = \phi(x) \tag{6.49}$$

by means of the identity (6.34), we wish to express the kinetic energy in terms of the infinitesimal generators of the group elements (r, λ, R). The infinitesimal generators of displacements r, λ have already been given by L_A and t_A respectively in equations (6.28) and (6.29) which with the interpretation (6.19) of t_{ij} are realized as vector fields on $\mathbb{R}^{3N}$. It remains therefore to choose a set of base fields for $O(N-3)\backslash O(N)$ which will serve as generators of infinitesimal displacements in R.

First observe that the group action

$$\phi(x) \overset{g}{\to} \phi(g^{-1}\cdot x), \quad g \in O(N) \tag{6.50}$$

of O(N) on a many-particle wave function $\phi(x)$ implies, by means of equation (6.34), the action

$$\psi(r, \lambda, R) \overset{g}{\to} \psi(r, \lambda, Rg) . \qquad (6.51)$$

The infinitesimal generators of this action are given by the realization of the o(N) Lie algebra

$$j_{mn} = \Sigma_{i=1}^{3} (x_{mi}p_{ni} - x_{ni}p_{mi}), \quad m, n = 1, \ldots, N . \qquad (6.52)$$

These are certainly vector fields tangent to the O(N) orbits in $\mathbb{R}^{3N}$. However there are too many of them. We require only 3N-6 base fields since this is the dimensionality of O(N-3)\O(N). In particular, we may eliminate at each point of a generic orbit, the infinitesimal generators of the little group O(N-3) which are realized as null vectors.

First we construct the vector fields corresponding to the infinitesimal generators of the left O(N) group action;

$$\Psi(R) \overset{g}{\to} \Psi(gR), \quad g \in O(N) . \qquad (6.53)$$

In parallel with the construction of L_A, eq. (6.28), we obtain

$$\mathcal{J}_{st} = \Sigma\, D^{1}_{sm}(R)\, D^{1}_{tn}(R)\, j_{mn} . \qquad (6.54)$$

A set of base fields for O(N-3)\O(N) is then obtained by restricting this set to the subset

$$\mathcal{J}_{sA}, \quad s = 1, \ldots, N, \quad A = 1, 2, 3, \ s > A ; \qquad (6.55)$$

i.e., we eliminate the infinitesimal generators ($\mathcal{J}_{\alpha\beta}$, α, β = 4,..., N) of the little group O(N-3).

Substituting the microscopic expression (6.41) for $D^{1}_{An}(R)$ we discover that

$$\mathcal{J}_{AB} = \Sigma\, D^{1}_{Am}(R)\, D^{1}_{Bn}(R)\, j_{mn} = - \mathcal{L}_{c} \qquad (6.56)$$

which does not surprise us in view of the identity (6.47). The remaining fields are given by

$$\mathcal{J}_{\alpha A} = - \lambda_A\, \Sigma_n\, D^{1}_{\alpha n}(R)\, \mathcal{D}^{1}_{Ai}(r)\, p_{ni}, \quad \begin{array}{l} \alpha = 4, \ldots, N \\ A = 1, 2, 3 \end{array} \qquad (6.57)$$

where to define $D^{1}_{\alpha n}(R)$ we must now specify R by picking a set of coset representatives for each O(N-3)R. A particular choice then defines a basis ($\mathcal{J}_{\alpha A}$) of intrinsic momentum operators to compliment the collective momenta (L_A, t_A, $\mathcal{L}_A$) but as we shall see, the resulting expression for the kinetic energy will not, or could it, depend on the choice.

Using the expansion (6.57) and the metric (6.11), we find that these intrinsic momenta are all orthogonal to the collective momenta and indeed to each other. We find

$$g(J_{\alpha A}, J_{\beta B}) = \delta_{\alpha\beta}\, \delta_{AB}\, \lambda_A^2 \ . \qquad (6.58)$$

Finally, as discussed in more detail in reference 8, the intrinsic kinetic energy to compliment the cm(3) kinetic energy of eq. (2.42) is found to be

$$T_{intr} = \sum_{\alpha,A} \frac{1}{2m\lambda_A^2} J_{\alpha A}^2 \ . \qquad (6.59)$$

One readily confirms that, because of the summation on α, this expression is independent of the choice of basis $(J_{\alpha A})$.

6.5 The Many-Fermion Hilbert Space

To construct the nuclear Hilbert space we consider the square integrable functions $L^2(\mathbb{R}^{3N})$ on $\mathbb{R}^{3N}$ and then select those functions of permutation symmetry that can be combined with spin-isospin functions of contragredient symmetry to form totally anti-symmetric states.

First consider the space of square integrable functions on $O(N-3)\backslash O(N)$. A complete set on O(N) is given by the matrix elements $D^{\omega}_{\sigma,\tau}$ of the O(N) unirreps where ω labels a unirrep and σ and τ label bases for the carrier space; i.e.,

$$D^{\omega}_{\sigma,\tau}(R) = \langle\omega, \sigma \mid U(R) \mid \omega, \tau\rangle \qquad (6.60)$$

where U denotes the unitary action of O(N) on the space. To obtain a complete set of functions on $O(N-3)\backslash O(N)$, we consider a basis which reduces the subgroup

$$[SO(3) \supset SO(2)] \times [O(N-3) \supset O(N-4) \supset \ldots] \subset O(N) \ .$$

Thus we set $\sigma \equiv \delta LK\Omega$ where δ is a multiplicity index, LK label the basis states for SO(3 ⊃ SO(2) and Ω labels the states of the O(N-3) chain. A complete set of functions on $O(N-3)\backslash O(N)$ is obtained by restricting the functions on O(N) to the subset that are invariant under O(N-3); i.e., the functions f having the property

$$f(R'\cdot R) = f(R), \quad R' \in O(N-3) \subset O(N), \quad R \in O(N) \ . \qquad (6.61)$$

Such a subset is simply obtained by selecting those D functions with $\Omega = 0$, where $\Omega = 0$ indexes the O(N-3) identity representation. Recall the parallel situation in which functions on $SO(2)\backslash SO(3)$ are obtained by restricting the SO(3) rotation matrices $\mathcal{D}^L_{KM}$ to the K = 0 subset. Thus a basis of square integrable functions on $O(N-3)\backslash O(N)$ is given by the set $\{D^{\omega}_{\delta LK0,\tau}\}$.

Looking at the coordinate decomposition (6.37) suggests that we consider the functions of (r, λ, R)

$$\mathcal{D}^{L}_{KM}(r)\ f^{L\mathcal{L}}_{K\mathcal{K}[\nu]}(\lambda)\ D^{\omega}_{\delta\mathcal{L}\mathcal{K}0,\tau}(R) \tag{6.62}$$

in parallel with eq. (2.30). However, as in the construction of the phenomenological Hilbert space, we must take into account the fact that $r \in SO(3)$ is only defined, by eq. (6.39), to within a rotation from the discrete subgroup $M \subset SO(3)$ that leaves λ invariant (cf. eq. (2.28)). Writing x_{ni} in the form (6.46), we see that

$$\begin{aligned} x_{ni} &= \sum_A G_{Ai}(m^{-1}\cdot\lambda\cdot mr)\ D^{1}_{An}(R) \\ &= \sum_A G_{Ai}(\lambda\cdot mr)\ D^{1}_{An}(mR)\ . \end{aligned} \tag{6.63}$$

Thus we need to symmetrize the functions (6.62) to make them invariant under the transformation $(r, \lambda, R) \to (mr, \lambda, mR)$, $m \in M$. In parallel with eq. (2.32) we obtain the basis functions for $\mathcal{L}^2(\mathbb{R}^{3N})$

$$\begin{aligned} \Psi^{\omega L}_{\delta\mathcal{L}\mathcal{K}[\nu]KM\tau}(r, \lambda, R) &= [1 + (-1)^{K+\mathcal{K}}]\ f^{L\mathcal{L}}_{K\mathcal{K}[\nu]}(\lambda) \\ &\times [\mathcal{D}^{L}_{KM}(r)\ D^{\omega}_{\delta\mathcal{L}\mathcal{K}0,\tau}(R) + (-1)^{L-K+\mathcal{L}-\mathcal{K}}\mathcal{D}^{L}_{-KM}(r) D^{\omega}_{\delta\mathcal{L}-\mathcal{K}0,\tau}(R)]. \end{aligned} \tag{6.64}$$

From among these $\mathcal{L}^2(\mathbb{R}^{3N})$ functions we must finally restrict to those of permissable permutation symmetry. Since the symmetric group S_N is a subgroup of O(N), all we have to do is choose a basis τ, for each unirrep ω, to reduce this subgroup. Note that a set of functions $D^{\omega}_{\sigma,\tau}$ can be constructed with σ and τ indexing completely unrelated basis states for the carrier space of the ω representation; i.e.,

$$D^{\omega}_{\sigma,\tau}(R) = \langle\Psi^{\omega}_{\sigma}|U(R)|\Phi^{\omega}_{\tau}\rangle\ . \tag{6.65}$$

Thus we set $\tau \equiv \alpha\lambda\mu$ where α is a multiplicity index, λ indexes a unirrep of $S_N \subset O_N$ and μ indexes a basis for this unirrep. One then simply selects those states having permutation symmetry given by Young tableaux of not more than four columns and combines them with spin-isospin wave functions of contragredient symmetry to span the space of totally anti-symmetric many-fermion states.

It is clear, by comparison with the phenomenological cm(3) results, that this basis reduces the many-fermion Hilbert space into irreducible cm(3) subspaces. Each cm(3) unirrep is characterized by the value of $\mathcal{L}$ which assumes integer values. In addition the basis reduces $O(N) \supset S_N$. Thus each cm(3) unirrep L occurs many times with different O(N) label ω, multiplicity index δ and permutation symmetry labels $\alpha\lambda$. Furthermore, the basis states are labelled by quantum numbers associated with the subgroup chain

$$\underset{L}{[CM(3)} \supset \underset{L}{SO(3)} \supset \underset{M}{SO(2)]} \times [\underset{\omega}{O(N)} \supset \underset{\lambda}{S_N]} \,. \qquad (6.66)$$

We recall, as discussed briefly in section 4, that the whole construction is trivially modified to separate out the many-fermion centre-of-mass motion. One simply replaces the many-fermion configuration space $\mathbb{R}^{3N}$ by the subspace $\mathbb{R}^{3(N-1)}$ of relative many-fermion configurations with the centre-of-mass subspace $\mathbb{R}^3$ removed. Correspondingly the Cartesian coordinates (x_{ni}; n = 1,..., N, i = 1, 2, 3) are replaced by a system of Cartesian coordinates ($\rho_{\nu i}$; ν = 1, ..., N-1, i = 1, 2, 3) for $\mathbb{R}^{3(N-1)}$ such as, for example Jacobi coordinates. Since the group O(N-1) contains the symmetric group S_N as a subgroup, the above construction of the many-fermion Hilbert space gives basis states labelled by the quantum numbers associated with the chain

$$\underset{L}{[CM(3)} \supset \underset{L}{SO(3)} \supset \underset{M}{SO(2)]} \times [\underset{\omega}{O(N-1)} \supset \underset{\lambda}{S_N]} \,. \qquad (6.67)$$

We also recall, as observed in section 4, the complimentarity of $Sp(3, \mathbb{R})$ and O(N-1) in the reduction of the many-fermion Hilbert space; i.e., we learned that the construction of a basis of many-fermion states such that they fall into irreducible subspaces of O(N-1) means that they simultaneously fall into irreducible subspaces of $Sp(3, \mathbb{R})$. In other words, the same label $\omega = (\omega_1, \omega_2, \omega_3)$ serves to classify both the O(N-1) unirreps. and the $Sp(3, \mathbb{R})$ unirreps. Furthermore it follows that we have constructed the many-fermion Hilbert space in a basis of states labelled by the quantum numbers

$$[\underset{\omega}{Sp(3, \mathbb{R})} \supset \underset{\delta\ L}{CM(3)} \supset \underset{L}{SO(3)} \supset \underset{M}{SO(2)]} \times [\underset{\omega}{O(N-1)} \supset \underset{\lambda}{S_N]} \qquad (6.68)$$

where δ is a multiplicity index.

7. SUMMARY AND CONCLUSIONS

We summarize what has been achieved so far and the outstanding problems.

First recall that in order to make the collective model more microscopically realizable it was necessary to extend it to the cm(3) model. This model in general has three more degrees of freedom than the Bohr-Mottelson model, namely the vortex rotations. The latter is a submodel realized in the $L = 0$ cm(3) representation. $L = 0$ representations occur in finite nuclei whenever the appropriate O(N-1) representation ω contains the identity

(L = 0, Ω = 0) representation of the SO(3) x O(N-4) subgroup. However $L \neq 0$ representations also occur and one must question which irreps. are likely to be the more important in the low-lying spectra of collective nuclei.

One also questions how good the cm(3) quantum number L is for physical states. The answer is probably that it is not so good. In particular the kinetic energy part of the Hamiltonian already does not conserve cm(3) symmetry. As has been shown in reference 7, pure L = 0 states correspond to irrotational flow collective motion. On the other hand rigid flow rotations, for example, correspond to mixtures of different L. Experimentally it is known that nuclear collective states have inertial parameters very different from either limit.

Some elementary cm(3) calculations with phenomenological potentials are currently being pursued. The major outstanding problem is the construction of a suitable set of basis functions $f^{LL}_{KK[\nu]}(\lambda)$; cf. eqs. (2.32) and (6.64). Note that the interest is not so much to do the most realistic calculations possible or to fit collective spectra, since the sp(3, $\mathbb{R}$) model is more general and simpler to apply, but to understand the kinds of spectra, the effects of vorticity and the properties that this purely collective model can have. We are also investigating the effects of mixing cm(3) irreps.

Microscopic cm(3) calculations are in principle possible with the Hamiltonian defined ideally simply as the restriction of the given microscopic Hamiltonian to the chosen irreducible cm(3) subspace. At present we do not know how to execute such calculations. We note however that Vanagas, as discussed in his notes, is having considerable success with a somewhat less ambitious objective. He defines his microscopic collective Hamiltonian as the O(N-1) scalar part of the fully microscopic Hamiltonian, where by scalar he means scalar with respect to both right and left group actions. In other words his collective Hamiltonian leaves not only the right quantum numbers (τ) of an O(N-1) labelled state (cf. eq. (6.60)) unchanged it also leaves the left labels $\sigma = \delta LK\Omega$ unchanged. Thus it preserves cm(3) symmetry. Clearly this Hamiltonian is not the largest possible component of the microscopic Hamiltonian that respects cm(3) symmetry. It is nevertheless of considerable interest.

Since the kinetic energy is in the sp(3, $\mathbb{R}$) algebra we expect the sp(3, $\mathbb{R}$) model to be more realistic. Thus one expects the cm(3) irreps. which belong to a single sp(3, $\mathbb{R}$) irrep. to be most strongly mixed. For this reason we have concentrated more on the sp(3, $\mathbb{R}$) model not only because it is more general but because it is more amenable to calculation.

The reason the sp(3, $\mathbb{R}$) model is both simpler in application and more likely to be valid is that it contains U(3) and SU(3) as subgroups and consequently it is naturally adapted to the shell model. Indeed, as we have shown in section 3, the independent particle harmonic oscillator shell model Hamiltonian is actually an element of the sp(3, $\mathbb{R}$) algebra. To date sp(3, $\mathbb{R}$) calculations have only been performed for ^{20}Ne with a phenomenological collective potential. Such calculations are presently possible for any light nucleus. We expect soon to be able to extend them to some highly deformed rare earth and actinide nuclei. The nuclei in question are the ones for which the asymptotic Nilsson model would normally apply. They are the nuclei in which the deformation potential is so large that the spin-orbit interaction can be included in perturbation theory. The problem here is primarily the calculation of SU(3) Clebsch-Gordon coefficient for the very large values of λ and μ that occur. Realistic applications of the sp(3, $\mathbb{R}$) model to the medium mass nuclei would appear to be further away since it is believed that for these nuclei the spin-orbit, tensor and pairing forces which are not in the sp(3, $\mathbb{R}$) enveloping algebra, play a major role.

The next step is to develop the facility to execute microscopic sp(3, $\mathbb{R}$) calculations. The microscopic sp(3, $\mathbb{R}$) Hamiltonian is defined to be the restriction of a given fully microscopic Hamiltonian to the chosen irreducible sp(3, $\mathbb{R}$) subspace. Techniques for doing this have been pioneered by Vasilevsky, Smirnov and Filippov and are already well advanced. The ultimate goal is to learn how to perform full shell model calculations in a Sp(3, $\mathbb{R}$) x $[O(N-1) \supset S_N]$ basis with mixing of sp(3, $\mathbb{R}$) unirreps. as found desirable to adequately take account of the spin-orbit, tensor, and pairing forces. Clearly a lot remains to be done but one can be optimistic that a lot of progress will be made in the microscopic realization of collective states within the next few years.

REFERENCES

1. A Bohr, Mat. Fys. Medd. Dan. Vid. Selsk. 26 (1952) 14; A. Bohr and B. Mottelson, Mat. Fys. Medd. Dan. Vid. Selsk. 27 (1953) 16.
2. G. Gneuss and W. Greiner, Nucl. Phys. A171 (1971) 449; R. Sedlmayer et al., Nucl. Phys. A232 (1971) 465.
3. G. Rosensteel and D.J. Rowe, Ann. Phys. (N.Y.) 96 (1976) 1.
4. G. Rosensteel and D.J. Rowe, Ann. Phys. (N.Y.) 123 (1979) 36.
5. G. Rosensteel and D.J. Rowe, Phys. Rev. Lett. 38 (1977) 10; "On the algebraic formulation of collective models III. The symplectic shell model of collective motion" (Ann. Phys. (N.Y.) to appear).
6. L. Weaver and L.C. Biedenharn, Nucl. Phys. A185 (1972) 1.
7. P. Gulshani and D.J. Rowe, Can. J. Phys. 54 (1976) 970.
8. D.J. Rowe and G. Rosensteel, J. Math. Phys. 20 (1979) 465; D.J. Rowe and G. Rosensteel, Ann. Phys. (N.Y.) 126 (1980) 198.
9. B. Buck, L.C. Biedenharn and R.Y. Cusson, Nucl. Phys. A317 (1979) 205; O.L. Weaver, R.Y. Cusson and L.C. Biedenharn, Ann. Phys. (N.Y.) 102 (1976) 493.
10. G. Rosensteel and D.J. Rowe, Ann. Phys. 104 (1977) 134.
11. M. Kashiwara and M. Vergne, Invent. Math. 44 (1978) 1.
12. M. Moshinsky and C. Quesne, J. Math. Phys. 11 (1970) 1631; 12 (1971) 1772; E. Chacon, M. Moshinsky and P. Winternitz, Kinam 1 (1979) 259.
13. D.J. Rowe, Nucl. Phys. A152 (1970) 273.
14. W. Zickendraht, J. Math. Phys. 12 (1971) 1663.

15. A.Y. Dzyublik, V.I. Ovcharenko, A.I. Steshenko and G.V. Filipov, Sov. J. Nucl. Phys. 15 (1972) 487.
16. V.S. Vasilevsky, Y.F. Smirnov and G.F. Filippov, "Generating function of a total oscillator basis with definite O(A-1) symmetry" (reprint).

THE MICROSCOPIC THEORY OF THE COLLECTIVE MOTION NUCLEI

V. Vanagas

Institute of Physics, Academy of
Sciences of the Lithuanian SSR,
Vilnus USSR

ISSN:0094-243X/81/710220-74$1.50

INTRODUCTION

These lecture notes are addressed to the reader, interested in the latest developments in the field of the application of algebraic methods in many-fermion system theory. The results presented here are based on the theory of the orthogonal and unitary groups. The very special feature of these groups is that their rank is proportional to the number of particles. This is the reason why the many-particle type properties of the equivalent fermion systems can be caught within one irreducible space of the group. The pioneering works in this field have been published about twenty years ago by Bargman and Moshinsky.[1 2]

These lecture notes are built up in two levels. The first one is for the beginners. We start with the brief review of the physical problem (the first two section). Then follows section 3, introducing the translational invariant variable and then sections 4,6,7 and 9 present the first part of the general theory exposed on the three -particle example. From the group-theoretical point of view the three- and four-particle systems has been studied in[3,4]. In the three-particle case we will use these results, specially adapted in [5] for the needs of the theory of the collective motions in nuclei.

On the first level very simple mathematics is used, only the orthogonal group in two dimensions and the symmetric group S_3 . In spite of this simplicity, on this example we are trying to explain a wide variety of questions: different chains of subgroups and the transformations between their bases, the technique of the explicit construction of the irreducible representations, the reflection properties and so on. We think it very easy to reproduce in all details the results of sections 4,6,7 and 9. In sections 5,8 and 10 general results are described; for the first acquaintance one can treate these sections as the brief summary of the original works.

The content of the second level is much more difficult. It begins at section 11. The results presented in sections 11-16 are destined for the specialists interested either in the microscopic theory of the collective motions in nuclei or in the applications of the special algebraic technique to the many-body problem in the quantum mechanics. More details on this subject, including the questions not discussed here, are given in[6] where the additional references to the original papers may be found.

1. PRELIMINARY REMARKS

These lecture notes will be devoted to the problem having a long history. The collective forms of motions in the atomic nuclei have been discovered a quarter of a century ago in well known works of J. Rainwater[7], A.Bohr and B. Mottelson[8,9]. The starting point of their theory is related to the classical picture of small oscillations in a noncompressible liquid drop. Taking into account only the quadrupole type surface vibrations and quantizing this picture, the rotational-vibrational model has been introduced. The Schrödinger equation for the Hamiltonian of this model has the following form:

$$H_{R-V}(\bar{\beta}\bar{\gamma})\psi_{R-V}(\bar{\beta}\bar{\gamma}) = E_{R-V}\psi_{R-V}(\bar{\beta}\bar{\gamma}) \tag{1.1}$$

where

$$H_{R-V} = -\frac{\hbar^2}{2B}\left[\sum_{s>s'=1}^{3}\frac{\hat{\mathcal{L}}^2_{s's}}{\mathcal{J}_{s's}(\bar{\beta}\bar{\gamma})} + \frac{1}{\bar{\beta}^4}\frac{\partial}{\partial\bar{\beta}}\bar{\beta}^4\frac{\partial}{\partial\bar{\beta}} + \frac{1}{\bar{\beta}^2}\frac{1}{\sin 3\bar{\gamma}}\frac{\partial}{\partial\bar{\gamma}}\sin 3\bar{\gamma}\frac{\partial}{\partial\bar{\gamma}}\right] + V_{R-V}(\bar{\beta},\cos 3\bar{\gamma}) \tag{1.2}$$

In (1.2)B ismass parameter, related to the density of the liquid drop. The variables $\bar{\beta}$ $\bar{\gamma}$ are the phenomenological collective variables, related to the intrinsic frme of reference, thus, according to the phenomenological picture, $\bar{\beta}$ and $\bar{\gamma}$ describe the shape of nucleus in the intrinsic coordinate system. In (1.2) $\mathcal{L}_{s's}$ denote the intrinsic angular momentum operators and $\mathcal{J}_{s's}$ -the following moments of inertia:

$$\mathcal{J}_{12} = 4\bar{\beta}^2 \sin^2\bar{\gamma}$$

$$\mathcal{J}_{13} = 4\bar{\beta}^2 \sin^2(\bar{\gamma} - \tfrac{2}{3}\Pi)$$

$$\mathcal{J}_{23} = 4\bar{\beta}^2 \sin^2(\bar{\gamma} - \tfrac{4}{3}\Pi) \tag{1.3}$$

In the rotational-vibrational model the collective potential energy V_{R-V} depends on $\bar{\gamma}$ in the form cos $3\bar{\gamma}$. This type of dependence follows from the space reflection symmetry requirements.

The expression of $V_{R-V}(\bar{\beta},\cos 3\bar{\gamma})$ in the phenomenological theory is unknown. Traditionally V_{R-V} is taken as some polynomial in $\bar{\beta}$ and cos $3\bar{\gamma}$. Many details of the potential energy surfaces of the polynomial type have been studied in[10].

The phenomenological model, based on the Schrödinger equation (1.1), has been widely used for studying of the properties of the deformed nuclei. In spite of the success, this model was never logically well-founded. The meaning of the variables $\bar{\beta}$ and $\bar{\gamma}$ is not always used carefully enough. They are quantum-mechanical variables, thus not observable in the states with definite energy. Nevertheless very often $\bar{\beta}$ and $\bar{\gamma}$ are treated as shape parameters. We will treat them strictly in the quantum mechanical sense, as some phenomenological variables.

Much more serious problems arise while trying to understand the microscopic meaning of the rotational-vibrational Hamiltonian, i.e. to get H_{R-V} from the microscopical picture. How is it possible to obtain H_{R-V} from the microscopical Hamiltonian H, what is the realationship between both of them? How many collective variables is it possible to introduce in the framework of the nonrelativistic quantum mechanics? Which is the collective part of the arbitrary potential of the nucleon-nucleon interaction? These and many other questions arise while trying to understand the microscopical quantum-mechanical meaning of the phenomenological Hamiltonian H_{R-V}

In spite of such inconsistency with the grounds of the quantum mechanics, the phenomenological rotational-vibrational model works surprizingly well. In some regions of the mass numbers many of the observed level properties can be explained reasonably using this phenomenological semi-classical picture.

Lately the rotational-vibrational model has been extended by taking into account the connection between collective and single-particle degrees of freedom and the theory of the structure of atomic nucleus based on this connection has been developed. The great success of this theory in interpreting various properties of nuclei is widely recognized to-day and in 1975 three physisists - A.Bohr, B.Mottelson and J.Rainwater have been awarded Nobel Prize for their contribution in that field.

However, these remarkable achievements give us no full satisfaction until the microscopic meaning of the collective forms of motion in atomic nuclei are fully explained. Making comments on this problem Eisenberg and Greiner write in their book[10]: "... In fact, it is one of the challenging problems in present day theoretical nuclear physics to pin down the true collective Hamiltonian of a nucleus".

We are going to take this challenge and obtain the collective Hamiltonian from the microscopical one. This is the main subject of these lecture notes.

2. THE MICROSCOPIC NUCLEAR HAMILTONIAN

Numerous papers, review articles and chapters of books have been devoted to the problem of the microscoscopical description of the collective motion in nuclei. The term "microscopic description" has been used in very different contexts. Thus we need the exact definition of the meaning we are going to put into these words. From the point of view of the nonrelativistic quantum mechanics the following chain of reasoning is natural. Suppose one has the microscopic nuclear Hamiltonian H, containing all the information needed about various forms of motion in nuclei. If only the collective forms of motion is being studied, one must extract from H its collective part, i.e. to get the following decomposition of H:

$$H = H_{coll} + H_{coll\text{-}intern} \qquad (2.1)$$

where H_{coll} is the pure collective term, depending on the collective variables only, and $H_{coll\text{-}intern}$ takes into account the interaction between the collective and the rest (let us call them the internal) degrees of freedom.

Because of the decomposition (2.1), every H gives birth to its' collective part H_{coll}, thus one has the Schrödinger equation

$$H_{coll}\,\psi_{coll} = E_{coll}\,\psi_{coll} \qquad (2.2)$$

The spectrum E_{coll} and eigenstates ψ_{coll} give the microscopic description of the purely collective features hidden in H. That very sense we put into the words "microscopic description".

As mentioned in section 1, the main task of these lecture notes is to obtain in the explicit form the decomposition (2.1). Before starting, let us recall general features of the microscopic nuclear Hamiltonian H.

We will present H written in the form, using the isospin formalizm. Let m_p and m_n be the mass of proton and neutron, $t^1_\eta(i)$ -spherical components of the isospin angular momentum of the i-th particle. Then the kinetic energy operator has the following expression:

$$H''_k = -\frac{\hbar^2}{4}\sum_{i=1}^{n}\Delta_i\left[\frac{1+2\,t^1_o(i)}{m_n}+\frac{1-2\,t^1_o(i)}{m_p}\right] \qquad (2.3)$$

where Δ_i is the Laplace operator for the i-th particle. From (2.3) one can see that we prescribe $2t^1_o(i)=-1$ for protons and $2t^1_o(1)=+1$ for neutrons; thus the total projection of the isospin quantum number M_T is

$$M_T = \frac{1}{2} (n_p - n_n) \tag{2.4}$$

where n_p and n_n are the numbers of protons and neutrons. The total number of nucleons is $n = n_p + n_n$. In (2.3) $\hbar^2/m_p$ = 41,5000 MeV·F^2 and $\hbar^2/m_n$ = 41,4428 MeV·F^2, thus $\hbar^2(1/m_p - 1/m_n)$ = 0,0572 MeV·F^2. There is no need to take into account the term with this small factor, so let us take $m_p = m_n = m$. Then instead of (2.3) one has:

$$H_k' = - \frac{\hbar^2}{2m} \sum_{i=1}^{n} \Delta_i \tag{2.5}$$

In order to reveal the translational invariant properties of the operator (2.5), it is convenient to rewrite H_k' in the following equivalent form:

$$H_k' = H_{c.m} + H_k \tag{2.6}$$

where

$$H_{c.m} = - \frac{\hbar^2}{2m} \frac{1}{n} \sum_{s=1}^{3} \frac{\partial^2}{\partial (R^s)^2} \tag{2.7}$$

and

$$H_k = - \frac{\hbar^2}{2m} \frac{1}{n} \sum_{i>j=1}^{n} \sum_{s=1}^{3} \frac{\partial^2}{\partial (x_i^s - x_j^s)^2} \tag{2.8}$$

In (2.7) R^s are the Cartesian components of the centrum-of-mass vector

$$\vec{R} = \frac{1}{n} \sum_{i=1}^{n} \vec{r}_i \tag{2.9}$$

and in (2.8) x_i^s are the Cartesian components of the single-particle position vector $\vec{r}_i$.

The first term in (2.6) is the plane wave kinetic energy operator, thus it may be omitted and the translational invariant expression (2.8) may be considered as the kinetic energy operator.

The potential energy terms in H consist of Coulomb and nucleon-nucleon interactions terms. The simplest Coulomb term is

$$H_{coul} = \frac{e^2}{4} \sum_{i>j=1}^{n} \frac{(1-2\, t_o^1(i))\,(1-2\, t_o^1(j))}{r_{ij}} \tag{2.10}$$

where $r_{ij} = |\vec{r}_i - \vec{r}_j|$ and $e^2 = 1,4400$ MeV.F.

The operator of the central nucleon-nucleon interaction has the following expression:

$$H_c = \sum_{i>j=1}^{n} \left[V_{cW}(r_{ij}) + V_{cM}(r_{ij}) P_{ij}^{(r)} + V_{cB}(r_{ij}) P_{ij}^{(\sigma)} + V_{cH} P_{ij}^{(\tau)} \right] \tag{2.11}$$

where $P_{ij}^{(r)}$, $P_{ij}^{(\sigma)}$, $P_{ij}^{(\tau)}$ are the exchange operators, acting on the space, spin and isospin variables correspondingly and V_{cW}, V_{cM}, V_{cB}, V_{cH} are the potentials of Wigner Majorana, Bartlett and Heizenberg terms.

The vectorial interaction H_v can be presented as follows:

$$H_v = \sum_{i>j=1}^{n} \left[V_{vo}(r_{ij}) + V_{v\tau}(r_{ij}) P_{ij}^{(\tau)} \right] \times \sum_{\eta} (-1)^{\eta} (s_{\eta}^1(i) + s_{\eta}^1(j))\, L^1_{-\eta}(\vec{r}^{\,o}_{ij}) \tag{2.12}$$

where $s_{\eta}^1(i)$ is the spherical component of the spin angular momentum of i-th particle and L_{η}^1 - the orbital angular momentum, depending on the components of the unit vector $\vec{r}^{\,o}_{ij}$.

The tensorial interaction H_t has the velocity-independent and velocity-dependent terms. In order to present both of them let us introduce the notations:

$$S_{\eta}^2(ij) = \sum_{\eta_1 \eta_2} s_{\eta_1}^1(i)\, s_{\eta_2}^1(j)\, C^{1\ 1\ 2}_{\eta_1 \eta_2 \eta} \tag{2.13}$$

and

$$L_{\eta}^2(\vec{r}^{\,o}_{ij}) = \sum_{\eta_1 \eta_2} L_{\eta_1}^1(\vec{r}^{\,o}_{ij})\, L_{\eta_2}^1(\vec{r}^{\,o}_{ij})\, C^{1\ 1\ 2}_{\eta_1 \eta_2 \eta} \tag{2.14}$$

where C are the Clebsh-Gordan coefficients. The tensorial part of H has the following expression:

$$H_t = \sum_{i>j=1}^{n} \left[V_{to}(r_{ij}) + V_{t\tau}(r_{ij}) P_{ij}^{(\tau)} \right] \sum_{\eta} (-1)^{\eta} S_{\eta}^{2}(ij)\, Y_{-\eta}^{2}(\vec{r}^{\,o}_{ij}) +$$

$$+ \sum_{i>j=1}^{n} \left[V'_{to}(r_{ij}) + V'_{t\tau}(r_{ij}) P_{ij}^{(\tau)} \right] \sum_{\eta} (-1)^{\eta} S_{\eta}^{2}(ij)\, L_{-\eta}^{2}(\vec{r}^{\,o}_{ij})$$

(2.15)

The Hamiltonian H consists of all the terms described above, i.e.

$$H = H_k + H_{coul} + H_c + H_v + H_t \tag{2.16}$$

and depends on the set of 10 potentials, namely $V_{cW}, V_{cM}, V_{cB}, V_{cH}$ - for the central term, V_{vo}, $V_{v\tau}$ - for the vectorial term and V_{to}, $V_{t\tau}$, V'_{to}, $V'_{t\tau}$ - for the tensorial term. These potentials are unknown and therefore we simply do not know what the real microscopic Hamiltonian looks like. Thus from the first sight the problem of projecting the collective part of H seems completely undetermined.

What should we do? If the explicit expression of H is unknown, the only way is to develop the microscopic theory of the collective motions in nuclei for the arbitrary H, i.e. for H with the arbitrary set of 10 potentials mentioned above. There are such examples in the theory of many-particle systems. One of them is the Hartree-Fock equations, suitable for the arbitrary potential $V(r_{ij})$; thus the Hartree-Fock method may be used in any case, when the single-particle field gives a good approximation. We will use the same point of view: the collective effects are undoubtedly important in nuclei and we are going to derive the dynamic equations for the collective forms of motion in case of arbitrary potentials.

Before passing to the next question let us make one more remark. Taking into account the expression (2.8) for the kinetic energy operator one can say, that every term in (2.16) may be treated as the two-particle operator and every one of them can be presented in the following form:

$$H(\kappa) = \sum_{i>j=1}^{n} \sum_{q} (-1)^{q}\, \hat{U}_{q}^{\kappa} (\sigma_i \sigma_j \tau_i \tau_j)\, \hat{W}_{-q}^{\kappa} (\vec{r}_i - \vec{r}_j)$$

(2.17)

In (2.17) κ represents the O_3^+ irreducible properties:
$\kappa = 0$ for the kinetic, Coulomb and central interaction,
$\kappa = 1$ for the vectorial term and $\kappa = 2$ for the tensorial term.

3. THE TRANSLATIONAL INVARIANT VARIABLES

In the shell-model calculations as well as in the theories based on the second quantization method the single-particle variables $\vec{r}_1, \vec{r}_2, \ldots \vec{r}_n$ are often used. These variables are in the space-fixed coordinate system, thus we encounter the problem of the elimination of the centrum-of-mass motion. In order to avoid this complication it is convenient to introduce the translational invariant variables.

The set of the translational-invariant variables is well known. That is the components of the Jacobi vectors, defined as

$$\vec{\rho}_1 = \frac{1}{\sqrt{2}} (\vec{r}_1 - \vec{r}_2)$$

$$\vec{\rho}_2 = \frac{1}{\sqrt{6}} (\vec{r}_1 + \vec{r}_2 - 2\vec{r}_3)$$

$$\vec{\rho}_3 = \frac{1}{\sqrt{12}} (\vec{r}_1 + \vec{r}_2 + \vec{r}_3 - 3\vec{r}_4) \tag{3.1}$$

and in general form

$$\vec{\rho}_i = \frac{1}{\sqrt{i(i+1)}} \left(\sum_{t=1}^{i} \vec{r}_t - i\, \vec{r}_{i+1} \right) \tag{3.2}$$

where $i = 1,2,\ldots,n-1$. The centrum-of-mass variables has been already introduced in (2.9). The normalization factor in (2.9), however, is not always convenient for use, because the coefficients of the linear combinations (3.2) and (2.9) form the non-orthogonal matrix. If instead of (2.9) the vector

$$\vec{\rho}_o = \frac{1}{\sqrt{n}} \sum_{i=1}^{n} \vec{r}_i \tag{3.3}$$

is introduced, then the transformation between the single-particle and Jacobi variables is orthogonal.

In the new variables it is possible to present every term of H. For example, in case of the three particles the term

$$V(|\vec{r}_1-\vec{r}_2|) + V(|\vec{r}_1-\vec{r}_3|) + V(|\vec{r}_2-\vec{r}_3|) \tag{3.4}$$

has the following form:

$$V(|\sqrt{2}\,\vec{\rho}_1|) + V(|\sqrt{\tfrac{3}{2}}\,\vec{\rho}_2+\tfrac{1}{\sqrt{2}}\,\vec{\rho}_1|) + V(|\sqrt{\tfrac{3}{2}}\,\vec{\rho}_2 - \tfrac{1}{\sqrt{2}}\,\vec{\rho}_1|) \tag{3.5}$$

The last expression is not very attractive and one can imagine how unhandy is the n particle Hamiltonian in this form.

This problem may be avoided if one takes the matrix representation for H. Let $\hat{Q}$ be the set of spin and isospin variables and let

$$\psi(\Gamma \mid \vec{\rho}_1, \vec{\rho}_2, \ldots, \vec{\rho}_{n-1}; \hat{Q}) \tag{3.6}$$

is some set of the antisymmetric wave functions labelled by quantum numbers Γ. On the antisymmetric wave functions every two-particle term of H gives the same contribution, one has 1/2 n(n-1) terms, thus

$$\langle \Gamma|H(\kappa)|\Gamma'\rangle =$$

$$= \frac{n(n-1)}{2} \sum_q (-1)^q \langle\Gamma|\hat{U}^{\kappa}_q(\sigma_{n-1}\sigma_n\tau_{n-1}\tau_n)\, \hat{W}^{\kappa}_{-q}(\vec{r}_{n-1}-\vec{r}_n)|\Gamma'\rangle \tag{3.7}$$

From the last expression one can see, that one needs the set of the translational invariant variables slighly different from those given in (3.2): the set (3.2) does not contain the vector $\vec{r}_{n-1}-\vec{r}_n$ one needs for the operator $\hat{W}(\vec{r}_{n-1}-\vec{r}_n)$ in (3.7). Let us modify the set of vectors (3.7) using the following transformation:

$$|\vec{\rho}_1,\ldots,\vec{\rho}_{n-3},\vec{\rho}_s,\vec{\rho}_a| = |\vec{\rho}_1,\ldots,\vec{\rho}_{n-3},\vec{\rho}_{n-2},\vec{\rho}_{n-1}| \times$$

$$\begin{vmatrix} e_{n-3} & 0 & 0 \\ 0 & \sqrt{\frac{n}{2(n-1)}} & -\sqrt{\frac{n-2}{2(n-1)}} \\ 0 & \sqrt{\frac{n-2}{2(n-1}} & \sqrt{\frac{n}{2(n-1)}} \end{vmatrix} \tag{3.8}$$

where e_{n-3} is (n-3) x (n-3) - dimensional unit matrix. From (3.2) and (3.8) one can easy find that

$$\vec{\rho}_a = \frac{1}{\sqrt{2}} (\vec{r}_{n-1} - \vec{r}_n) \tag{3.9}$$

thus $\vec{\rho}_a$ is the variable, one needs in order to calculate the matrix elements (3.7). The subscript a of $\vec{\rho}_a$ indicates, that $\vec{\rho}_a$ is antisymmetric with respect to the permutation of the single-particle variables $\vec{r}_{n-1}$ and $\vec{r}_n$.

4. THE NEW TYPE OF VARIABLES. THREE-PARTICLE CASE

Our aim is to extract the collective part H_{coll} from H. In order to achieve it one needs exact definitions. What meaning are we going to put into the words "the collective part of H"? If one has decided to start with the microscopic Hamiltonian H, there is no need to use the picture of the noncompressible liquid drop. For the time being the reader must forget about the Bohr-Mottelson theory. We need to find some new conceptions, otherwise we shall not be able to explain the microscopic meaning of the phenomenological theory.

The idea as how to define H_{coll} can be grasped from the properties of the space variables. As a rule, there are special variables adapted to every problem in quantum mechanics. For example, in case of the particle, moving in the central field, the only natural variables are spherical ones. In this case it would be unwise to use the Cartesian variables. For the same reason it is highly unlikely, that the variables, discussed in the previous section, are suitable for the disclosure of the collective properties of the nuclei.

The idea as how to choose the new type of variables can be explained on the following simple example. Let us take the polar coordinates

$$x_1 = \rho \cos \theta$$

$$x_2 = \rho \sin \theta \tag{4.1}$$

with $0 \leq \theta < 2\Pi$. In order to clear up the group-theoretical meaning of the expression (4.1) let us take the rotation matrix in the two-dimensional space:

$D^{(1_2)}(\theta) =$

	1	2
1	$\cos\theta$	$\sin\theta$
2	$-\sin\theta$	$\cos\theta$

(4.2)

The polar coordinates (4.1) may be easily presented in the following form:

$$x_i = \rho \, D^{(1_2)}_{2i}(-\theta) \tag{4.3}$$

where i=1,2 and $D^{(1_2)}_{i'i}$ are the matrix elements given in (4.2).

The last expression indicates, that angle variables, used in mathematics, are somehow related to the matrix elements of the rotation matrices. This conclusion may be confirmed by the next, more complicated example. Let us take the rotation matrix in three-dimensional space:

$$D^{(1_3)}(\theta'\theta''\theta''') = \begin{vmatrix} c' & s' & o \\ -s' & c' & o \\ o & o & 1 \end{vmatrix} \begin{vmatrix} 1 & o & o \\ o & c'' & s'' \\ o & -s'' & c'' \end{vmatrix} \begin{vmatrix} c''' & s''' & o \\ -s''' & c''' & o \\ o & o & 1 \end{vmatrix} \tag{4.4}$$

In (4.4) the abbreviation is introduced $c' \equiv \cos\theta'$, $s'' \equiv \sin\theta''$ etc. The Euler angles $\theta'\theta''\theta'''$ take the values $o \leqslant \theta' < 2\pi$, $o \leqslant \theta'' < \pi$, and $o \leqslant \theta''' < 2\pi$. The product of matrices in (4.4) gives:

$$D^{(1_3)}(\theta'\theta''\theta''') = \begin{vmatrix} c'c'''-s'c''s''' & c's'''+s'c''c''' & s's'' \\ -s'c'''-c'c''s''' & -s's'''+c'c''c''' & c's'' \\ s''s''' & -s''c''' & c'' \end{vmatrix} \tag{4.5}$$

Now, in the complete analogy with (4.3), let us take

$$x_i = \rho \, D^{(1_3)}_{3i}(-\theta''\theta''') \tag{4.6}$$

where i=1,2,3. Using the explicit expression for $D^{(1_3)}$

one has

$$x_1 = \rho \sin \theta'' \quad \sin \theta'''$$

$$x_2 = \rho \sin \theta'' \quad \cos \theta'''$$

$$x_3 = \rho \cos \theta'' \tag{4.7}$$

i.e. (4.6) means the usual change of the Cartesian variables to the spherical ones.

Now let us concentrate our attention on the generalization of (4.6) in the case of the variables ρ_i^s with the indices of two types,- the vector index i=1,2,...n-1 and the Cartesian index s=1,2,3. This generalization in the case of n-1 quasiparticles (we say quasiparticles, because the centrum-of-mass motion has been eliminated, thus there are no single-particle variables) moving in the three-dimensional space, has been proposed in[11,12] and was also used in the less explicit form in[13]. In this section we will continue study the three-particle example. In this case there are two Jacobi vectors $\vec{\rho}_1$, $\vec{\rho}_2$ and thus six variables ρ_i^s (i=1,2; s=1,2,3). The formula changing them into the new ones is composed using the matrix elements of the matrices $D^{(1_2)}$ and $D^{(1_3)}$. One can change the variables ρ_i^s into the new ones putting

$$\rho_i^s = \sum_{s_o=2,3} \rho^{(s_o)} D^{(1_3)}_{s_o s} (\theta'\theta''\theta''') D^{(1_2)}_{s_o-1\; i} (\theta) \tag{4.8}$$

or in the explicit form

$$\rho_1^1 = \rho^{(2)} D^{(1_3)}_{21} D^{(1_2)}_{11} + \rho^{(3)} D^{(1_3)}_{31} D^{(1_2)}_{21}$$

$$\rho_1^2 = \rho^{(2)} D^{(1_3)}_{22} D^{(1_2)}_{11} + \rho^{(3)} D^{(1_3)}_{32} D^{(1_2)}_{21}$$

$$\rho_1^3 = \rho^{(2)} D^{(1_3)}_{23} D^{(1_2)}_{11} + \rho^{(3)} D^{(1_3)}_{33} D^{(1_2)}_{21}$$

$$\rho_2^1 = \rho^{(2)} D^{(1_3)}_{21} D^{(1_2)}_{12} + \rho^{(3)} D^{(1_3)}_{31} D^{(1_2)}_{22}$$

$$\rho_2^2 = \rho^{(2)} D^{(1_3)}_{22} D^{(1_2)}_{12} + \rho^{(3)} D^{(1_3)}_{32} D^{(1_2)}_{22}$$

$$\rho_2^3 = \rho^{(2)} D^{(1_3)}_{23} D^{(1_2)}_{12} + \rho^{(3)} D^{(1_3)}_{33} D^{(1_2)}_{22}$$

(4.9)

In (4.8) and (4.9) $\rho^{(2)}$, $\rho^{(3)}$ are the radial variables, taking the values $0 \leqslant \rho^{(2)}, \rho^{(3)} < +\infty$. The matrix elements of the matrices $D^{(1_2)}$ and $D^{(1_3)}$ are given in (4.2) and (4.5).

The formula (4.8) is not complete for our purposes. For further generalization let us discuss the group-theoretical meaning of the expression (4.8). Four of the new variables, namely θ and $\theta'\theta''\theta'''$ are the elements of the rotation groups O_2^+ and O_3^+. Let us take the group O_2^+. If $\hat{g}^+$ is an arbitrary element of O_2^+, the action of $\hat{g}^+$ on (4.8) gives:

$$\hat{g}^+ \rho_i^s = \sum_{s_o=2,3} \rho^{(s_o)} D^{(1_3)}_{s_o s}(\theta'\theta''\theta''') D^{(1_2)}_{s_o-1\, i}(\theta g^+) =$$

$$= \sum_{i'} \sum_{s_o=2,3} \rho^{(s_o)} D^{(1_3)}_{s_o s}(\theta'\theta''\theta''') D^{(1_2)}_{s_o-1\, i'}(\theta) D^{(1_2)}_{i' i}(g^+)$$

(4.10)

thus

$$\hat{g}^+ \rho_i^s = \sum_{i'} \rho_{i'}^s D^{(1_2)}_{i' i}(g^+)$$

(4.11)

i.e. the action of $\hat{g}^+$ transforms the Jacobi variables ρ_i^s in the usual way. Now let us discuss the permutation properties of $\vec{\rho}_1$ and $\vec{\rho}_2$. The permutation p_{12} gives

$$p_{12}\, |\vec{\rho}_1, \vec{\rho}_2| = p_{12}\, \left| \frac{1}{\sqrt{2}} (\vec{r}_1 - \vec{r}_2), \frac{1}{\sqrt{6}} (\vec{r}_1 + \vec{r}_2 - 2\,\vec{r}_3) \right| =$$

$$= \left| \frac{1}{\sqrt{2}} (\vec{r}_2 - \vec{r}_1), \frac{1}{\sqrt{6}} (\vec{r}_2 + \vec{r}_1 - 2\vec{r}_3) \right| = |\vec{\rho}_1\ \vec{\rho}_2| \begin{vmatrix} -1 & 0 \\ 0 & 1 \end{vmatrix}$$

(4.12)

and similar

$$p_{23}\, |\vec{\rho}_1\ \vec{\rho}_2| = |\vec{\rho}_1\ \vec{\rho}_2| \begin{vmatrix} \frac{1}{2} & \frac{\sqrt{3}}{2} \\ \frac{\sqrt{3}}{2} & -\frac{1}{2} \end{vmatrix}$$

(4.13)

From (4.11), (4.12) and (4.13) one can see, that the action of p_{12} and p_{23} can not be represented by the rotation (4.2). Really, the determinant of the matrix (4.2) is equal to +1 while the determinants of the matrices in (4.12) and (4.13) are equal to -1.

This is very inconvenient from the point of view of the group theory. It is worth to extend the rotation group O_2^+ to the orthogonal group O_2, i.e. to the rotation group with reflections. This is easily done, introducing the reflection group, consisting of two elements, unit element and the reflection element. In two-dimensional space, spanned by Jacobi vectors $\vec{\rho}_1$ and $\vec{\rho}_2$ we will define the action of the elements $\hat{\sigma}$ of the reflection group as follows:

$$\hat{\sigma} \, | \vec{\rho}_1 \; \vec{\rho}_2 | = | \vec{\rho}_1 \; \vec{\rho}_2 | \begin{vmatrix} 1 & 0 \\ 0 & (-1)^\sigma \end{vmatrix} \tag{4.14}$$

where $\sigma = 0$ for the unit element and $\sigma = 1$ for the reflection element. Now all the elements $\hat{g}$ of the orthogonal group O_2 may be presented like follows:

$$g \in \begin{vmatrix} 1 & 0 \\ 0 & (-1)^\sigma \end{vmatrix} \begin{vmatrix} \cos\theta & \sin\theta \\ -\sin\theta & \cos\theta \end{vmatrix} = \begin{vmatrix} \cos\theta & \sin\theta \\ (-1)^{\sigma+1}\sin\theta & (-1)^\sigma \cos\theta \end{vmatrix} \tag{4.15}$$

One may easily check, that the matrices in (4.12) and (4.13) are particular cases of (4.15), when $\sigma = 1$ and

$$\theta = \theta_j = (5-2j)\,\frac{\pi}{3} \tag{4.16}$$

where j=1 in the case of p_{12}, and j=2 in the case of p_{23}. Thus the action of the permutation is the inner operation of the orthogonal group O_2.

Now it is clear, how the expression (4.8) must be generalized (this generalization in the case n-1 quasiparticles moving in three-dimensional case has been proposed in[14]) in order to take into account the permutation properties. We have:

$$\rho_i^s = \sum_{s_o=2,3} \rho^{(s_o)} \, D^{(1_3)}_{s_o s} (\theta'\theta''\theta''') \, D^{(1_2)}_{s_o-1\; i} (\sigma\;\theta) \tag{4.17}$$

where the argument $\sigma\ \theta$ indicates dealing with matrices (4.15).

5. THE NEW TYPE OF VARIABLES. GENERAL CASE

In this section we present new variables and differential operators in new variables for the general case of the r quasiparticles moving in r_o -dimensional space. More details on these questions may be found in[6].

We need some convenient notations. We will use r_o and r dimensional matrices, so, let us denote them by $D^{(1 r_o)}$ and $D^{(1 r)}$. We are going to use all columns but often only a few rows of some matrices D. In order to enumerate them, let us introduce the special index s_o and let $s_o=1,2,\ldots,r_o$ if $r_o \leqslant r$ and $s_o=1+r_o-r, 2+r_o-r,\ldots, r_o$ if $r_o \geqslant r$. In both cases we will use the min (r_o,r) last rows of both matrices $D^{(1 r_o)}$ and $D^{(1 r)}$.

One can replace the $r_o r$ variables ρ_i^s $(i=1,2,\ldots, r;\ s=1,2,\ldots, r_o)$ by the new ones of one puts

$$\rho_i^s = \sum_{s_o} \rho^{(s_o)}\ D^{(1 r_o)}_{s_o s}(G^+)\ D^{(1 r)}_{r-r_o+s_o, i}\ (\sigma\ q^+) \tag{5.1}$$

where the radial variables $\rho^{(s_o)}$ take values $0 \leqslant \rho^{(s_o)} < +\infty$. The number of $\rho^{(s_o)}$ is min (r_o,r).

In order to get the explicit expression of the matrix elements in (5.1) let us introduce the following rotation matrix, acting in the a - dimensional space:

$$T^{(t)}_{p-1\,p}(\theta^{(t)}_p) = \begin{array}{c|ccc|c|c|ccc}
 & 1 & \ldots & & p-1 & p & p+1 & \ldots & a \\
\hline
1 & 1 & & 0 & & & & & \\
\cdot & & \ddots & & & & & & \\
\cdot & & & & 0 & 0 & & 0 & \\
\cdot & 0 & & 1 & & & & & \\
p-2 & & & & & & & & \\
\hline
p-1 & & & 0 & c_p^{(t)} & s_p^{(t)} & & 0 & \\
\hline
p & & & 0 & -s_p^{(t)} & c_p^{(t)} & & 0 & \\
\hline
p+1 & & & & & & 1 & & 0 \\
\cdot & & & & & & & \ddots & \\
\cdot & & & 0 & 0 & 0 & & & \\
\cdot & & & & & & & & \\
a & & & & & & 0 & & 1
\end{array} \tag{5.2}$$

where $p = 2,3,\ldots,a$; $c_p^{(t)} \equiv \cos\ \theta_p^{(t)}$ and $s_p^{(t)} \equiv \sin\ \theta_p^{(t)}$.

Let us introduce the product of matrices (5.2)

$$T(a,b)=\prod_{\kappa=t-b+1>1}^{t}\left[\prod_{t=a}^{2} T^{(a+\kappa-t)}_{\kappa-1,\kappa}(\theta^{(a+\kappa-t)}_{\kappa})\right] \qquad (5.3)$$

where a = 2,3,..., and b = 1,2,...,a-1. The variables θ takes the values $0\leqslant \theta_p < \Pi$, if $p>2$ and $0\leqslant \theta_2 < 2\Pi$, if p = 2. The operators T in (5.3) are ordered in the following way: T with the largest t and κ are on the left side. The square parenthesis in (5.3) indicate, that at first one must take the product with respect to t and afterwards with respect to κ. Let us take some examples of the operator (5.3):

$$T(2,1)=T^{(2)}_{12}(\theta^{(2)}_{2})$$

$$T(3,2)=T^{(2)}_{12}(\theta^{(2)}_{2})T^{(3)}_{23}(\theta_3^{(3)})T^{(3)}_{12}(\theta^{(3)}_{2})$$

$$T(4,3)=T^{(2)}_{12}(\theta^{(2)}_{2})T^{(3)}_{23}(\theta^{(3)}_{3})T^{(4)}_{34}(\theta^{(4)}_{4})T^{(3)}_{12}(\theta^{(3)}_{2})T^{(4)}_{23}(\theta^{(4)}_{3}) \times T^{(4)}_{12}(\theta^{(4)}_{2}) \qquad (5.4)$$

thus in case of b=a-1, the expression (5.3) gives the parametrization of the rotation group O^+_a . If b=a-2, we have:

$$T\ (3,1) = T^{(3)}_{23}(\theta_3^{(3)})\ T^{(3)}_{12}(\theta_2^{(3)})$$

$$T\ (4,1)= T^{(4)}_{34}(\theta_4^{(4)})\ T^{(4)}_{23}(\theta_3^{(4)})\ T^{(4)}_{12}(\theta_2^{(4)}). \qquad (5.5)$$

The last expression gives the partial parametrization of the rotational group. In the group theory such type of parametrization is called the factor space $O^+_1\diagdown O^+_3$ and $O^+_1\diagdown O^+_4$. In general, the expression (5.3) gives the parametrization of the factor space $O^+_{a-b}\diagdown O^+_a$ depending on 1/2a(a-1)-1/2 (a-b)(a-b-1) variables. These questions are closely related to the theory of the induced representations. The realationship between this theory and the problems of the collective motion in nuclei have been disclosed in[14], where also the principal group-theoretical problems, related to the microscopical description fo the collective degrees of freedom have been solved (for details see also [15,16]).

The explicit expression of the matrix $D^{(1_r)}$ is

$$D^{(1_r)}\ (\sigma\, q^+) = \sigma_r\, D^{(1_r)}\ (q^+) \tag{5.6}$$

In (5.6) σ_r is the following r - dimensional matrix

$$\sigma_r = \begin{array}{|c|c|} \hline e_{r-1} & 0 \\ \hline 0 & (-1)^\sigma \\ \hline \end{array} \tag{5.7}$$

where e_{r-1} is the (r-1)-dimensional unit matrix and $\sigma = 0,1$. In (5.6)

$$D^{(1_r)}\ (q^+) = \begin{cases} T\ (r,r_o)\ , \text{ if } r_o < r \\ T\ (r,r-1), \text{ if } r_o \geqslant r \end{cases} \tag{5.8}$$

thus q^+ denotes the set of $1/2\ r_o(2r-r_o-1)$ variables θ, if $r_o \leqslant r$ and the set of $1/2\ r\ (r-1)$ variables θ, if $r_o \geqslant r$. The expression of $D^{(1_{r_o})}$ in (5.1) is:

$$D^{(1_{r_o})}\ (G^+) = \begin{cases} T\ (r_o,r), \text{ if } r < r_o \\ T\ (r_o,r_o-1), \text{ if } r \geqslant r_o, \end{cases} \tag{5.9}$$

thus G^+ denotes the set of $1/2\ r_o\ (r_o-1)$ variables θ, if $r_o \leqslant r$ and the set of $1/2r(2r_o-r-1)$ variables θ, if $r_o \geqslant r$. Taking into account the number of variables $\rho^{(s_o)}$ one can see, that in the right side of (5.1) there are always $r_o r$ continuous variables and one discrete variable. In particular, if $r_o = 1$, s_o takes only one value $s_o=1$, $D^{(1_1)}=1$ and from (5.1),

$$\rho_i^1 = \rho^{(1)}\ D_{ri}^{(1_2)}\ (\sigma\, q^+) \tag{5.10}$$

In the case of $\sigma = 0$ this formula can be easily related to the expressions changing r Cartesian variables into spherical ones. In the case of r=2,3,(5.10) gives (4.3) and (4.6). If $r_o=3$ and r=2, (5.10) coincides with (4.8).

Besides terms, depending on ρ_i^s, the Hamiltonian H contains terms depending on derivatives $\frac{\partial}{\partial\rho}$. Thus, in order to get H in new variables, one needs to rewrite $\frac{\partial}{\partial\rho}$ in the variables $\rho^{(s_o)}$, G^+ and q^+. This complicated question (one encounters the problem of inverting $r_o r$-dimensional Jacobi matrix) in the case of $r_o=3$ has been solved in[18] and generalized in[6] for the arbitrary r_o.

Without goint into details we will present the final expression:

$$\frac{\partial}{\partial\rho_i^s} = \sum_{s'_o} D^{(1_{r_o})}_{s'_o s_o} D^{(1_r)}_{r-r_o+s'_o,i} \frac{\partial}{\partial\rho^{(s'_o)}} - \sum_{s'}\sum_{i'} D^{(1_{r_o})}_{s' s_o} D^{(1_r)}_{i' i} \hat{\mathcal{K}}_{(s'i')} -$$

$$- \sum_{s'_o > s''_o} \left[\frac{\rho^{(s'_o)}}{(\rho^{(s'_o)})^2 - (\rho^{(s''_o)})^2} D^{(1_{r_o})}_{s'_o s_o} D^{(1_r)}_{r-r_o+s''_o,i} + \right.$$

$$\left. + \frac{\rho^{(s''_o)}}{(\rho^{(s'_o)})^2 - (\rho^{(s''_o)})^2} D^{(1_{r_o})}_{s''_o s_o} D^{(1_r)}_{r-r_o+s'_o,i} \right] \hat{\mathcal{J}}_{r-r_o+s'_o,r-r_o+s''_o} -$$

$$- \sum_{s'_o > s''_o} \left[\frac{\rho^{(s'_o)}}{(\rho^{(s'_o)})^2 - (\rho^{(s''_o)})^2} D^{(1_{r_o})}_{s''_o s_o} D^{(1_r)}_{r-r_o+s'_o,i} + \right.$$

$$\left. + \frac{\rho^{(s''_o)}}{(\rho^{(s'_o)})^2 - (\rho^{(s''_o)})^2} D^{(1_{r_o})}_{s'_o s_o} D^{(1_r)}_{r-r_o+s''_o,i} \right] \hat{\mathcal{L}}_{s'_o s''_o} \tag{5.11}$$

In (5.11) s_o take the same values as in (5.1). Matrix elements of the matrices $D^{(1_{r_o})}$ and $D^{(1_r)}$ depend respectively on G^+ and σq^+. Operators $\hat{\mathcal{J}}$ and $\hat{\mathcal{L}}$ are the left-shift infinitesimal operators for $O^+_{r_o}$ and O^+_r. In the second term of (5.11) $i'=1,2,\dots,r-r_o$; $s'=1,2,\dots,r_o$, if $r>r_o$ and $s'=1,2,\dots r_o-r$; $i'=1,2,\dots,r$ if $r<r_o$. The operator $\hat{\mathcal{K}}$ in (5.11) has the following expression:

$$\hat{\mathcal{K}}_{(s'i')} = \begin{cases} (\rho^{(s')})^{-1}\, \hat{\mathcal{J}}_{r-r_o+s',i}\,, & r>r_o \\ (\rho^{(r_o-r+i')})^{-1}\, \hat{\mathcal{L}}_{r_o-r+i',s'} & r<r_o \\ 0 & r=r_o \end{cases} \tag{5.12}$$

The operators $\hat{\mathcal{J}}_{ii'}$ (as well as the operators $\hat{\mathcal{L}}_{s_o s'_o}$ satisfy the following commutation relations:

$$\left[\hat{\mathcal{J}}_{i_1 i_2}, \hat{\mathcal{J}}_{i_3 i_4}\right] = \hat{\mathcal{J}}_{i_1 i_3}\,\delta(i_2 i_4) + \hat{\mathcal{J}}_{i_2 i_4}\,\delta(i_1 i_3) - $$

$$- \hat{\mathcal{J}}_{i_1 i_4}\,\delta(i_2 i_3) - \hat{\mathcal{J}}_{i_2 i_3}\,\delta(i_1 i_4) \qquad (5.13)$$

The details of the definitions of $\hat{\mathcal{J}}$ and $\hat{\mathcal{L}}$ are given in[6].

Using (5.11) it is possible to get the following expression of the $r_o r$ dimensional Laplace operator:

$$\Delta = \sum_{is} \frac{\partial^2}{\partial(\rho_i^s)^2} = \sum_{s'_o} \frac{\partial^2}{\partial(\rho^{(s'_o)})^2} + |r-r_o| \sum_{s'_o} \frac{1}{\rho(s'_o)} \frac{\partial}{\partial\rho(s'_o)} +$$

$$+ 2 \sum_{s'_o > s''_o} \frac{1}{(\rho^{(s'_o)})^2 - (\rho^{(s''_o)})^2} \left[\rho^{(s'_o)} \frac{\partial}{\partial\rho(s'_o)} - \rho^{(s''_o)} \frac{\partial}{\partial\rho(s''_o)}\right] +$$

$$+ \sum_{s'i'} (\hat{\mathcal{K}}_{(s'i')})^2 + \sum_{s'_o > s''_o} \frac{(\rho^{(s'_o)})^2 + (\rho^{(s''_o)})^2}{((\rho^{(s'_o)})^2 - (\rho^{(s''_o)})^2)^2} \left[(\hat{\mathcal{J}}_{r-r_o+s'_o,\, r-r_o+s''_o})^2\right.$$

$$\left. + (\hat{\mathcal{L}}_{s'_o s''_o})^2\right] + 4 \sum_{s'_o > s''_o} \frac{\rho^{(s'_o)}\, \rho^{(s''_o)}}{((\rho^{(s'_o)})^2 - (\rho^{(s''_o)})^2)^2} \hat{\mathcal{J}}_{r-r_o+s'_o,\, r-r_o+s''_o} \times$$

$$\times \hat{\mathcal{L}}_{s'_o s''_o} \qquad (5.14)$$

In the case of $r_o=3$ this expression has been obtained in[11-13]. In order to explain the microscopic meaning of the phenomenological rotational-vibrational model, we will need the special case of the expression (5.14), when the infinitesimal operators of the group O_r act in the O_r scalar space. One can trace the origin of terms in (5.11) (for details see[6]) and obtain , that in O_r scalar space (5.14) reduce to

$$\Delta^{(o)} = \sum_{s'_o} \frac{\partial^2}{\partial(\rho^{(s'_o)})^2} +$$

$$+ \sum_{s'_o > s''_o} \frac{1}{(\rho^{(s'_o)})^2 - (\rho^{(s''_o)})^2} \left[\rho^{(s'_o)} \frac{\partial}{\partial \rho^{(s'_o)}} - \rho^{(s''_o)} \frac{\partial}{\partial \rho^{(s''_o)}} \right] +$$

$$+ \sum_{s'_o > s''_o} \frac{(\rho^{(s'_o)})^2 + (\rho^{(s''_o)})^2}{((\rho^{(s'_o)})^2 - (\rho^{(s''_o)})^2)^2} (\hat{\mathcal{L}}_{s'_o\, s''_o})^2 \qquad (5.15)$$

This operator does not depend on the number of particles.

This brief review of the general results ends this section and we return to the simplified version of the theory, -the three-particle case.

6. THE COMPLETE SET OF FUNCTIONS FOR THE ANGULAR VARIABLES. THREE-PARTICLE CASE

In order to continue our discussion we need a complete set of functions, depending on the Euler angles as well as the variables $\sigma\ \theta$. The set, depending on $\theta'\theta''\theta'''$, is given by the Wigner functions

$$D^L_{KM}\ (\theta'\theta''\theta''') \qquad (6.1)$$

labelled by the quantum number L of the total space angular momentum and its projection M in the space-fixed coordinate system as well as the projection K in the body-fixed coordinate system. Thus we need only the set of functions, depending on the elements $\sigma\ \theta$ of the orthogonal group O_2.

In order to construct this set let us start with well known one-dimensional matrices of the irreducible representations of the rotation group O_2^+:

$$D^{(m)}\ (\theta) = e^{im\theta} \qquad (6.2)$$

where m=o, ± 1, ± 2,...

The O_2 irreducible matrices may be obtained from (6.2) taking into account the reflection properties. These matrices may be constructed either in the basis, labelled by the quantum numbers of the chain

$$O_2 \supset O_2^+ \qquad (6.3)$$

or in the basis of the chain

$$O_2 \supset O_1' \dotplus O_1'' \qquad (6.4)$$

where O_1' and O_1'' are the reflection groups, isomorphic to the one-dimensional orthogonal group. The groups O_1' and O_1'' are equivalent, but we suppose that O_1' acts on the com ponents of the vector $\vec{\rho}_1$, while O_1'' acts on the components of $\vec{\rho}_2$.

The O_2 irreducible matrices would be easily got if one knows the reflection operator in the basis of both chains mentioned above. Thus let us start with this ques tion. Let $\hat{S}$ be the reflection matrix in the basis (6.3). This operator must anticommute with the infinitesimal operator

$$\hat{I} = -i \frac{\partial}{\partial\theta} \qquad (6.5)$$

(because the reflection changes the sing of θ). The second condition, put on $\hat{S}$, is that the square of $\hat{S}$ is equal to the unit operator (because two reflections give an identity). Thus we have

$$\hat{S}\hat{I} = -\hat{I}\hat{S}$$

$$\hat{S}^2 = e \qquad (6.6)$$

In order to get the matrix representation for $\hat{I}$ let us act with (6.5) on the functions $e^{im\theta}$:

$$\hat{I}\, e^{im\theta} = m\, e^{im\theta} \qquad (6.7)$$

Thus the matrix elements of $\hat{I}$ in the basis spanned by (6.2) are

$$(\hat{I})_{mm'} = \delta(mm')m \qquad (6.8)$$

where $m=0,\pm1,\pm2,$ ... Let us take the "box" with definite m. In the case of $m=0$, $(\hat{I})_{00} = 0$ and the solution of (6.6) is ±1. If $m \neq 0$, then (6.6) gives

$$\begin{pmatrix} s_{11} & s_{12} \\ s_{21} & s_{22} \end{pmatrix}\begin{pmatrix} m & o \\ o & -m \end{pmatrix} - \begin{pmatrix} m & o \\ o & -m \end{pmatrix}\begin{pmatrix} s_{11} & s_{12} \\ s_{21} & s_{22} \end{pmatrix} = 0$$

$$\begin{pmatrix} s_{11} & s_{12} \\ s_{21} & s_{22} \end{pmatrix} \begin{pmatrix} s_{11} & s_{12} \\ s_{21} & s_{22} \end{pmatrix} = \begin{pmatrix} 1 & 0 \\ 0 & 1 \end{pmatrix} \tag{6.9}$$

One can easily check, that the general solution of (6.9) is

$$S^{(m)} = \begin{array}{c|c|c} & m & -m \\ \hline m & 0 & e^{i\varphi} \\ \hline -m & e^{-i\varphi} & 0 \end{array} \tag{6.10}$$

with an arbitrary φ . Let us take $\varphi = \Pi$ and use the variable σ introduced in the previous section. Then for both elements σ of the reflection group we have

$$S^{(o)}(\sigma) = (\pm 1)^{\sigma} \tag{6.11}$$

in case of m=0 and

$$S^{(m)}(\sigma) = \begin{array}{c|c|c} & m & -m \\ \hline m & 1-\sigma & \sigma \\ \hline -m & \sigma & 1-\sigma \end{array} \tag{6.12}$$

in case of $m \neq 0$.

The O_2 irreducible matrix in the basis (6.4) is the product of (6.12) and (6.2):

$$D^{(\omega)}(\sigma\theta) = \begin{pmatrix} 1-\sigma & \sigma \\ \sigma & 1-\sigma \end{pmatrix} \begin{pmatrix} e^{im\theta} & 0 \\ 0 & e^{-im\theta} \end{pmatrix} = \begin{pmatrix} (1-\sigma)e^{im\theta} & \sigma e^{-im\theta} \\ \sigma e^{im\theta} & (1-\sigma)e^{-im\theta} \end{pmatrix} \tag{6.13}$$

where ω denotes the O_2 irreducible representation and takes the values $\omega = 1,2,3,...$ We will see later that it is convenient to introduce for ω the following notation:

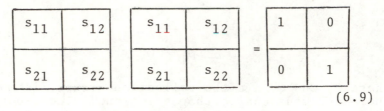

$\omega \equiv k \pmod 3$. This notation means, that $\omega=3a+k$, where $a = 0,1,2,\ldots$ and for every $a, k = 0,1,2$.

The product (6.2) with m=0 and (6.11) gives two one-dimensional representations of the group O_2, the scalar representation

$$D^{(0)}(\sigma\theta) = 1 \tag{6.14}$$

and the pseudoscalar representation

$$D^{(0)*}(\sigma\theta) = (-1)^{\sigma} \tag{6.15}$$

Both (6.14) and (6.15) do not depend on θ, thus, as a matter of fact, they give the irreducible representations of the reflection group.

Let us summarize. The set of the O_2 irreducible representations consists of the infinite number of two-dimensional matrices, labelled by positive integer ω. These representations split on the chain $O_2 \supset O_2^+$ into two non-equivalent one-dimensional O_2^+ representations (6.2), labelled by $m=\pm\omega$. There are also only two one-dimensional O_2 representations labelled by (o) and (o)*. On the chain $O_2 \supset O_2^+$ both of them reduce to the scalar O_2^+ representation, i.e. to (6.2) with m=0.

Now we are going to analize the properties of the basis for the chain (6.4). This chain is canonical and its branching rules are well known. The orthogonal group O_1 is isomorphic to the reflection group, thus O_1 has only two representations (6.14) and (6.15). Let us introduce a special notation $\bar{\omega}$ and let $\bar{\omega} = 0$ in case of the scalar representation (6.14) and $\bar{\omega} = 1$ in case of the pseudoscalar representation (6.15). The branching rules on the chain (6.4) are as follows: ω = (o) contains scalar for both O_1' and O_1''. If ω = (o)*, then $\bar{\omega}$ = o for O_1' and $\bar{\omega} = 1$ for O_1'' (it depends on our definition. We defined the reflection operator in (4.14) as the operator, acting on the vector $\vec{\rho}_2$). If $\omega \neq$ (o), $\omega \neq$ (o)* (in this case ω is a two-dimensional matrix), the quantum numbers of the groups O_1' and O_1'' take the values, given in the Table I, where for the sake of convenience, the previous cases are also included.

Table I. The branching rules on the chain (6.4).

$O_2 \supset$	O_1' +	O_1''
$\omega = (o)$	$\bar{\omega} = o$	$\bar{\omega} = o$
$\omega = (o)^*$	$\bar{\omega} = o$	$\bar{\omega} = 1$
ω - even	$\bar{\omega} = o$	$\bar{\omega} = o$
$\omega \neq (o), \omega \neq (o)^*$	$\bar{\omega} = 1$	$\bar{\omega} = 1$
ω - odd	$\bar{\omega} = o$	$\bar{\omega} = 1$
$\omega \neq (o), \omega \neq (o)^*$	$\bar{\omega} = 1$	$\bar{\omega} = o$

From the Table I one can see, that the irreducible representation of O_1'' is uniquely defined by the representation of O_1' and by the partity of ω. Thus the matrix of the subgroup $O_1' + O_1''$ containing in D (in other words -the reflection operator in the basis labelled by the chain (6.4)) has the following expression:

$$\bar{S}(\omega) = \begin{array}{c|c|c} \bar{\omega} \backslash \bar{\omega}' & 1 & 0 \\ \hline 1 & (-1)^{\omega+1} & 0 \\ \hline 0 & 0 & (-1)^{\omega} \end{array} \tag{6.16}$$

The rows and columns of this matrix are labelled by the quantum numbers $\bar{\omega}$ of the group O_1'.

We have got already the reflection operator in the basis (6.3) (the matrix (6.10) with $\varphi=\Pi$) and in the basis (6.4) (the matrix (6.16)), thus the unitary transformation between (6.10) and (6.16) gives the transformation between the both bases. In order to obtain it one must solve the matrix equation

$$\mathcal{M}^{-1}\,\bar{S}^{(\omega)}\mathcal{M} = S^{(\omega)} \tag{6.17}$$

It is easy to check, that the general solution of (6.17) is

$$\mathcal{M}^{(\omega)} = \begin{array}{c|c|c} \bar{\omega} \diagdown m & +|m| & -|m| \\ \hline 1 & \frac{1}{\sqrt{2}}\, e^{i\psi''} & \frac{(-1)^{\omega+1}}{\sqrt{2}}\, e^{i(\psi''+\Pi)} \\ \hline 0 & \frac{1}{\sqrt{2}}\, e^{i\psi'} & \frac{(-1)^{\omega}}{\sqrt{2}}\, e^{i(\psi'+\Pi)} \end{array} \tag{6.18}$$

with arbitrary ψ' and ψ''. The matrix elements of (6.18) may be presented in the following form:

$$\mathcal{M}^{(\omega)}_{\bar{\omega}m} = \frac{1}{\sqrt{2}}\, e^{i\left[(1-\bar{\omega})\psi' + \bar{\omega}\psi'' + \frac{1}{2}(1-\mathrm{sign}\, m)(\omega-\bar{\omega}+1)\Pi\right]} \tag{6.19}$$

In (6.2) $\omega \neq (o)$, $\omega \neq (o)^*$ and

$$\mathrm{sign}\ x = \begin{cases} 1, & \text{if } x > 0 \\ 0, & \text{if } x = 0 \\ -1, & \text{if } x < 0 \end{cases} \tag{6.20}$$

For the sake of simplicity let us take in (6.19) $\psi' = \frac{\Pi}{2}$ and $\psi'' = 0$. In case of $\omega = (o)$ and $\omega = (o)^*$ we define:

$$\mathcal{M}^{(o)} = +1$$

$$\mathcal{M}^{(o)*} = e^{i\frac{\Pi}{2}} \tag{6.21}$$

Now for the matrix elements of D^{ω} in the basis labelled by the chain (6.4) one has:

$$D^{\omega}_{\bar{\omega}\,\bar{\omega}'}(\sigma\theta) = \sum_{mm'} \left[\mathcal{M}^{(\omega)}_{\bar{\omega}\,m}\right]^{-1} D^{\omega}_{mm'}(\sigma\theta)\, \mathcal{M}^{(\omega)}_{m'\bar{\omega}'} \tag{6.22}$$

and using (6.2) and (6.19)

$$D^{\omega}(\sigma\theta) = \begin{array}{c|c|c} \bar\omega \backslash \bar\omega' & 1 & 0 \\ \hline 1 & (-1)^{\sigma(\omega+1)}\cos\omega\theta & (-1)^{\sigma(\omega+1)}\sin\omega\theta \\ \hline 0 & (-1)^{\sigma\omega+1}\sin\omega\theta & (-1)^{\sigma\omega}\cos\omega\theta \end{array} \quad (6.23)$$

The unitary transformation with one-dimensional matrices (6.21) does not change (6.14) and (6.15), thus

$$D^{(o)}_{oo}(\sigma\theta) = 1$$

$$D^{(o)*}_{11}(\sigma\theta) = (-1)^{\sigma} \quad (6.24)$$

The matrices (6.23) and (6.24) present the final result needed for our analysis.

7. THE PHYSICAL BASIS. THREE-PARTICLE CASE

The matrix elements in (6.23) satisfy the usual orthogonality relation

$$\frac{d\omega_1}{2}\sum_{\sigma}\int_{0}^{2\Pi}\frac{d\theta}{2\Pi}\overset{*}{D}{}^{\omega_1}_{\bar\omega_1\bar\omega'_1}(\sigma\theta)\,D^{\omega_2}_{\bar\omega_2\bar\omega'_2}(\sigma\theta) =$$

$$= \delta(\omega_1\omega_2)\,\delta(\bar\omega_1\bar\omega_2)\,\delta(\bar\omega'_1\bar\omega'_2) \quad (7.1)$$

as well as the completeness condition

$$\sum_{\omega\bar\omega\bar\omega'}\frac{d\omega}{4\Pi}\overset{*}{D}{}^{\omega}_{\bar\omega\bar\omega'}(\sigma\theta)\,D^{\omega}_{\bar\omega\bar\omega'}(\sigma'\theta') = \delta(\sigma\sigma')\,\delta(\theta-\theta') \quad (7.2)$$

Thus they present a complete set of functions, depending on the elements of the group O_2. Every function, depending on $\rho^{(2)}\rho^{(3)}$, $\theta'\theta''\theta'''$, $\sigma\theta$ and characterized by quantum numbers LM, $\omega\bar\omega$, may be presented as follows:

$$\psi(\Gamma\ L\ M\ \omega\bar\omega \mid \rho^{(2)}\rho^{(3)},\ G_3^{+},\sigma\theta) =$$

$$= \sum_{k\bar{\omega}'} \Theta(\Gamma L K \omega \bar{\omega}' | \rho^{(2)} \rho^{(3)}) D^{L}_{KM} (G_3^{+}) D^{\omega}_{\bar{\omega}', \bar{\omega}} (\sigma\nu) \quad (7.3)$$

where G_3^{+} denotes elements of the group O_3^{+}, i.e. $G_3^{+} \equiv \theta'\theta''\theta'''$. In (7.3) Θ are the decomposition coefficients depending on $\rho^{(2)}$ $\rho^{(3)}$ and also on a set of additional quantum numbers Γ. In (7.3) we exposed quantum numbers $LM, \omega\bar{\omega}$ needed for the future analysis. Coupling LM with spin quantum numbers $S\ M_S$ we get the total angular momentum J and its projection M_J.

We have not yet discussed the symmetry properties. The nucleus is a system of the equivalent fermions, thus one needs to assure the Pauli principle requirements. In order to be able to do that, let us transform the set of function (7.3) to the new basis, labelled by quantum numbers of the physical chain of groups, namely the chain

$$O_2 \supset S_3 \supset S_2 \quad (7.4)$$

where S_3 and S_2 are the symmetric groups, acting correspondingly on $\vec{r}_1, \vec{r}_2, \vec{r}_3$ and $\vec{r}_1, \vec{r}_2$. Thus the transformation matrix $A^{(\omega)}$ with matrix elements

$$A^{(\omega)}_{\bar{\omega}\ ,\lambda\bar{\lambda}} \quad (7.5)$$

is needed. The rows in (7.5) are labelled by quantum numbers of the chain (6.4), i.e. by the O_1 irreducible representation, while the columns -by quantum numbers of the chain (7.4), i.e. by the S_3 irreducible representation λ(λ takes values [3], [21] , [111]) and by the S_2 irreducible representation $\bar{\lambda}$ ($\bar{\lambda}$ takes values [2] , [11]). If matrix elements (7.5) are known, then the transformation

$$\psi\ (\Gamma\ L\ M\ \omega\lambda\bar{\lambda}) = \sum_{\bar{\omega}} \psi(\Gamma L M\ \omega\bar{\omega})\ A^{(\omega)}_{\bar{\omega},\ \lambda\bar{\lambda}} \quad (7.6)$$

gives basis functions with definite permutation symmetry.

The branching rules on the chain (7.4) reducing O_2 representations on the symmetric group S_3 may be easily obtained. The characters $\chi^{(\omega)}$ of (6.23) and (6.24) are

$$\chi^{(\omega)}\ (\sigma\theta) = D^{\omega}_{11}(\sigma\theta) + D^{\omega}_{oo}\ (\sigma\theta) =$$

$$= (-1)^{\sigma\omega}\ (1+(-1)^{\sigma})\ \cos\ \omega\theta \quad (7.7)$$

if $\omega \neq (o)$, $\omega \neq (o)^*$ and

$$\chi^{(o)}(\sigma\theta) = D^{(o)}_{oo}(\sigma\theta) = 1$$

$$\chi^{(o)*}(\sigma\theta) = D^{(o)}_{11}(\sigma\theta) = (-1)^{\sigma} \qquad (7.8)$$

We know already how S_3 is imbedded in O_2. In section 4 it has been show, that (4.15) gives S_3 irreducible matrix for p_{12}, if $\sigma = 1$, $\theta = \Pi$ and (4.15) gives S_3 irreducible matrix for p_{23}, if $\sigma = 1$, $\theta = 1/3\ \Pi$. Thus for S_3 element $p_{12} \cdot p_{23}$ one has $\sigma = o$, $\theta = (-1 + 1/3 + 2)\Pi$ and for the identity element e of S_3 — $\sigma = 0$, $\theta = 0$. The elements e, p_{12}, $p_{12} \cdot p_{23}$ represent all classes of S_3 and for each of them using (7.7) and (7.8) one can calculate the character $\chi^{(\omega)}(p)$ of the S_3 representation $D^{(\omega)}(p)$. For example from (7.7) one has

$$\chi^{(\omega)}(e) = 2$$

$$\chi^{(\omega)}(p_{12}) = 0$$

$$\chi^{(\omega)}(p_{12} \cdot p_{23}) = 2 \cos \frac{4}{3}\omega\Pi \qquad (7.9)$$

In general this representation is a reducible one. The branching rules on the chain (7.4) may be obtained from the following wellknown expression

$$N(\lambda) = \frac{1}{6}\Big[\chi^{(\omega)}(e)\ \chi^{\lambda}(e) + 3\chi^{(\omega)}(p_{12})\chi^{(\lambda)}(p_{12}) + 2\chi^{(\omega)}(p_{12} \cdot p_{23})\chi^{(\lambda)}(p_{12} \cdot p_{23}) \Big] \qquad (7.10)$$

where $N(\lambda)$ is a number of λ in ω and $\chi^{(\lambda)}$ is the character of S_3 irreducible representation λ .

The expression (7.10) gives the following rules: $\omega = (o)$ contains $\lambda = [3]$; $\omega = (o)^*$ contains $\lambda = [111]$; $\omega = (3a+k)$ contains $[3] + [111]$, if $k = 0$ and $[21]$, if $k = 1,2$. Thus on the chain (7.4) only O_2 representation $\omega = (3a)$ splits into two S_3 irreducible representations.

Now, when S_3 irreducible structure of the matrix (6.23) is known (there is no need to consider (6.24) - they are both already S_3 irreducible representations), the structure of the transformation matrix between the two bases, labelled by chains (6.4) and (7.4), is also known. Taking into account this structure one gets the following equation:

$$(A^{(\omega)})^{-1} D^{(\omega)}(\sigma=1,\theta_j)\, A^{(\omega)} = \begin{cases} D^{\lambda}(p_{jj+1}), & \text{if } \omega \neq (3a) \\ D^{[3]}(p_{jj+1}) \dot{+} D^{[111]}(p_{jj+1}), & \text{if } \omega = (3a) \end{cases} \qquad (7.11)$$

where j = 1,2. In (7.11) D^{λ} are S_3 irreducible matrices and λ takes the values, given by the branching rules. Up to the phase factors the equation (7.11) gives the matrix elements (7.3). As the solutions of (7.11) one can choose the following expression:

$$A^{(\omega)}_{\bar{\omega},\, \lambda\bar{\lambda}} = (-1)^{[(k+\bar{\omega})\,\bar{\omega}\,]\,\mathrm{sign}\, a} \qquad (7.12)$$

for all $\omega = 3a + k$, as well as for $\omega = (o)$ and $\omega = (o)^*$ ($a=k=o$, $\bar{\omega}=0$, if $\omega = (o)$ and $a=k=o$, $\bar{\omega} = 1$, if $\omega = (o)^*$). Let us mention, that according to the branching rules, $\bar{\lambda}$ is always uniquely defined by ω, δ, λ and this is the reason, why $A^{(\omega)}$ is always one-dimensional matrix.

Finally, using (7.6) one gets

$$\psi(\Gamma LM\ \omega\lambda\bar{\lambda} \,|\, \rho^{(2)}\rho^{(3)},\ G_3^{+},\ \sigma\theta) =$$

$$= \sum_{K\bar{\omega}} \Theta(\Gamma LK\ \omega\bar{\omega} \,|\, \rho^{(2)}\ \rho^{(3)})\ D^{L}_{KM}(G_3^{+})\ D^{\omega}_{\bar{\omega},\ \lambda\bar{\lambda}}(\sigma\theta) \qquad (7.13)$$

where

$$D^{\omega}_{\bar{\omega},\lambda\bar{\lambda}}(\sigma\theta) = D^{\omega}_{\bar{\omega}',\bar{\omega}}(\sigma\theta)\ (-1)^{[(k+\bar{\omega})\bar{\omega}]\,\mathrm{sign}\, a} \qquad (7.14)$$

The set of functions (7.13) has a definite permutation symmetry, thus it is adapted to assure the Pauli principle requirements.

8. THE COMPLETE SET OF FUNCTIONS FOR THE ANGULAR VARIABLES. GENERAL CASE

In this section we will briefly present the results of the general case of n particles moving in three-dimensional space. This generalization does not concern the Euler angles $G_3^{\pm} \equiv \theta'\theta''\theta'''$; the set of functions (6.1) suits the arbitrary number of particles, thus we need only to describe the functions depending on 3(n-3) continuous variables q^+ and the reflection σ (see section 5). The complete set of such a type of functions in the physical basis has been introduced in[14] (see also [15-17,6]). Let us denote it as

$$D^{\omega}_{\nu^\circ,\alpha\lambda\eta}(\sigma\ q^+) \tag{8.1}$$

and let us discuss the meaning of the quantum numbers in (8.1). The quantum number ω in the case of n-1>6 stands for three numbers, i.e. $\omega \equiv (\omega_1\omega_2\omega_3)$, where $(\omega_1\omega_2\omega_3)$ labels the O_{n-1} irreducible representation (if n-1=6,5, $\omega \equiv (\omega_1\omega_2)$; if n-1=4,3, $\omega \equiv (\omega_1)$).

The symmetric group S_n is imbedded into O_{n-1} on the chain $O_{n-1} \supset S_n$; the mathematical aspects of this question have been cleared up in[19]. The chain $O_{n-1} \supset S_n$ has been used in[20] (see also [21]) in order to construct the antisymmetric oscillator wave functions. On this chain one is encountered with the multiplicity problem, thus in (8.1) the multiplicity index α is introduced. The basis of the S_n irreducible representation λ is labelled by η . Thus the columns of the matrix elements (8.1) have the same structure as in (7.5). The only difference is, that the groups and their irreducible representations are much more complicated.

The new features appear in the labelling of the rows. The rows in (8.1) are labelled by the quantum numbers of the following canonical chain of the orthogonal groups:

$$O_{n-1} \supset O_{n-2} \supset O_{n-3} \cdots \supset O_1 \tag{8.2}$$

One need only a small part of the quantum numbers of this chain, namely the set of numbers which ends up with the O_{n-4} scalar. This circumstance is connected with the properties of the set of varibles q^+. As it was mentioned in section 5, the set q^+ parametrizes the factor-space $O^+_{n-4} \diagdown O^+_{n-1}$ and this is the reason, why one needs only the O_{n-4} scalar basis. Thus the quantum numbers ν° in (8.1) denote the following "small" canonical set of three numbers

$$\nu^{\circ} \equiv \left\backslash \begin{array}{ccccc} \bar{\omega}_1 & & \bar{\omega}_2 & & \bar{o} \\ & \bar{\bar{\omega}} & & \bar{\bar{o}} & \\ & & \bar{\bar{\bar{o}}} & & \end{array} \right/ \tag{8.3}$$

where $(\bar{\omega}_1\ \bar{\omega}_2)$, $(\bar{\bar{\omega}}_1)$ and $(\bar{\bar{\bar{0}}})$ correspondingly denote the irreducible representations of the groups O_{n-2}, O_{n-3} and O_{n-4}. In the future we shall need this basis, therefore let us introduce the notations $\bar{\omega} \equiv (\bar{\omega}_1\ \bar{\omega}_2)$ and $\bar{\bar{\omega}} \equiv (\bar{\bar{\omega}}_1)$. In case of n-3 in the chain (8.2) only the O_{n-2} subgroup has the meaning, thus ν° means $\bar{\omega}$, already discussed in section 6. Instead of the canonical chain (8.2) one can use other chain, for example the chain

$$\begin{array}{ccc} O_{n-1} \supset & O_3 & \dot{+}\ O_{n-4} \\ & \cup & \\ & O_3^+ & \\ & \cup & \\ & O_2^+ & \end{array} \tag{8.4}$$

In th case instead of ν° one has the quntum numbers $\beta_\nu\ \ell_\nu\ m_\nu$, where ℓ_ν and m_ν are the quasi-momentum and its projection in the abstract three-dimensional space (not connected with the three-dimensional space we are living in), spanned by the Jacobi vectos $\vec{\rho}_{n-3}$, $\vec{\rho}_{n-2}$, $\vec{\rho}_{n-1}$ and β_ν is the multiplicity index for the chain (8.4). These quantum numbers have been introduced in[14]. Another possibility is to introduce the chain[16]:

$$\begin{array}{ccc} O_{n-1} \supset & O_3 & \dot{+}\ O_{n-4} \\ & \cup & \\ & \bar{S}_3 & \end{array} \tag{8.5}$$

where $\bar{S}_3$ is symmetric group, permutating the Jacobi vectors $\vec{\rho}_{n-3}$, $\vec{\rho}_{n-2}$, $\vec{\rho}_{n-1}$ (not to be mixed with the group S_3 , permutating the single-particle vectors).

We need the generalization of the formula (7.14) and this question is not a trivial one. In the case of n-1=2 there was only one continuous varible and due to it the matrix $A^{(\omega)}$ is one-dimensional. In the general case one has 3(n-3) varibles q^+ and it is difficult to obtain the transformation matrix between two bases, labelled by the quantum numbers of the chains

$$O_{n-1} \supset O_{n-2} \supset O_{n-3} \supset \dots \supset O_1$$

$$O_{n-1} \supset S_n \supset S_{n-1} \supset \dots \supset S_2 \qquad (8.6)$$

The only way is to develope the recurrent technique. This technique has been proposed in[22]. The final results are as follows. One must take the set of variables and decompose it into two subsets $\sigma\bar{q}^{+}\sigma_1^{-1}$ and $\sigma_1\, q_1^{+}$ where $\bar{q}^{+}$ consists of the 3(n-9) variables and q_1^{+} consists of three variables (we are not goint to discuss this question in details, see[6,17]. The meaning of the set q_1^{+} will be explained in section 10). One also must introduce the transformation between the following two chains:

$$O_{n-1} \supset O_{n-2} \supset S_{n-1}$$

$$O_{n-1} \supset S_n \supset S_{n-1} \qquad (8.7)$$

The matrix elements of this transformation according to Shur lemma are diagonal with respect to the S_{n-1} irreducible representation $\bar{\lambda}$ and do not depend on its basis. Thus the matrix elements needed may be labelled as follows:

$$A^{(\omega\ \bar{\lambda})}_{\bar{\omega}\ \bar{\alpha},\alpha\lambda} \qquad (8.8)$$

where $\bar{\omega}$ and λ denote the O_{n-2} and S_n irreducible representations; α and $\bar{\alpha}$ are the multiplicity indexes on the chains $O_{n-1} \supset S_n$ and $O_{n-2} \supset S_{n-1}$.

In[14] (see also [6,17]) it has been shown, that the set of functions (8.1) may be decomposed in the following way:

$$D^{\omega}_{\nu^\circ,\alpha\lambda\bar{\lambda}\bar{\eta}}(\sigma\, q^{+}) =$$

$$= \sum_{\bar{\omega}\,\bar{\alpha}\,\bar{\nu}^\circ} D^{\omega}_{\nu^\circ,\ \bar{\omega}\ \bar{\nu}^\circ}(\sigma\bar{q}^{+}\sigma_1^{-1})\, D^{\bar{\omega}}_{\bar{\nu}^\circ,\alpha\bar{\lambda}\bar{\eta}}(\sigma_1 q_1^{+})\, A^{(\omega\ \bar{\lambda})}_{\bar{\omega}\ \bar{\lambda},\alpha\ \lambda} \qquad (8.9)$$

where $\bar{\nu}^\circ$ denotes the set of quantum numbers (8.3) for the chain $O_{n-3} \supset O_{n-4} \supset O_{n-5}$. In the three particle case O_{n-2} is the reflection group, but in the argument of function D there are two reflections σ and σ_1^{-1}, thus in this case $D^{\omega}=1$, $A^{(\omega)}$ is one-dimensional and (8.9) reduces to (7.14)

The decomposition (8.9) is the fractional-parentage type decomposition and (8.8) may be called the fractional-parentage coefficients. Those coefficients are, of course essentially different from those used in the shell model calculations. It is convenient to introduce the following bilinear forms[22]

$$Q_{\omega\,\bar{\omega},\omega'\,\bar{\omega}}(\bar{\alpha}\,\bar{\lambda};\alpha\lambda,\alpha'\lambda') = \sum_{\bar{\alpha}} A^{(\omega\,\bar{\lambda})}_{\bar{\omega}\bar{\alpha},\alpha\lambda}\, A^{(\omega'\,\bar{\lambda})}_{\bar{\omega}\bar{\alpha},\alpha'\lambda'} \tag{8.10}$$

The quantity Q is the group-theoretical analogue of the generalized one-particle density matrix in quantum mechanics.

9. THE O_2 IRREDUCIBLE DECOMPOSITION OF THE HAMILTONIAN. THREE PARTICLE CASE

In this section we will make the last step towards our goal - to project H_{coll} from H. We are going to decompose H in the series of the O_2 irreducible operators, i.e. to get the decomposition

$$H = \sum_{\chi} H^{(\chi)} \tag{9.1}$$

where χ denotes the O_2 irreducible representation. We will perform this decomposition partially in the matrix and partially in the coordinate representation.

We use a very simple idea. Let us explain it by example. Suppose one has a particle, moving in the non-central field with an arbitrary potential $V(xyz) \equiv V(r\theta\psi)$ and one is looking for the O_3^+ irreducible decomposition of V in order to get the central part of V. Let us introduce the following matrix representation for V:

$$\langle \ell m | V(r) | \ell' m' \rangle =$$

$$= \int_0^{\pi} \sin\theta \, d\theta \int_0^{2\pi} d\psi \, \mathcal{Y}^{*\ell}_{m}(\theta\psi)\, V(r\theta\psi)\, \mathcal{Y}^{\ell'}_{m'}(\theta\psi) \tag{9.2}$$

where $\mathcal{Y}^{\ell}_{m}$ are the spherial harmonics. Let us say that one needs the matrix elements (9.2) in some many-particle problem, i.e. one needs the following linear combinations:

$$\sum_{\ell\ell' mm'} \langle \ell m|V|\ell' m'\rangle \; Q_{\ell m,\ell' m'} \tag{9.3}$$

The meaning of the coefficients Q in (9.3) is not important for this example. Using the orthogonality properties of the Clebsch-Gordan coefficients

$$\sum_{\kappa\eta} \frac{d_\kappa}{2\ell+1} C^{\kappa}_{\eta}{}^{\ell'}_{m'}{}^{\ell}_{m} \; C^{\kappa}_{\eta}{}^{\ell'}_{m'_1}{}^{\ell}_{m_1} = \delta(m'm'_1)\;\delta(mm_1) \tag{9.4}$$

one can easily present (9.3) in the form

$$\sum_{\substack{\ell\ell' \\ mm'm_1m'_1}} \langle \ell m|V(r)|\ell' m'\rangle \times$$

$$\Big(\sum_{\kappa\eta} \frac{d_\kappa}{2\ell+1} C^{\kappa}_{\eta}{}^{\ell'}_{m'}{}^{\ell}_{m} \; C^{\kappa}_{\eta}{}^{\ell'}_{m'_1}{}^{\ell}_{m_1}\Big) \; Q_{\ell m_1,\ell' m'_1} =$$

$$= \sum_{\substack{\ell\ell' \\ \kappa\eta}} \mathrm{J}_{\ell m\kappa\eta} \; Q_{\ell\ell'\kappa\eta} \tag{9.5}$$

where

$$\mathrm{J}_{\ell\ell'\kappa\eta}(r) = \sum_{mm'} \frac{d_\kappa}{2\ell+1} \langle \ell m|V(r)|\ell' m'\rangle \; C^{\kappa}_{\eta}{}^{\ell'}_{m'}{}^{\ell}_{m} \tag{9.6}$$

and

$$Q_{\ell m\kappa\eta} = \sum_{m_1m'_1} Q_{\ell m_1,\ell' m'_1} \; C^{\kappa}_{\eta}{}^{\ell'}_{m'_1}{}^{\ell}_{m_1} \tag{9.7}$$

The integrals (9.6) depend on the O_3^{+} irreducible representation index κ, thus they give the $O_3^{\pm}$ irreducible decomposition of the interaction $V(r\theta\psi)$. As long as one is looking for the $O_3^{\pm}$ scalar part of the expression (9.3), one must simply take only the term in (9.5) with $\kappa=0$. In this case from (9.6) and (9.2), taking into account the well known formula

$$\mathcal{Y}^{\ell}_{m}(\theta\psi) = \sqrt{\frac{2\ell+1}{4\Pi}}\; D^{\ell}_{om}(o-\theta-\psi) \qquad (9.8)$$

one has

$$\mathcal{J}_{\ell\ell' oo}(r) = \delta(\ell\ell') \sum_{m} \frac{1}{2\ell+1} \langle \ell m|\; V(r)\,|\ell m\rangle =$$

$$= \sum_{m} \frac{\delta(\ell\ell')}{4\Pi} \int_{o}^{\Pi} \sin\theta\, d\theta \int_{o}^{2\Pi} d\psi\; \overset{*}{D}{}^{\ell}_{om}(0-\theta-\psi) V(r\theta\psi)\; D^{\ell}_{om}(0-\theta-\psi) \qquad (9.9)$$

Taking the sum over m in (9.9) one has

$$\mathcal{J}_{\ell\ell' oo}(r) = \delta(\ell\ell')\, \frac{1}{4\Pi} \int_{o}^{\Pi} \sin\theta\, d\theta \int_{o}^{2\Pi} d\psi\; V(r\theta\psi) \qquad (9.10)$$

thus the 0^{+}_{3} scalar part of the expression (9.3) is

$$\sum_{\ell\ell' mm'} \langle \ell m|\; V(r)\,|\ell'\; m'\rangle \quad Q_{\ell m,\ell' m'} \Bigg|_{0^{+}_{3}\ \text{scalar}} =$$

$$= \delta(\ell\ell') \sum_{\ell} \mathcal{J}_{\ell\ell oo}(r) \sum_{m} Q_{\ell m,\ell m} \qquad (9.11)$$

In this example the relationship of the variable r, we've used in (9.2), with the many-particle system variables and the meaning of the coefficients Q in (9.3) has not been discussed. Thus this example does illustrate the next steps we are going to perform, but it does not illustrate the previous steps, we've done already.

Now let us return to the three particle case and let us construct the antisymmetric wave functions (3.6) by coupling the space functions (7.12) with the spin-isospin functions. In case of the three particles the latter has the following expression:

$$\Phi\;(\;\tilde{\lambda}\;\tilde{\tilde{\lambda}}\;S\;M_S\;T\;M_T|\;\tilde{Q}\;), \qquad (9.12)$$

where $\tilde{\lambda}\ \tilde{\tilde{\lambda}}$ denote the S_3 and S_2 irreducible representations, conjugated with those in (7.12). Using the Clebsch-Gordan coefficients for the S_3 group one gets:

$$\psi(\Gamma\ (LS)\ JM_J TM_T \lambda |\ \rho^{(2)}\ \rho^{(3)},\ G_3^+, \sigma\theta\ ;\ \tilde{\tilde{Q}}\ ) =$$

$$= \sum_{\substack{\eta M M_S \\ K\bar{\omega}'}} \Theta(\Gamma LK\ \omega\bar{\omega}' |\ \rho^{(2)}\ \rho^{(3)})\ D^L_{KM}(G_3^+) D^{\omega}_{\bar{\omega}',\bar{\omega}}\ (\sigma\theta)\ A^{(\omega)}_{\bar{\omega},\ \lambda\bar{\lambda}}\ \times$$

$$\times\ \ \Phi(\tilde{\lambda}\tilde{\tilde{\lambda}} SM_S TM_T\ |\ \tilde{Q})\ C^{\lambda\tilde{\lambda}\ [111]}_{\bar{\lambda}\tilde{\tilde{\lambda}}\ [11]}\ C^{L\ \ S\ \ J}_{M\ M_S\ M_J} \qquad (9.13)$$

All the Clebsch-Gordan coefficients for the group S_3 used in (9.13), have the following expression

$$C^{[3]\ \ [111]\ \ [111]}_{[2]\ \ [11]\ \ [11]} = C^{[111]\ \ [3]\ \ [111]}_{[11]\ \ [2]\ \ [11]} = 1$$

$$C^{[21]\ \ [21]\ \ [111]}_{[2]\ \ [11]\ \ [11]} = -C^{[21]\ \ [21]\ \ [111]}_{[11]\ \ [2]\ \ [11]} = \frac{1}{\sqrt{2}} \qquad (9.14)$$

Now, when the Pauli principle is assured, the formula (3.7) may be used, thus instead of $H(\kappa)$, given in (2.17), only the operator

$$\sum_q (-1)^q\ \hat{U}^{\kappa}_q\ (\sigma_2\sigma_3\tau_2\tau_3)\ \hat{W}^{\kappa}_{-q}\ (\sqrt{2}\ \vec{\rho}) \qquad (9.15)$$

is needed. In general $\hat{W}^{\kappa}_{-q}$ depends on derivatives, but in order not to be bogged down with the details (we will present the general result in the next section) let us take the simplest case when $\hat{W}$ depends only on $|\vec{\rho}_a|$ and $\hat{U} = 1$, i.e. $\hat{W}$ is the potential of the Wigner term in the central interaction (2.11). Let us use the formula (4.7) for the variables $\vec{\rho}_s$, $\vec{\rho}_a$. Then taking into account the orthogonality properties of D^L and D^{ω} one gets

$$\rho_a^2 = (\rho^{(2)})^2 \sin^2\theta + (\rho^{(3)})^2 \cos^2\theta \qquad (9.16)$$

thus

$$V(\sqrt{2}\,\rho_a) = V\left(\sqrt{2(\rho^{(2)})^2 \sin^2\theta + 2(\rho^{(3)})^2\cos^2\theta}\right) \tag{9.17}$$

Now let us introduce the integrals

$$\mathcal{J}^{(\bar{\omega}_o\bar{\omega}'_o)}_{\omega\bar{\omega},\omega'\bar{\omega}}(\rho^{(2)}\rho^{(3)}) =$$

$$= \frac{1}{2}\sqrt{d_\omega d_{\omega'}}\sum_\sigma \int_0^{2\pi} d\theta\, \overset{*}{D}{}^\omega_{\bar{\omega}\bar{\omega}_o}(\sigma\theta)\, V(\rho^{(2)}\rho^{(3)})\, D^{\omega'}_{\bar{\omega}'\bar{\omega}'_o}(\sigma\theta) \tag{9.18}$$

Those integrals are still O_2 reducible ones (compare (9.18) with (9.2)). In order to calculate the matrix elements of the operator (9.17) on the basis (9.13) one needs the matrices D^ω with columns, labelled by $\lambda\,\bar{\lambda}$, thus, according (7.14) and (7.12)

$$\mathcal{J}^{(\bar{\omega}_o\,\bar{\omega}'_o)}_{\omega\bar{\omega},\omega'\bar{\omega}}(\rho^{(2)}\rho^{(3)})\; Q_{\omega\bar{\omega},\omega'\bar{\omega}} \tag{9.19}$$

is needed. In (9.19)

$$Q_{\omega\bar{\omega},\omega'\bar{\omega}}(\bar{\lambda};\lambda,\lambda') = A^{(\omega)}_{\bar{\omega},\lambda\bar{\lambda}}\; A^{(\omega')}_{\bar{\omega},\lambda'\bar{\lambda}} \tag{9.20}$$

(compare (9.19) with (9.3); because $A^{(\omega)}$ is one-dimensional, there is no sum in (9.19)). Now if one repeats all the steps made between the formulae (9.3) and (9.7) using the Clebsch-Gordan coefficients

$$C\begin{matrix}\chi & \omega' & \omega\\ \bar{\chi} & \bar{\omega}' & \bar{\omega}\end{matrix} \tag{9.21}$$

for the group O_2 and coupling with them the quantum numbers of O_2, labelling the rows of the matrices D^ω, one easily gets the following matrix elements of the operator (9.17) on the basis (7.13):

$$\langle\bar{\omega}_o\omega\;\lambda\bar{\lambda}\;|V|\;\bar{\omega}'_o\omega'\lambda\;\bar{\lambda}'\rangle =$$

$$= \delta(\bar{\lambda}\ \bar{\lambda}') \sum_{\chi} \mathcal{J}^{(\bar{\omega}_o \bar{\omega}'_o)}_{\omega\omega'\chi} Q_{\omega\omega'\chi} \tag{9.22}$$

In (9.10) the following notations are introduced:

$$Q_{\omega\omega'\chi} = \sum_{\bar{\omega}} Q_{\omega\bar{\omega},\omega'\bar{\omega}} C \begin{matrix} \chi & \omega' & \omega \\ 0 & \bar{\omega} & \bar{\omega} \end{matrix} \tag{9.23}$$

and

$$\mathcal{J}^{(\bar{\omega}_o \bar{\omega}'_o)}_{\omega\omega'\chi} (\rho^{(2)} \rho^{(3)}) =$$

$$\sum_{\bar{\omega}} \frac{d_\chi}{d_\omega} \mathcal{J}^{(\bar{\omega}_o\ \bar{\omega}'_o)}_{\omega\bar{\omega},\omega'\bar{\omega}} C \begin{matrix} \chi & \omega' & \omega \\ o & \bar{\omega} & \bar{\omega} \end{matrix} \tag{9.24}$$

There is only the O'_1 scalar basis of χ in (9.24), because O'_1 does not act on the Jacobi vector $\vec{\rho}_a$. The Clebsch-Gordan coefficients (9.21) are very simple, and there is no problem to calculate them. The rules of coupling χ with ω' to ω are as follows:

$$(o) \times (\omega) = (\omega)$$

$$(o)^* \times (\omega) = \begin{cases} (\omega), & \text{if } \omega \neq (o), \omega \neq (c)^* \\ (o)^*, & \text{if } \omega = (o) \\ (o), & \text{if } \omega = (o)^* \end{cases} \tag{9.25}$$

and

$$(\omega) \times (\omega) = (o) + (o)^* + (2\omega)$$

$$(\omega_1) \times (\omega_2) = (|\omega_1 - \omega_2|) + (\omega_1 + \omega_2), \text{ if } \omega_1 \neq \omega_2 \tag{9.26}$$

In (9.26) all the representations ω, ω_1 and ω_2 are two dimensional.

The term in (9.22) with the definite χ gives the O_2 irreducible component of the interaction (9.17), presented in the matrix representation with respect to the variable θ and in the coordinate representation with respect to the variables $\rho^{(2)}$ $\rho^{(3)}$. In order to get the matrix representation with respect to all variables, one must just take the expectation values of the χ component of (9.17) on the set of the arbitrary radial functions Θ, used in the decomposition (9.13). Thus (9.17) gives the O_2 irreducible decomposition of the Wigner interaction. In particular, if one is interested only in the O_2 scalar part of (9.17), one takes $\chi = 0$ and, similarly to (9.10), one has

$$V(\sqrt{2}\,|\vec{\rho}_a|)\Big|_{O_2\ \text{scalar}} = \frac{1}{2\pi}\int_0^{2\pi} d\theta \;\; V\left(\sqrt{2(\rho^{(2)})^2\sin^2\theta + 2(\rho^{(3)})^2\cos^2\theta}\right) \tag{9.27}$$

In the section 11 we will prove, that the last expression gives the collective part of the Wigner interaction.

10. THE O_{n-1} IRREDUCIBLE DECOMPOSITION OF THE HAMILTONIAN. GENERAL CASE

In the general case due to (3.7), the matrix elements are needed only for the operator $W^{\kappa}_{-q}\ (\sqrt{2}\,\vec{\rho}_a)$ (this time we are not simplyfying this operator, therefore in general it depends on the vector $\vec{\rho}_a$ and derivatives with respect to the components of $\vec{\rho}_a$). We will take (8.1) as the basis functions and we will use (5.1) and (5.11) in order to change ρ^s_a and $\frac{\partial}{\partial \rho^s_a}$ by new variables. It is essential, that the last column of $D^{(1_{n-1})}$ in (5.1) (for ρ^s_a only the last column is needed) depends only on the three angles and has the following expression:

$$D^{(1_{n-1})}_{n-4,n-1}(\bar{q}^+) = \sin\theta_3 \sin\theta_2 \;\sin\theta_1$$

$$D^{(1_{n-1})}_{n-3,n-1}(\bar{q}^+) = \cos\theta_3 \sin\theta_2 \;\sin\theta_1$$

$$D^{(1_{n-1})}_{n-2,n-1}(\bar{q}^+) = \cos\theta_2 \sin\theta_1$$

$$D^{(1_{n-1})}_{n-1,n-1}(\bar{q}^+) = \cos\theta_1 \tag{10.1}$$

Thus the set of the variables $\bar{q}^+$ used in (8.9) consists of three variables used in (10.1), i.e. $\bar{q}^{+} \equiv \theta_3\ \theta_2\ \theta_1$. Evidently, every $\hat{W}$ be rewritten in the new variables, thus

$$\hat{W}^{\kappa}_{-q}\ (\sqrt{2}\,\vec{\rho}_a) =$$

$$= \hat{W}^{\kappa}_{-q}\ (\rho^{(s_o)},\ \frac{\partial}{\partial\rho(s_o)},\ G_3^+,\ \hat{\mathcal{L}}_{s_o\ s_o'},\ \sigma\bar{q}^{+},\ \hat{\mathcal{J}}_{s_o\ s_o'})$$

(10.2)

Acting with (10.2) on the matrices $D^{(\omega)}$ in the canonical basis one may convert the polynoms depending on $\hat{\mathcal{J}}_{s_o s_o'}$ into their matrices, thus

$$\hat{W}\ D^{\omega}_{\nu^\circ\nu}(q) = \sum_{\omega'\nu^{\circ\prime}} \langle\omega\nu^\circ|\hat{W}|\omega'\nu^{\circ\prime}\rangle\ D^{\omega'}_{\nu^{\circ\prime}\nu}\ (q) \qquad (10.3)$$

Now the right side of (10.3) does not depend on $\hat{\mathcal{J}}$, thus every matrix element of W in (10.3) with respect to the varibles q is essentially the same kind of operator, as of the previous section. Therefore all the proof is analogous, so we present only the final result. This result has been obtained in[23,6].

The matrix elements of the operator (10.2) in the basis (8.1) may be presented in the following form:

$$\langle\ \nu^\circ\omega\alpha\lambda\eta|\hat{W}|\nu^{\circ\prime}\ \omega'\ \alpha'\ \lambda'\ \eta'\ \rangle =$$

$$= \delta\ (\eta\eta')\ \sum_{\chi\beta}\ \mathrm{J}^{(\nu^\circ\nu^{\circ\prime})}_{\omega\omega'\chi\beta}\ \ Q_{\omega\omega'\chi\beta} \qquad (10.4)$$

where

$$Q_{\omega\omega'\chi\beta} = \sum_{\bar{\omega}}\ Q_{\omega\ \bar{\omega},\omega'\ \bar{\omega}}\ C^{\chi}_{\bar{o}}\ \ {}^{\omega'}_{\bar{\omega}}\ \ {}^{\beta}\ \ {}^{\omega}_{\bar{\omega}} \qquad (10.5)$$

and

$$\mathrm{J}^{(\underline{\nu}^\circ\nu^{\circ\prime})}_{\omega\bar{\omega},\omega'\bar{\omega}}\ (\rho^{(s_o)},\ \frac{\partial}{\partial\rho(s_o)},\ G_3^+,\ \hat{\mathcal{L}}_{s_o s_o'}) =$$

$$= \frac{\sqrt{d_\omega d_{\omega'}}}{d_{\bar{\omega}}} \quad \frac{1}{2} \sum_{\substack{\sigma\bar{\nu}^\circ \\ \omega''\nu^{\circ''}}} \int d\tau\,(\bar{q}^+)\ \overset{*}{D}{}^{\omega}_{\nu^\circ,\bar{\omega}\bar{\nu}^\circ}(\sigma\bar{q}^+) \quad \times$$

$$\times \quad \langle\, \omega'\nu^{\circ'} \mid \hat{W}(\sigma\bar{q}^+) \mid \omega''\ \nu^{\circ''} \,\rangle \quad D^{\omega''}_{\nu^{\circ''},\bar{\omega}\bar{\nu}^\circ}(\sigma\bar{q}^+) \tag{10.6}$$

In (10.4)-(10.6) β is the multiplicity index in the direct product $\chi\times\omega'\to\beta\omega$; $Q_{\omega\bar{\omega},\omega'\bar{\omega}}$ - the matrix elements (8.10) of the density matrix, $d\tau(\bar{q}^+)$ - the following volume element. If $n>5$:

$$d\tau(\bar{q}^+) = \frac{\Gamma\left(\frac{n-1}{2}\right)}{\Pi\sqrt{\Pi}\,\Gamma\left(\frac{n-4}{2}\right)} \sin^{n-5}\theta_3 \ \sin^{n-4}\theta_2 \ \sin^{n-3}\theta_1 d\theta_3 d\theta_2 d\theta_1 \tag{10.7}$$

and

$0 \leqslant \theta_3,\ \theta_2,\ \theta_1 < \Pi$. If $n=5$:

$$d\tau(\bar{q}^+) = \frac{1}{2\Pi^2} \ \sin\theta_2 \ \sin^2\theta_1 \ d\theta_3 \ d\theta_2 \ d\theta_1 \tag{10.8}$$

and

$0 \leqslant \theta_3 < 2\Pi$; $0 \leqslant \theta_2, \theta_1 < \Pi$. If $n=4$:

$$d\tau(\bar{q}^+) = \frac{1}{4\Pi} \sin\theta_1 \ d\theta_2 \ d\theta_1 \tag{10.9}$$

and $0 \leqslant \theta_2 < 2\Pi$; $0 \leqslant \theta_1 < \Pi$. In (10.7) Γ is gamma function.

Similarly to the three particle case, (10.4) gives the O_{n-1} irreducible decomposition of the arbitrary operator $\hat{W}(\sqrt{2}\,\vec{\rho}_a)$. In order to get the matrix representation for the O_{n-1} irreducible term χ in (10.4), the expectation values of this term on the set of functions

$$\sum_{K} \ \Theta\left({}^{\Gamma\omega LKM}_{\nu^\circ} \mid \rho^{(s_o)} \right) D^L_{KM}(G_3^+) \tag{10.10}$$

must be calculated. One may also get the matrix representation for the two-particle spin-isospin operator $\overset{\times}{U}{}^{\kappa}_{q}$ in

(2.17), calculating its expectation values on the following set of the n-particle spin-isospin wave functions

$$\Phi\ (\tilde{\lambda}\tilde{\eta}\tilde{\alpha}\ S\ M_S\ T\ M_T \mid \tilde{Q}\ ) \qquad (10.11)$$

where $\tilde{\alpha}$ is the multiplicity index. Then, composing those results we get the matrix representation for the O_{n-1} irreducible component $H^{(\chi)\beta}$ of H, averaged on the set of the antisymmetric wave functions

$$\psi\ \left(\begin{matrix} \Gamma\omega(LS)\ JM_J \\ TM_T\ \tilde{\alpha}\ (\lambda\tilde{\lambda}J)\,[1^n] \end{matrix} \middle| \rho^{(s_o)},\ G_3^+\ ,\ \sigma q^+;\ \tilde{Q}\ \right) =$$

$$= \sum_{\substack{\nu^\circ K \\ MM_S\eta}} \Theta \left(\begin{matrix} \Gamma\omega LMK \\ \nu^\circ \end{matrix} \middle| \rho^{(s_o)} \right) D^L_{KM}\ (G_3^+)\ D^\omega_{\nu^\circ,\alpha\lambda\eta}\ (\sigma\ q^+)\ \times$$

$$\times\ \ \Phi(\tilde{\lambda}\tilde{\eta}\tilde{\alpha}\ SM_S\ TM_T \mid \tilde{Q})\ C^{L}_{M}\ {}^{S}_{M_S}\ {}^{J}_{M_J}\ C^{\lambda}_{\eta}\ {}^{\tilde{\lambda}}_{\tilde{\eta}}\ {}^{[1^n]}_{\eta_{[1^n]}} \qquad (10.12)$$

In (10.12) $[1^n]\ \eta_{[1^n]}$ is the antisymmetric representation of S_n and its basis.

We have proved the existence of the O_{n-1} irreducible decomposition of the arbitrary Hamiltonian by help of the explicit construction (it was not only the abstract proof, - one can use the final expression for calculations). But if one has proven the existence of $H(\kappa)$ acting in the Hilbert space in a particular representation, one has proven by the same token its existence in the arbitrary (thus also in the coordinate) representation[25]. That means that one has got the following decomposition[24,23,6]

$$H = H^{(o)} + \sum_{\chi \neq o,\ \beta} H^{(\chi)\beta} \qquad (10.13)$$

where the first term is the O_{n-1} scalar part of H.

11. THE MICROSCOPIC COLLECTIVE HAMILTONIAN

In this section we will discuss the first term of the decomposition (10.13). The important feature of $H^{(o)}$ is that $H^{(o)}$ depends only on the variables $\rho^{(s_o)}$, $G_3^\pm$. Therefore in order to clear up the physical meaning of

$H^{(o)}$ one must understand the meaning of the variables $\rho^{(s_o)}$, G_3^{+}. What is their most important feature that differs them from the rest of 3(n-3) variables q^{+} ?

We have no intention to treat $\rho^{(s_o)}$, G_3^{+} and q^{+} geometrically, because in quantum mechanics the real density distribution, strictly speaking is not connected with the varibles, used in the Schrödinger equation. The geometric picture always has some phenomenological roots. Some quantum mechanics criteria must be found in order to make clear distinction between the sets of varibles $\rho^{(s_o)}$, $G_3^{\pm}$ and q^{+}. In[15] the following definition has been proposed: the variables are called collective variables, if they are scalars with respect to the symmetric group, acting on the single particle variables. The variables are non collective (let us call them the internal variales) in case they are not scalars with respect to the symmetric group, mentioned above.

From the general expression (5.1) it is clear that the varibles $\rho^{(s_o)}$, G_3^{+} are scalars, thus according to our definition, they are collective variables. In fact the transformations of the group O_{n-1} act only on q^{+} (see the example (4.11)), the symmetric group S_n in the subgroup of O_{n-1}, thus S_n acts only on q^{+}. That gives the proof, that $\rho^{(s_o)}$, $G_3^{\pm}$ are the S_n scalars.

The question remains, whether all varibles q^{+} are not scalars with respect to the action of S_n. In[15] has been proved that all of them are not scalar of S_n and that there is no possibility to reduce their number, without failing to assure the Pauli principle requirements if the total number of variables is 3(n-1). Thus, according to our definition, the set q^{+} consists of the noncollective variables and now one gets clear quantum-mechanical distinction between the variables $\rho^{(s_o)}$, G_3^{+} and q^{+}. Thus we have the following theorem[15,6]:

THEOREM. Without violating the requirements of microscopic translational invariance and antisymmetry properties of the wave function of an isolated fermion system consisting of r+1 equivalent particles moving in the r_o dimensional space, one may introduce N collective variables, where $N=1/2\ r_o(r_o+1)$, if $r_o \leq r$ and $N=rr_o - 1/2\ r\ (r+1)$, if $r_o \geq r$. The remaining $r_o r-N$ continuous variables and one reflection variable are essentially noncollective (internal) ones.

The important point of this theorem is that one is treating protons and neutrons as equivalent particles, i.e. that one is using the isospin formalizm. If protons and neutrons are treated as different particles, the nucleus is a system consisting of two types of fermion subsystems. For this version of the theory one has N_p for protons and N_n for neutrons and also r_o variables for the relative motion of both subsystems, thus in this case one has the total number of the collective variables $N_p+N_n+r_o$.

The definition of the collective and the internal variables we've just described makes the physical meaning of $H^{(o)}$ clear. The O_{n-1} scalar part of H is its collective part, i.e. $H^{(o)}= H_{coll}$. The irreducible properties of the decomposition (10.13) guarantee that H_{coll} consists of the whole part of H, which depends on the collective variables only. Every additional term, i.e. the therms in (10.13) with $\chi \neq (o)$, takes into account the interaction between collective and internal degrees of freedom. Thus, taking into account the physical meaning of the terms in (10.13) we've just explained, let us rewrite (10.14) in the following form

$$\Sigma\ H(\kappa) = \Sigma\ H_{coll}(\kappa) + \Sigma\ H_{coll-intern}(\kappa) \tag{11.1}$$

where Σ means, that all the term of H given in (2.16) are taken and κ denotes their O_3^+ irreducible properties. Now the decomposition (2.1) became constructive because

$$H_{coll} = \Sigma\ H_{coll}(\kappa) \tag{11.2}$$

and

$$H_{coll-intern} = \Sigma\ H_{coll-intern}(\kappa) \tag{11.3}$$

In[23] it has been proved that the O_{n-1} irreducible term $H_{coll}(\kappa)$ in the matrix representation (i.e. on the set of functions (10.12)) has the following expression:

$$< \begin{matrix} \Gamma(LS)JM_J \\ TM_T\tilde{\alpha}\omega\alpha \end{matrix} (\lambda\tilde{\lambda})\ [1^n] \left| H^{(\kappa\kappa)o}_{coll} \right| \begin{matrix} \Gamma'\ (L'S')JM_J \\ T'\ M_T\tilde{\alpha}'\omega'\alpha' \end{matrix} (\lambda'\tilde{\lambda}')\ [1^n]> =$$

$$= \frac{n(n-1)}{2}\delta(\omega\omega')\ \delta(\alpha\alpha')\ \delta(\lambda\lambda')\ \delta(TT')(-1)^{L+S'+J}\sqrt{(2L+1)(2S+1)}\times$$

$$\begin{Bmatrix} L & S & J \\ S' & L' & \kappa \end{Bmatrix} <\tilde{\lambda}\tilde{\alpha}STM_T||\ \hat{U}^{\kappa}\ ||\tilde{\lambda}\tilde{\alpha}'\ S'\ T\ M_T> \times$$

$$< \Gamma L\omega||\ \hat{W}^{\kappa}_{coll}\ ||\ \Gamma'\ L'\ \omega\ > \tag{11.4}$$

The reduced matrix elements in (11.4) have been calculated using the Wigner-Eckart theorem in the following form:

$$<\ell m \mid T_q^{\kappa} \mid \ell' m'> = <\ell \| T^{\kappa} \| \ell'> C_q^{\kappa} \begin{matrix} \ell' \\ m' \end{matrix} \begin{matrix} \ell \\ m \end{matrix} \tag{11.5}$$

The explicit expression for the reduced matrix elements $<\| \hat{U}^{\kappa} \|>$ has been obtained in[26] and their numerical values are known in all regions of the mass numbers, including the heavy deformed nuclei. Thus in order to get the numerical matrix of the H_{coll} one must calculate only the reduced matrix elements $<\| \hat{W}^{\kappa}_{coll} \|>$ on the set of functions (10.10), using as $\hat{W}^{\kappa}_{coll}$ the first term in (10.4) with $\chi = (o)$. In this case (10.5) reduces to

$$Q_{\omega\omega'}\ (o) = \delta(\omega\omega')\ \delta(\alpha\alpha')\ \delta(\lambda\lambda') \tag{11.6}$$

and there is no need for the matrix elements (8.8). From (10.6) one has the following expressions for the W_{coll}:

$$\left[\hat{W}^{\kappa}_{coll}\right]_{\nu^{o\prime}\nu^{o}} = \frac{1}{2} \sum_{\sigma} \int d\tau(\bar{q}^{+}) <\omega\nu^{o\prime} | W(\sigma\bar{q}^{+}) | \omega\nu^{o}> \tag{11.7}$$

in general case,

$$\hat{W}^{\kappa}_{coll} = \frac{1}{2} \sum_{\sigma} \int d\tau\ (\bar{q}^{+})\ \hat{W}\ (\sigma\bar{q}^{+}) \tag{11.8}$$

in case $\hat{W}^{\kappa}$ does not depend on the infinitesimal operators of the group O_{n-1} and

$$W^{(o)}_{coll} \equiv V_{coll} = \int d\tau(\bar{q}^{+}) V\left(\sqrt{2(\rho^{(1)})^2 c_3^2 s_2^2 s_1^2 + 2(\rho^{(2)})^2 c_2^2 s_1^2 + 2(\rho^{(3)})^2 c_1^2} \right) \tag{11.9}$$

in case of the arbitrary central interaction potential $V(\sqrt{2}\ |\vec{\rho}_a|)$. In (11.9) $c_3 \equiv \cos\theta_3$ etc. and $d\tau(\bar{q}^{+})$ is the volume elements (10.7)-(10.9).

From (11.4) one can see, that $\Sigma H_{coll}^{(\kappa)}$ is diagonal with respect to the quantum numbers ω, α, λ, T and does not depend on α. Thus the collective part of the arbitrary H acts within some "box" of the Hilbert space, labelled by $\omega\lambda T$, and the Schrödinger equation for (11.2) is

$$(\Sigma H^{(\omega\lambda T)}_{coll}(\kappa))\psi^{(\omega\lambda T)}_{coll} = \mathcal{E}_{coll}\ (J M_T\ \omega\lambda T)\ \psi^{(\omega\lambda T)}_{coll} \tag{11.10}$$

In (11.10) the integrals of motion $JM_J\Pi$ (Π is the parity) are also exposed. The equation (11.10) describes all collective degrees of freedom in nuclei, when the isospin formalizm is used. The general features of the collective phenomena in the two component fermion system (i.e. in the case of the theory without the isospin formalizm) have been discussed in[27,28]. The equation (11.10) was obtained in[24,23].

The physical picture described by $\psi_{coll}^{(\omega\lambda T)}$ is the microscopic collective model of nucleus, thus we have completed the general formulation of the microscopic theory of the collective motion in nuclei and now a lot of details must be worked out. But it is, of course, not the topic for these lecture notes. In the few following sections we will discuss only some questions. But before starting let us have a look at what a kind of collective potentials may be got from the potentials of the nucleon-interaction.

12. THE MICROSCOPIC COLLECTIVE POTENTIALS IN THE CASE OF THE CENTRAL INTERACTION

Formula (11.9) gives the collective part V_{coll} of the arbitrary potential $V(\sqrt{2}|\vec{\rho}_a|)$. In this section, following[29], we will discuss the dependence of V_{coll} on the shape of $V(\sqrt{2}|\vec{\rho}_a|)$. One of the simplest types of the potentials, often used in the theory of heavy deformed nuclei, is the multipole type interaction:

$$V^a(|\vec{r}_{n-1}-\vec{r}_n|) = (|\vec{r}_{n-1}-\vec{r}_n|)^{2a} \tag{12.1}$$

where a=1,2,3... In order to get its collective part, according to (11.9), we must find the integral

$$V^a_{coll} = \int d\tau(\bar{q}^+)\left(2(\rho^{(1)})^2 c_3^2 s_2^2 s_1^2 + 2(\rho^{(2)})^2 c_2^2 s_1^2 + 2(\rho^{(3)})^2 c_1^2\right)^a \tag{12.2}$$

The integration gives the following expression:

$$W_a(\rho^{(1)}\rho^{(2)}\rho^{(3)}) \equiv V^a_{coll} =$$

$$= \frac{(n-3)!!(2a)!!}{(2a+n-3)!!} \sum_{k+\ell+m=a} \frac{(2k-1)!!(2\ell-1)!!(2m-1)!!}{k!\,\ell!\,m!} (\rho^{(1)})^{2k}(\rho^{(2)})^{2\ell}(\rho^{(3)})^{2m} \tag{12.3}$$

where k,ℓ,m are the nonnegative integers. If k=0 and n=3 the formula (12.3) also gives V^a_{coll} in the three-particle case. In particular, from (12.3) we get:

$$W_1 = \frac{2}{n-1} \sum_{s_o} (\rho^{(s_o)})^2$$

$$W_2 = \frac{4}{(n+1)(n-1)} \left(3 \sum_{s_o} (\rho^{(s_o)})^4 + 2 \sum_{s'_o > s_o} (\rho^{(s'_o)})^2 (\rho^{(s_o)})^2\right) \qquad (12.4)$$

Let us introduce[6,17]

$$\rho^2 = (\rho^{(1)})^2 + (\rho^{(2)})^2 + (\rho^{(3)})^2$$

$$u = \frac{1}{\sqrt{6}} ((\rho^{(1)})^2 + (\rho^{(2)})^2 - 2(\rho^{(3)})^2) = \beta \cos \gamma$$

$$v = \frac{1}{\sqrt{2}} ((\rho^{(1)})^2 - (\rho^{(2)})^2) = \beta \sin \gamma \qquad (12.5)$$

and let us discuss the symmetry properties of the expression (12.3). This expression is scalar with respect to the action of the symmetric group $\bar{\bar{S}}_3$, permuting the variables $\rho^{(s_o)}$. From (12.5) one may find, that the action of the elements $e; p_{12}; p_{23}; p_{12} \cdot p_{23}; p_{23} \cdot p_{12}$ and $p_{23} \cdot p_{12} \cdot p_{23}$ on the angle γ (this is the only variable that $\bar{\bar{S}}_3$ acts on) correspondingly is γ; $(-\gamma)$; $(-\gamma + 2/3\,\Pi)$; $(\gamma + 2/3\Pi)$; $(\gamma - 2/3\Pi)$ and $(-\gamma - 2/3\Pi)$. Thus, being the scalar of $\bar{\bar{S}}_3$, the expression (12.3) must depend on γ in the form $\cos 3\gamma$. In the new variables, (12.3) has the following form:

$$W_a(\rho^2, \beta, \cos 3\gamma) =$$

$$= \frac{(n-3)!!(2a+1)!!}{(2a+n-3)!!} \left(\frac{2}{3} \rho^2\right)^a \sum_{k=0}^{a} \frac{a!}{k!(a-k)!} \left(-\sqrt{6} \frac{\beta}{\rho^2}\right)^k B_k(\cos 3\gamma) \qquad (12.6)$$

where

$$B_k (\cos 3\gamma) = \sum_{\ell=0}^{[k/2]} (\cos\gamma)^{k-2\ell} (\sin\gamma)^{2\ell} \times$$

$$\times \sum_{t=\ell}^{[k/2]} \sum_{m=o}^{k-2t} \frac{(-1)^{t+\ell+m}\, 3^{2t+m-\ell} k!\; (2t+2m-1)!!}{2^t \ell!\; m!\; (t-\ell)!\; (k-2t-m)!(4t+2m+1)!!} \qquad (12.7)$$

By direct calculations it is possible to ckeck that (12.7) depends only on the cos 3γ . In particular $B_o=1, B_1=0$, $5\ B_2=1$, $35\ B_4=3$ and

$$B_3 = \frac{2}{35} \cos 3\gamma$$

$$B_5 = \frac{4}{77} \cos 3\gamma$$

$$B_6 = \frac{45}{1001} + \frac{8}{1001} (\cos 3\gamma)^2$$

$$B_7 = \frac{6}{143} \cos 3\gamma \qquad (12.8)$$

Let us introduce the ratios $u_o=u/\rho^2$ and $v_o=v/\rho^2$. Taking into account, that the radial variables $\rho(s_o)$ take the values $0 \leqslant \rho(s_o) < \infty$, from (12.5) we get, that u_o, v_o take the values in the equiangular triangle, the angles of which are in the points with the coordinates (u_o, v_o) equal to $(-2/\sqrt{6}, o)$, $(1/\sqrt{6}, 1/\sqrt{2})$ and $(1/\sqrt{6}, -1/\sqrt{2})$; thus for the variable β one has $0 \leqslant \beta^2 \leqslant 2/3\ \rho^4$. Using the expressions

$$\left(\frac{\beta}{\rho^2}\right)^2 = u_o^2 + v_o^2$$

$$\left(\frac{\beta}{\rho^2}\right)^3 \cos 3\gamma = u_o(u_o^2 - 3\ v_o^2) \qquad (12.9)$$

and introducing

$$W'_a = \frac{(2a+n-3)\,!!}{(2a+1)!!\ (n-3)!!} \left(\frac{3}{2\rho^2}\right)^a W_a \qquad (12.10)$$

from (12.6) one gets

$$W'_1 = 1$$

$$W'_2 = 1+\frac{6}{5}(u_o^2 + v_o^2)$$

$$W'_3 = 1+\frac{18}{5}(u_o^2 + v_o^2) - \frac{12\sqrt{6}}{35} u_o(u_o^2-3v_o^2)$$

$$W'_4 = 1+\frac{36}{5}(u_o^2+v_o^2) - \frac{48\sqrt{6}}{35} u_o(u_o^2-3v_o^2) + \frac{108}{35}(u_o^2+v_o^2)^2 \qquad (12.11)$$

The graphical illustration of the functions (12.11) is given in Fig.1. The shaded area shows the domain for the variables u_o, v_o . All the functions W_2', W_3' and W_4' have a common minimum when $u_o=v_o=o$. Because of the symmetry, connected with the dependence of W_a' on γ in the form cos 3 γ , the functions W_2', W_3', W_4' have three identical maxima for the values of (u_o,v_o) in the points, mentioned above. Fig. 1 shows that the maxima of W_a' noticably increases with a.

Let us take a more realistic type of $V(\sqrt{2}|\vec{\rho}_a|)$ and for the beginning let us project the collective part from the Gaussian potential

$$V^\alpha(r) = V_o\ e^{-\alpha r^2} \tag{12.12}$$

In the three-particle case, using (9.17), one gets:

$$V^\alpha_{coll}(\rho,\beta_o) = V_o\ e^{-\alpha\rho^2}\ I_o(\alpha\beta_o) \tag{12.13}$$

where I_o is the modified Bassel function (see e.g.[35]), satisfying the relation $I_o(-x)=I_o(x)$. In (12.13)

$$\rho^2 = (\rho^{(2)})^2 + (\rho^{(3)})^2$$

$$\beta_o = (\rho^{(2)})^2 - (\rho^{(3)})^2 \tag{12.14}$$

thus β_o takes the values $|\beta_o| \leqslant \rho^2$.

In the general n-particle case starting with (11.9) one can prove that the collective part of the potential (12.12) is

$$V^\alpha_{coll}(\rho,\beta,\gamma) = V_o e^{-(\frac{2}{3}\alpha\ \rho^2)} \sum_{k=0}^{\infty} \frac{(2k+1)!!\,(n-3)!!}{k!\,(2k+n-3)!!}\left(\frac{2\sqrt{2}}{\sqrt{3}}\alpha\beta\right)^k$$

$$\times\ \Phi\left(\frac{n}{2}-2,\frac{n-1}{2}+k,\ \frac{2}{3}\alpha\ \rho^2\right) B_k(\cos 3\gamma) \tag{12.15}$$

where Φ is the confluent hypergeometric function and B_k the polynoms (12.7). In the case of n=4

$$\Phi\left(0,\ \frac{3}{2} + k,\ \frac{2}{3}\ \alpha\ \rho^2\right) = 1 \tag{12.16}$$

thus in n=4 (12.15) depends on ρ^2 only in the exponential form.

The wide variety of potentials may be presented in terms of the Gaussians, including some realistic ones, thus (12.13) and (12.15) give the possibility to study various cases. As an example let us take the set of the potentials given by the expression

$$V(r) = V_0 \left(e^{-\alpha_1 r^2} - e^{(\alpha_2-\alpha_1) r_0^2} e^{-\alpha_2 r^2} \right) \tag{12.17}$$

In (12.17) the parameter r_0 is introduced, because $V(r_0)=0$. It is also convenient to introduce the new parameters V_1, r_1 and κ instead of α_1, α_2 and V_0. They have the following meaning: the minimum value of $V(r)$ is V_1 and $V(r_1)=V_1$. The parameter κ is a ratio of α_2 and α_1, i.e. $\kappa = \alpha_2/\alpha_1$. In the new parameters

$$V(r)=V_1 \frac{\kappa^{\frac{r^2}{(1-\kappa)(r_1^2-r_0^2)}} - \kappa^{\frac{r_0^2+\kappa(r^2-r_0^2)}{(1-\kappa)(r_1^2-r_0^2)}}}{\kappa^{\frac{r_1^2}{(1-\kappa)(r_1^2-r_0^2)}} - \kappa^{\frac{r_0^2+\kappa(r_1^2-r_0^2)}{(1-\kappa)(r_1^2-r_0^2)}}} \tag{12.18}$$

and

$$\alpha_1 = \frac{\ln \kappa}{(1-\kappa)(r_0^2 - r_1^2)} \tag{12.19}$$

The expression (12.18) is invariant with respect to the change of κ into $1/\kappa$, thus it is enough to take $0 < \kappa \leqslant 1$. Let $V_1=-5$, $r_1=5$ and $\kappa=1$. For those values of V_1, r_1 and κ the dependence of $V(r)$ on r is shown in Fig. 2. This figure also shows the dependence of the shape of $V(r)$ on r_0. In Fig. 3 the dependence of $V(r)$ on κ is illustrated in case of $V_1=-5$, $r_1=2{,}5$ and $r_0=2{,}5$. From Fig. 2 and Fig. 3 one can see, that the asymptotic behaviour of $V(r)$, when $\to \infty$, strongly depends on r_0 and κ.

From (12.12) and (12.13) we have the following expression for the collective part of the potential (12.18):

$$W^{(3)}(r_0,r_1,V_1,\kappa;\rho,\beta_0) = V_0 \left[\kappa^{\frac{\rho^2}{(1-\kappa)(r_1^2-r_0^2)}} I_0\left(\frac{\ln\kappa}{(1-\kappa)(r_1^2-r_0^2)} \beta_0\right) - \kappa^{\frac{r_0^2+\kappa(\rho^2-r_0^2)}{(1-\kappa)(r_1^2-r_0^2)}} I_0\left(\frac{\kappa \ln\kappa}{(1-\kappa)(r_1^2-r_0^2)} \beta_0\right) \right] \tag{12.20}$$

where

$$V_0=V_1 \left[\kappa^{\frac{r_1^2}{(1-\kappa)(r_1^2-r_0^2)}} - \kappa^{\frac{r_0^2+\kappa(r_1^2-r_0^2)}{(1-\kappa)(r_1^2-r_0^2)}} \right]^{-1} \tag{12.21}$$

The graphical illustration of the function (12.20) in the variables $\rho^{(2)}$, $\rho^{(3)}$ is shown in Fig. 4-6, where the numerical values of the parameters V_1, r_o, r_1 and κ are also given. Because of the symmetry of $W^{(3)}$ with respect to the interchange of $\rho^{(2)}$ and $\rho^{(3)}$, in Fig. 4-6 the shape of $W^{(3)}$ is presented only in the case of $\rho^{(3)} \geq \rho^{(2)}$. As it was already mentioned, on the line $\rho^{(2)} = \rho^{(3)}$ the shape of $W^{(3)}$ and nucleon-nucleon interaction $V(r)$, one has started with, is identical. On this line, when $\rho^{(2)} = \rho^{(3)} = 5/\sqrt{2}$, $W^{(3)}$ has the only minimum. If $\rho^{(3)} > \rho^{(2)}$ the shape of $W^{(3)}$ tends to diverge from the shape of $V(r)$ and this tendency depends on the parameters of the potential $V(r)$. The dependence of W on $\rho^{(2)}$, when $\rho^{(3)} > 5/\sqrt{2}$ is fixed, also has the minima; one may see it in Fig. 4-6, when $\rho^{(3)}=10$. The set of those minima forms the hollow between the line $\rho^{(2)} = \rho^{(3)}$ and the axis $\rho^{(3)}$. Different shapes of $W^{(3)}$ with various values of r_o and κ are well noticable when $\rho^{(2)}=0$. On Fig. 4-6 one may see, that $W^{(3)}(\rho^{(2)}=0, \rho^{(3)})$ changes the sign, while in Fig. 5 $W^{(3)}(\rho^{(2)}=0, \rho^{(3)})$ takes the positive values only.

13. THE COMPLETE SET OF THE RADIAL BASIS FUNCTIONS

We have not yet discussed the method of calculation of the reduced matrix elements $\langle \| V_{coll} \| \rangle$ in (11.4). The spectrum of H_{coll} is not necessarily a discrete one. Thus in general one must use not only the square integrable set of the basis functions in (10.12). Taking interest only in bound states, one can span the Hilbert space for H_{coll} on the discrete spectrum basis functions. One of the simplest examples of the set of square integrable functions, often used in the calculations, is the harmonic oscillator wave functions. The Hamiltonian of the $3(n-1)$ dimensional harmonic oscillator possesses high symmetry. It is the first Casimir operator of the unitary group $U_{3(n-1)}$. Thus, spanning the Hilbert space on the eigenstates of this Hamiltonian, the powerful algebraic technique for calculations may be exploited.

Let us take the Schrödinger equation for the $3(n-1)$ -dimensional oscillator Hamiltonian:

$$\frac{1}{2}\left(-\Delta + \rho^2\right)\psi_{oscill} = \left(E + \frac{3}{2}(n-1)\right)\psi_{oscill} \tag{13.1}$$

and let us solve it in the collective and internal variables. The Laplace operator in the new variables has the expression (5.14) and depends on the operators $\mathcal{L}$ and $\mathcal{J}$ in the body-fixed coordinate system. Therefore one must take the solutions of (13.1) in the form (10.12), act with $\mathcal{L}$ and $\hat{\mathcal{J}}$ on the matrix elements D^{L}_{KM} and $D^{\omega}_{\nu,\nu'}$ and use the formula (10.3). One gets the complicated system of the differential equations and there is no hope to solve it in the general n particle case. In[30] the solutions of the equation (13.1) have been obtained in the indirect way, using the group-theoretical technique. In case of $n > 3$ the final expression has the following form (for details see[6]):

$$\Theta\left(\begin{matrix} E\ \gamma & L & K \\ \alpha_o\omega & \bar{\omega} & \bar{\bar{\omega}} \end{matrix}\ \bar{\bar{\bar{o}}}\ \middle|\ \rho^{(x)}\ \rho^{(y)}\ \rho^{(z)}\right) =$$

$$= \sum_{\substack{E_x E_y E_z \\ \omega_x \omega_y \omega_z}} R_{E_x\omega_x}(\rho^{(x)})\, R_{E_y\omega_y}(\rho^{(y)})\, R_{E_z\omega_z}(\rho^{(z)})\, B_{E_x\omega_x E_y\omega_y E_z\omega_z} \tag{13.2}$$

where

$$B = \sum_{\substack{E_{xy}\,\omega_{xy} \\ \alpha_{xy}\bar{\omega}_x\bar{\omega}_y}} (d\omega_x\, d\omega_y\, d\omega_z)^{1/2}\, D^{(\omega_x\ \bar{o})}_{\bar{o}\ \bar{\omega}_x}\, D^{(\bar{\omega}_x\ \bar{\bar{o}})}_{\bar{\bar{o}}\ \bar{\bar{\omega}}}\, D^{(\omega_y\ \bar{o})}_{\bar{o}\ \omega_y} \times$$

$$\times\, C^{E_x\ \ \ E_y\ \ \ E_{xy}}_{\omega_x\bar{\omega}_x\bar{\bar{\omega}}\ \ \omega_y\bar{\omega}_y\bar{\bar{o}}\ \ \alpha_{xy}\ \omega_{xy}\bar{\omega}\bar{\bar{\omega}}}\ C^{E_{xy}\ \ \ E_z\ \ \ E}_{\alpha_{xy}\omega_{xy}\bar{\omega}\bar{\bar{\omega}}\ \ \omega_z\bar{o}\bar{\bar{o}}\ \ \alpha_o\omega\bar{\omega}\bar{\bar{\omega}}} \times$$

$$M^{(E)}_{E_{xy}\,E_x,\gamma\,L\,K} \tag{13.3}$$

The functions $R_{E_{s_o}\omega_{s_o}}(\rho^{(s_o)})$, $(s_o = x,y,z)$ in (13.2) are the radial wave functions for the (n-1)-dimensional oscillator, E_{s_o} and ω_{s_o} correspondingly denote U_{n-1} irreducible representation $[E_{s_o}0\ldots0]$ and O_{n-1} irreducible representation $(\omega_{s_o}0\ldots0)$. In (13.2) and (13.3) E denotes U_3 and U_{n-1} irreducible representation $[E_1\ E_2\ E_3]$, α_o and γ - the multiplicity indexes on the chains $U_{n-1}\supset O_{n-1}$ and $U_3\supset O_3^+$. The meaning of ω, $\bar{\omega}$, $\bar{\bar{\omega}}$ and $\bar{\bar{o}}$ has already been discussed in section 8. In (13.3) C are the Clebsch-Gordan coefficients for the group U_{n-1} with the basis labelled by the quantum numbers of the chain

$$U_{n-1}\supset O_{n-1}\supset O_{n-2}\supset O_{n-3}\supset O_{n-4} \tag{13.4}$$

The matrix $M^{(E)}$ in (13.3) transforms the canonical basis of the chain $U_3 \supset U_2 \supset U_1$ into the basis of the chain $U_3 \supset O_3^{+} \supset O_2^{+}$. In the explicit form this matrix has been found in[31] and in another form in[32]. The factor D in (13.3) has the following expression[30]:

$$D^{(\omega\bar{\bar{o}})}_{\bar{o}\;\bar{\bar{\omega}}} = (-1)^{\omega-\bar{\omega}} \sum_{m\bar{m}} M^{(\omega\bar{o})}_{\bar{o}\;m} D^{(m\bar{\bar{o}})}_{\bar{o}\;\bar{m}} M^{(\omega\bar{m})}_{m\;\bar{\bar{\omega}}} M^{(\bar{\omega}\;\bar{\bar{o}})}_{\bar{m}\;\bar{\bar{o}}} \tag{13.5}$$

where $M^{(\omega\bar{m})}$ are very simple one or two dimensional matrices, similar to the matrix (6.18). Their explicit expression is given in[33]. In (3.5)

$$D^{(m\bar{\bar{o}})}_{\bar{o}\;\bar{m}} = \frac{1+(-1)^{m-\bar{m}}}{2} (-1)^{\frac{m+\bar{m}}{2}} \frac{2^{\bar{m}}\,\Gamma(1/2(m+\bar{m}+n-3))}{\Gamma(\frac{1}{2}(n-3))(\frac{1}{2}(m-\bar{m}))!} \times$$

$$\times \sqrt{\frac{m!\,(m-\bar{m})!\,(n-4)!\,(\bar{m}+n-5)!\,(2\bar{m}+n-4)!}{\bar{m}!\,(m+\bar{m}+n-4)!\,(m+n-4)!}} \tag{13.6}$$

The functions (13.2) are orthogonal in the following sense[30]:

$$\sum_{K\bar{\omega}\bar{\bar{\omega}}} \frac{1}{(2L+1)d_{\omega}} \int d\tau(\rho^{(s_o)})\, \Theta^{*}\left(\begin{smallmatrix}E\gamma LK \\ \alpha_o\;\omega\bar{\omega}\bar{\bar{\omega}}\bar{\bar{o}}\end{smallmatrix} \middle|\, \rho^{(s_o)}\right) \times$$

$$\times\ \Theta\left(\begin{smallmatrix}E'\gamma' LK \\ \alpha'_o\;\omega\bar{\omega}\bar{\bar{\omega}}\bar{\bar{o}}\end{smallmatrix} \middle|\, \rho^{(s_o)}\right) = \delta(EE')\,\delta(\gamma\gamma')\,\delta(\alpha_o\alpha'_o) \tag{13.7}$$

where[11]:

$$d\tau(\rho^{(s_o)}) = \frac{2^{2n-y}[\Gamma(\frac{1}{2}(n-1))]^2}{3\,\Gamma(n-3)} |(\rho^{(x)})^2-(\rho^{(y)})^2||(\rho^{(y)})^2 -$$

$$(\rho^{(z)})^2||(\rho^{(z)})^2-(\rho^{(x)})^2|(\rho^{(x)}\rho^{(y)}\rho^{(z)})^{n-4}d\rho^{(x)}d\rho^{(y)}d\rho^{(z)} \tag{13.8}$$

The normalization factor in (13.8) has been obtained in[30] taking into account the condition

$$\int d\tau(\rho^{(s_o)})\, e^{-\rho^2} = 1 \tag{13.9}$$

All the factors in (13.3) are knowx except the Clebsch-Gordan coefficients, thus the problem of constructing (13.2) reduces to the calculation of a special type of the Clebsch-Gordan coefficients for the chain (13.4).

In the three particle case one has only two variables $\rho^{(s_o)}$. This leads to the simplification in (13.3). In this case the Clebsch-Gordan coefficients in (13.3) have the expression

$$C^{[E_1' \; E_1''][E_2' \; E_2''][E' \; E'']}_{\omega_1 \bar{\omega}_1 \; \omega_2 \bar{\omega}_2 \; \omega \bar{\omega}} = \sum_{m_1 m_2} (-1)^{\frac{1}{2}(E_1'-2E_1''+m_1+E_2'-2E_2''+m_2)} \times$$

$$M^{(\omega_1)}_{\bar{\omega}_1 m_1} M^{(\omega_2)}_{\bar{\omega}_2 m_2} (M^{(\omega)})^{-1}_{m\bar{\omega}} \; e^{i(F(m_1E_1'')+F(m_2E_2'')-F(mE''))} \times$$

$$\times \; C^{\frac{1}{2}(E_1'-E_1'') \; \frac{1}{2}(E_2'-E_2'') \; \frac{1}{2}(E'-E'')}_{\frac{1}{2}m_1 \; \frac{1}{2}m_2 \; \frac{1}{2}m} \tag{13.10}$$

where $M^{(\omega)}$ has the expression (6.19) or (6.21). On the right side of (13.10) C are the Clebsch-Gordan coefficients of the group SU_2, well known in the theory of the angular momentum. In (13.10) F denotes the following phase factor:

$$F(mE) = \left[\psi_o' + (E\,\Pi - (|m| -1)\,\psi)\frac{(1-\text{sign } m)}{2} \right] |\text{sign } m| +$$

$$+ \psi_o(1-\text{sign } |m|) \tag{13.11}$$

with $\psi_o' = (1-|m|)\frac{\Pi}{2}$, $\psi = \Pi$ and $\psi_o = -\Pi\, E_2$. In case of n=3 the functions (13.2) are orthogonal in the sense of (13.7), with the volume element[30]:

$$d\tau(\rho^{(S_o)}) = |(\rho^{(2)})^2 - (\rho^{(3)})^2| \rho^{(2)} \rho^{(3)} d\rho^{(2)} d\rho^{(3)} \tag{13.12}$$

The phase factor in (13.10) is adapted to the following standard expression for the radial wave function of the p dimensional harmonic oscillator:

$$R^{(p)}_{\varepsilon\ell}(r)=\sqrt{\frac{2(\frac{1}{2}(\varepsilon-\ell))!}{\Gamma(\frac{1}{2}(\varepsilon+\ell+p))}}\, e^{-\frac{1}{2}r^2} \sum_{s=0}^{\frac{1}{2}(\varepsilon-\ell)} (-1)^s \quad \times$$

$$\times \quad \frac{\Gamma(\frac{1}{2}(\varepsilon+\ell+p))}{s!\,(\frac{1}{2}(\varepsilon-\ell)-s)!\ \Gamma(\ell+s+\frac{1}{2}p)}\, r^{2S+\ell} \tag{13.13}$$

In (13.13) $\varepsilon = 0,1,2,...$ and $\ell = \varepsilon,\varepsilon-2,...,0$ or 1.

14. THE MICROSCOPIC MEANING OF THE PHENOMENOLOGICAL VARIABLES

We have not yet discussed the correspondence between the microscopic theory of the collective motion in nuclei and the phenomenological rotational-vibrational model. The first question is whether our definition of the collective variables given the possibility to explain the microscopic meaning the phenomenological variables β γ used in the Bohr-Mottelson approach. In this section we will see, that the answer to this question is positive.

Traditionally the phenomenological variables are always treated as the variables, connected with the surface oscillations. Let us give them another interpretation. In order to do that, let us introduce the following bilinear forms of the Jacobi variables:

$$q^{ss'}=\sqrt{\frac{2}{1+\delta(ss')}} \sum_{i=1}^{n-1} \rho_i^s\ \rho_i^{s'} \tag{14.1}$$

Substituting in (14.1) the expression (5.1) for ρ_i^s, one has

$$q^{ss'}=\sqrt{\frac{2}{1+\delta(ss')}} \sum_{s_o} (\rho^{(s_o)})^2 D^{(1_3)}_{s_o s}(G_3^+) D^{(1_3)}_{s_o s'}(G_3^+) \tag{14.2}$$

thus one can see, that $q^{\underline{s}\,\underline{s}'}$ are functions of the microscopic collective variables $\rho^{(\underline{s}_o)}$ and G_3^+. Following[34], let us decompose (14.2) into the O_3^+ irreducible terms. Using the Clebsch-Gordan coefficients in the Cartesian basis one has:

$$q_\eta^\kappa = \sum_{\underline{s}' \geqslant \underline{s}} q^{\underline{s}\,\underline{s}'} C^{1\ \ 1\ \ \kappa}_{\underline{s}\ \ \underline{s}'\ \ q} = \sum_{\eta'} p_{\eta'}^\kappa D^\kappa_{\eta'\eta}(G_3^+) \tag{14.3}$$

where $\kappa = 0,2$ and

$$p_\eta^\kappa = \sum_{\underline{s}_o} (\rho^{(\underline{s}_o)})^2 C^{1\ \ 1\ \ \kappa}_{\underline{s}_o\ \underline{s}_o\ \eta} \tag{14.4}$$

Using the numerical values of the Clebsch-Gordan coefficients in (14.4), given in (7.12)[6], one can see that there are only three variables p_q^κ, namely $p_0^0 = \frac{1}{\sqrt{3}}\rho^2$, $p_{11}^2 = u$ and $p_{11}^2 = v$, where ρ^2, u and v have the expression (12.5)

Let us introduce the Laplace operator in the variables $q^{\underline{s}\,\underline{s}'}$:

$$\Delta_q = \sum_{s' \geqslant s=1}^{3} \frac{\partial^2}{\partial (q^{\underline{s}\,\underline{s}'})^2} = \frac{\partial^2}{\partial X^2} + \frac{\partial^2}{\partial Y^2} + \frac{\partial^2}{\partial Z^2} + 2\left(\frac{1}{X-Y}+\frac{1}{X-Z}\right)\frac{\partial}{\partial X} +$$

$$+ 2\left(\frac{1}{Y-Z}+\frac{1}{Y-X}\right)\frac{\partial}{\partial Y} + 2\left(\frac{1}{Z-X}+\frac{1}{Z-Y}\right)\frac{\partial}{\partial Z} + \frac{2\hat{\mathcal{L}}_{12}^2}{(X-Y)^2} + \frac{2\hat{\mathcal{L}}_{13}^2}{(X-Z)^2} + \frac{2\hat{\mathcal{L}}_{23}^2}{(Y-Z)^2}, \tag{14.5}$$

where $X \equiv (\rho^{(1)})^2$, $Y \equiv (\rho^{(2)})^2$, $Z = (\rho^{(3)})^2$. In the variables (12.5) the last expression has the following form[17]:

$$\Delta q = 3\frac{\partial^2}{\partial (\rho^2)^2} + 4\left(\sum_{s>s'=1}^{3} \frac{\hat{\mathcal{L}}_{s's}(G_3^+)}{\mathcal{J}_{s's}(\beta\gamma)} + \frac{1}{\beta^4}\frac{\partial}{\partial\beta}\beta^4\frac{\partial}{\partial\beta} + \right.$$

$$\left. + \frac{1}{\beta^2}\frac{1}{\sin 3\gamma}\frac{\partial}{\partial\gamma}\sin 3\gamma\frac{\partial}{\partial\gamma}\right) \tag{14.6}$$

where

$$\mathcal{J}_{12} = 4\beta^2 \sin^2 \gamma$$
$$\mathcal{J}_{13} = 4\beta^2 \sin^2 (\gamma - \tfrac{2}{3}\Pi)$$
$$\mathcal{J}_{23} = 4\beta^2 \sin^2 (\gamma - \tfrac{4}{3}\Pi) \tag{14.7}$$

Let us take the potential energy term discussed in section 12. Then we get the following Hamiltonian;

$$H(q^{ss'}) = -\frac{\hbar^2}{2}\Delta_q + V_{coll}(\rho^2, \beta, \cos 3\gamma) \tag{14.8}$$

The equivalence of the both (14.8) and (1.2) Hamiltonians is clearly seen, than we can say, that we derived the rotational-vibrational Hamiltonian (thus the phenomenological rotational-vibrational model) without referring to the surface oscillations (the "breething mode" variable ρ^2 has been suppressed in the Bohr-Mottelson theory).

That is not yet the microscopical foundation of the phenomenological approach, because the kinetic energy operator (14.5) is not yet related to the microscopic operator (5.14) and also because the potential energy term in (14.8), according to the general theory, may be used only on the set of the antisymmetric wave functions. In this section the question of the Pauli principle requirements has not yet been raised. Thus we just only explain the microscopic meaning of the phenomenological variables $\bar{\beta}, \bar{\gamma}$ - they are the bilinear forms of $\rho^{(s_0)}$. Those results also bespeak the clear link to the microscopic theory.

We will present the microscopic foundation of the rotational-vibrational model in the next section. In order to do that, we need the explicit construction of the Hilbert space for the Hamiltonian (14.8). Let us introduce the complete set of the square integrable functions depending on the variables $q^{ss'}$. There are six of them, thus let us take the following Schrödinger equation for the Hamiltonian of the six-dimensional harmonic oscillator:

$$\frac{1}{2}(-\Delta_q + \sum_{s \geqslant s'} (q^{ss'})^2)\psi(\varepsilon) = (\varepsilon + 3)\psi(\varepsilon) \tag{14.9}$$

The solution $\psi(\varepsilon)$ of (14.9) in the group-theoretical language may be treated as the basis of the U_6 irreducible representation $\varepsilon \equiv [\varepsilon 00000]$. In order to label this basis various chains of subgroups[34] may be employed, for example, the chains

$$U_6 \supset O_5 \supset O_5^+ \supset O_3^+ \tag{14.10}$$

$$U_6 \supset O_6 \supset O_5 \supset O_3^+ \tag{14.11}$$

or the chain

$$U_6 \supset U_3 \supset O_3^+ \qquad (14.12)$$

In the case of the last chain one has the following branching rules: if ε is the irreducible representation of U_6, then one can enumerate all the U_3 irreducible representations contained in ε by counting all the partitions of the number 2ε into no more than three parts, i.e. $2\varepsilon = E_1+E_2+E_3$, where $E_1 \geqslant E_2 \geqslant E_3$ and all the numbers E_1, E_2, E_3 are even. The representations $[E_1\ E_2\ E_3]$ obtained in this way denote all U_3 irreducible representations contained in ε. For example, the U_6 representation $[3]_6$ contains $[600]$, $[420]$ and $[222]$ U_3 representations.

Taking into account those branching rules let us denote the basis functions for the six-dimensional oscillator Hamiltonian as

$$\Theta(\varepsilon[E_1'\ E_2'\ E_3']\gamma\ LK\ \ (\rho^{(s_o)^2})) \qquad (14.13)$$

The matrix of the Hamiltonian (14.8) may be calculated on this set of functions and thus the eigenfunctions of (14.8) for the bound states of Hamiltonian (14.8) may be found. In other words, the Hilbert space for the Hamiltonian (14.8) may be spanned on the set of functions (14.13). Let us denote this space as $\mathcal{R}_{phen}$. The set of functions (14.13) or the equivalent set, labelled by the quantum numbers of the chain $U_6 \supset U_5 \supset O_5 \supset O_3^+$ or its subchain has been constructed in[36-39].

15. THE MICROSCOPIC FOUNDATION OF THE ROTATIONAL-VIBRATIONAL MODEL

In order to find the guiding idea as to how to explain the microscopic meaning of the rotational-vibrational model, let us examin the transformation properties of the variables $q^{s\bar{s}'}$ with respect to the left-shift operators $\hat{T}_L$ of the group O_{n-1}. If g is an arbitrary element of this group, one has:

$$\hat{T}_L(g)\ \rho_i^s = \sum_{s_o} \rho^{(s_o)} D^{(1_3)}_{s_o s}(G_3^+) D^{(1_{n-1})}_{n-4+s_o i}(g^{-1}\sigma\ q^+) \qquad (15.1)$$

From (5.1) and (15.1) one can easily get that

$$\hat{T}_L\ (g)\ \ q^{s\bar{s}'} = q^{ss'} \qquad (15.2)$$

i.e., the variables $q^{ss'}$ are the left-shift (as well as the right-shift) scalars of the group O_{n-1}. It means that $q^{ss'}$ do not "feel" the existence of the group O_{n-1} in the same maner as, for example, the radial variable r, being O_3^+ scalar, does not "feel" the action of the group O_3^+. In other words, in the Hilbert space $\mathcal{R}_{phen}$ spanned on the basis functions (14.13), every action of the group O_{n-1} is completely "frozen"; thus in order to establish the relationship with the microscopic theory, we must find some "frozen' subspace in the Hilbert space $\mathcal{R}^{(\omega)}_{micr}$, spanned on the set of functions (13.2). The action of O_{n-1} in the space $\mathcal{R}^{(\omega)}_{micr}$ may be "frozen" if $\omega = (o)$ is put in (13.2). One can expect that in the case of $\omega = (o)$ the set of quantum numbers in (13.2) is isomorphic to the set in (14.13).

This prediction is immediately confirmed by checking the branching rules on the chain $U_{n-1} \supset O_{n-1}$. Those rules are well known[19] and only the following statement is needed here: the U_{n-1} irreducible representation $[E_1\ E_2\ E_3]$ contains the O_{n-1} scalar representation, if and only if all E_1, E_2, E_3 are even, and in this case the multiplicity of the O_{n-1} scalar representation $\omega = (o)$ in $[E_1\ E_2\ E_3]$ is equal to one.

Thus we have cleared up the point that the subset of the functions (13.2) with $\omega = (o)$, namely the subset

$$\Theta\left(\begin{matrix} E[E_1 E_2 E_3]\gamma & L & K \\ \omega = (o) & & \end{matrix} \Big| \rho(s_o)\right) \tag{15.3}$$

is characterized by the set of quantum numbers, isomorphic to the set in (14.13). (That does not mean, that the functions (13.2) and (14.13) are the same; see[6]) This isomorphizmus allows us to make the following statement[6]:

THEOREM. The Hilbert space $\mathcal{R}^{(o)}_{micr}$ spanned on the set of function (15.3) is isomorphic to the Hilbert space $\mathcal{R}_{phen}$, spanned on the set of function (14.13).

At the bottom of this theorem rests the algebraic properties. One can see that there are some problems connected with the domains for the variables $q^{ss'}$; for example, $0 \leqslant q^{ss} < +\infty$ thus the eigenvalue equation for the six-dimensional oscillator, we have used here, is not the usual one. From the algebraic point of view those details are not so important and may be corrected with the help of the normalization factors.

The theorem stated above establishes the relationship between two spaces $\mathcal{R}^{(o)}_{micr}$ and $\mathcal{R}_{phen}$ and provides the clue for the microscopic foundation of the rotational-vibrational model.

How to make use of the isomorphism between the spaces $\mathcal{R}^{(o)}_{micr}$ and $\mathcal{R}_{phen}$? Let us have a look at the Schrödinger equation (11.10). If one puts $\omega = (o)$ in this equation, one must also put $\lambda = [n]$ (because of the chain $O_{n-1} \supset S_n$). But in case of $n > 4$ the Pauli principle does not allow the S_n scalar states. Thus, strictly speaking, in the sense of the kinematically correct microscopic theory in the subspace with $\omega = (o)$ the Schrödinger equation (11.10) is not defined. Thus there is a contradiction. On the one hand, using $q^{ss'}$ as the variables we were able to get the equation (14.8), i.e. the Schrödinger equation used in the phenomenological theory, but this equation follows from the microscopic theory only if the Pauli principle is violated. Thus the phenomenological model is defined in the nonphysical subspace $\mathcal{R}^{(o)}_{micr}$ of the space $\mathcal{R}^{(\omega)}_{micr}$ (let us mention, that it is a well known set of ω, allowed by the Pauli principle for details see[6]).

In order to explain the phenomenological theory we have no choice but to violate the Pauli principle. For the collective Hamiltonian in case of $\omega = (o)$ one must slightly modify the expression (11.4), as it was obtained for the antisymmetric wave functions. After this modification (for details see[6]), using (13.2) as the radial basis functions, from (11.4) in case of $\omega = (o)$ we have:

$$\left\langle \begin{matrix} E\ \gamma(L\underset{\sim}{S})JM_J \\ T\ M_T\ \alpha(o)\lambda \end{matrix} \middle| H^{(\kappa\kappa)o}_{coll} \middle| \begin{matrix} E'\gamma'(L'S')JM_J \\ T\ M_T\ \tilde{\alpha}'(o)\tilde{\lambda} \end{matrix} \right\rangle =$$

$$= (-1)^{L+S'+J} \sqrt{(2L+1)(2S+1)} \begin{Bmatrix} L & S & J \\ S' & L' & \kappa \end{Bmatrix} \times$$

$$\times \langle \tilde{\lambda}\tilde{\alpha}STM_T || \sum_{i>j=1}^{n} U^{\kappa}(\sigma_i\sigma_j\tau_i\tau_j) || \tilde{\lambda}\tilde{\alpha}'S'TM_T \rangle \times$$

$$\times \langle [E_1E_2E_3]\gamma L(o) || \hat{W}^{\kappa}_{coll} || [E'_1E'_2E'_3]\gamma' L'(o) \rangle \qquad (15.4)$$

In the particular case, if H_{coll} does not depend on the spin and isospin variables the last expression reduces to

$$\frac{n(n-1)}{2} \langle [E_1 E_2 E_3] \gamma\ L(o) || \hat{W}_{coll} || [E_1' E_2' E_3'] \gamma'\ L(o) \rangle \tag{15.5}$$

For the kinetic energy one can use either (15.5) or the expression (5.15).

The microscopic theory has been developed for the arbitrary H, thus it may be concluded, that for the arbitrary Hamiltonian (14.8) acting in the space $\mathcal{R}$ phen the equivalent microscopic collective Hamiltonian (11.2) (introduced at the cost of the violating of the Pauli principle) acting in the space $\mathcal{R}^{(o)}_{micr}$ may be found. This prove, based on the explicit constructions described above, has been given in[6].

16. THE ROLE OF U_6 IN THE COLLECTIVE PHENOMENA. DZHOLOS-JANSSEN-DÖNAU AND ARIMA-IACHELLO APPROACH

In two previous sections the microscopic foundation of the rotational-vibrational model has been presented. This interpretation allowed us to introduce in a natural way the unitary group U_6 acting on the variables $q^{ss'}$[34]. In the works of Dzholos-Janssen-Dönau[40] and later in the papers of Arima-Iachello[41,42], the unitary group U_6 has been also introduced in connection with the conception of the active bosons, created by the coupled pairs of fermions. The group U_6, used in this approach is isomorphic to U_6, acting on the states in sd -shell[43,44], thus the formal analogy has been established as well as the physical picture proposed, stating that s- and d -bosons create some kind of the collective features, similar to those in the Bohr-Mottelson theory. The numerical results confirm this opinion, but the relationship between both phenomenological approaches have not yet been established.

Thus at present there are two very different physical conceptions, connected with the role of U_6 in the problem of the collective phenomena in nuclei. The space variables $q^{ss'}$ (see their definition in (14.1) are bilinear functions of all quasi-particle variables. Thus if one supposes, that U_6 acts of $q^{ss'}$, that means that all the quasi-particles are engaged in the collective motion. Consequently this conception does not suppose the existence of the active and passive nucleon formations in nuclei. That is the main difference between the two approaches, proposed in[34] and[40-42].

Another difference becomes clear if one is trying to understand the microscopic meaning of the approaches, based on the transformation properties of the group U_6. As it was already described in the previous sections, if one uses the variables $q^{ss'}$, the group U_6 provides the complete set of functions depending on them. This interpretation gives a clear distinction between the dynamical problem (to solve the Schrödinger equation for the Hamiltonian (14.8)) and the technical problem (to do that using the matrix representation of the Hamiltonian (14.8) by the help of the U_6 irreducible basis (14.13)). The calculations can be carried out by using the algebraic technique. Instead of the variables $q^{ss'}$ the Euler angles and the variables $(\rho^{(s}o^{)})^2$ (see (14.3)) may be used and $(\rho^{(s}o^{)})^2$ may be expressed in terms of the creation and annihilation operators, and thus the collective part of the multipole interaction (12.3) may be rewritten in them (or even the collective part of more sophisticated potentials, presented in the section 12) and finally the collective Hamiltonian in terms of generators of the noncompact symplectic group $S_p(12,R)$ introduced in[34] may be obtained. Restricting the interaction of the Hamiltonian (14.8), presented in this form to the irreducible space of the unitary group U_6, embedded in $S_p(12,R)$ on the chain

$$S_p(12,R) \supset U_6 \tag{16.1}$$

and taking either vibrational or rotational type basis, labelled by the quantum numbers of the chains (14.10), (14.11) or (14.12) one gets an approximate group-theoretical description of the eigenstates and eigenvalues of the Hamiltonian (14.8), acting within the U_6 irreducible space $\varepsilon=[\varepsilon 00000]$ with some definite ε. Technically this restriction can be achieved by taking into account only terms of H, composed from the multilinear forms of the infinitesimal operators of the group U_6. Similarly one can restrict the action of H either within the definite U_3 irreducible space (rotational limit) or within the O_5 irreducible space (vibrational limit).

We have not yet discussed the SU_3 states $(\lambda\eta)$, used in the interacting boson model calculations. The question to be answered in whether those states are allowed by the Pauli principle. For example, the active boson states used in[42] are (24,0), (20,2) in the case of $^{156}_{64}Gd_{92}$ and (34,0), (30,2) in the case of $^{170}_{68}Er_{102}$. The notation of the SU_3 irreducible representation $(\lambda\eta)$ used in[42] is related with the notation of the U_3 irreducible representation $[E_1E_2E_3]$ in (13.2) in the following way: $E_1 = \lambda + \eta + x$, $E_2 = \eta + x$, $E_3 = x$, where x is an arbitrary nonnegative integer. Thus the total number of oscillator quanta used in[42] is 3x+24 in the case of $^{156}_{64}Gd_{92}$ and 3x+34 in the case of $^{170}_{68}Er_{102}$.

From the results, obtained in[45] it follows, that in the region of $141 \leqslant n \leqslant 182$ the smallest number of oscillator quanta, allowed by the Pauli principle, is $5n_n + 4n_p - 210$, where n_n and n_p are the numbers of neutrons and protons. Thus 3x+24 and 3x+34 must be correspondingly equal to 506 and 572, which is impossible, because x is a nonnegative integer. Thus the states, used in[42], are not allowed by the Pauli principle (One may argue, that it is not necessary to take the smallest allowed number of oscillator quanta. But this argument has no sense, because there is no physical reason to ignore the lowest allowed U_3 states). The correct SU_3 states, obtained by Sabaliauskas[46] in the case of $^{170}_{68}Er_{102}$ are (4,38), (7,35) (10,32), etc. Those states belong to the two times excited SU_3 configuration $(o)^4\ (1)^{12}\ (2)^{24}\ (3)^{40}\ (4)^{58}\ (5)^{32}$.

Thus we came to the conclusion, that accepting the conception, proposed in[34,6] let us to explain the microscopic role of the group U_6. It also gives the constructive method to get the collective Hamiltonian from the microscopic one. Therefore we insist that the conception of active bosons should yield to the interpretation of the role of U_6 in the collective phenomena, proposed in[34,6].

17. CONCLUDING REMARKS

The problems associated with the collective motions in nuclei are too many-sided to be described in these lecture notes. We discussed only some aspects of the general theory, connected with the key definitions and the explicit construction of the microscopical collective Hamiltonian. The applications of the general theory raise new questions, the answer to which must be given in the future. Let us mention some of the problems.

The first question to be discussed concerns the properties of the eigenstates of H_{coll}. What new effects are hidden in the microscopical equation (11.10) in comparison with the features of the phenomenological theory. The complete answer to this question may be given only if ψ_{coll} is known, however some of the properties of the eigenstates of H_{coll} may be seen from the structure of the Hamiltonian H_{coll}.

The first new feature of the spectrum of H_{coll} concerns the existence of the collective states in the odd mass nuclei; states of such type are not allowed in the phenomenological theory. The well known physical picture, treating the deformed odd mass nucleus as the sys tem, consisting of the deformed core and an extra particle, moving outside the core is no longer needed in the microscopic theory. From the point of view of the existence of the pure collective states, there is no difference between the even and the odd mass nuclei; in both cases the Hamiltonian H_{coll} has the eigenstates. The only difference is the eigenvalue distribution. The spectrum of H_{coll} is similar to the traditional rotational-vibrational picture only in special cases. Often, especially in the odd mass nuclei, the new picture arise, not known in the phenomenological theory. The level distribution depends on the shape of the potentials of the nucleon-nucleon interactions as well as on the presence of the vectorial and tensorial terms in H_{coll}.

As it was mentioned at the very beginning of these lecture notes, the potentials of the nucleon-nucleon interaction are not known and because of it, the theory is adapted to the arbitrary Hamiltonian H. This feature, being valuable for its generality, raises some problems. One of them is connected with the "inertia" of the results with respect to the details of the nucleon-nucleon interaction. If this interaction is changed how much will very the spectrum and expectation values of various operators calculated in the eigenstates of the Hamiltonian H_{coll}?

This problem is too difficult to be solved in a general way. The Hamiltonian H_{coll} depends on three variables (if the vectorial and tensorial terms are suppressed, i.e. if H_{coll} acts in the "boxes" with $\lambda = [4...4]$ or $\lambda = [4...42...2]$), or on six variables (if the vectorial and tensorial terms are not suppressed). Thus, the problem of the exact solution of the Schrödinger equation for H_{coll} is very difficult. In order to be able to make the qualitative analysis it is desirable to make some simplifications.

Those simplifications may be made in a few directions. It is possible to take a very simple interaction (in section 13 we demonsrrated the solutions of H_{coll} in the case of the potential of the harmonic oscillator), or study only a few nucleon systems (for example, if n=3, one may diagonalize the matrix of H_{coll}, calculated on the basis functions (13.10)). There is one more direction One may try to find the following decomposition of the H_{coll}:

$$H_{coll} = H^{\circ}_{coll} + H'_{coll} \qquad (17.1)$$

where H°_{coll} is much simpler than H_{coll}. If (17.1) is known and if it is possible to solve the Schrödinger equation for H°_{coll}:

$$H^{\circ}_{coll} \ \psi^{\circ}_{coll} = E^{\circ}_{coll} \ \psi^{\circ}_{coll} \tag{17.2}$$

then we have ψ°_{coll} and E°_{coll}. If the decomposition (17.1) is properly guessed, we may hope, that ψ°_{coll} and E°_{coll} give the zeroth order approximation for the eigenstates and eigenvalues of H_{coll}.

The decomposition of the type (17.1) has been proposed in[47] (see for details[6]), using the following idea. We know, that H_{coll} has the meaning of the O_{n-1} scalar part of H. Let us put O_{n-1} into U_{n-1} i.e. let us consider the chain U_{n-1} O_{n-1}. This chain gives the possibility to get the decomposition (17.1), extracting from H_{coll} its U_{n-1} scalar part H°_{coll}. The symmetry of H°_{coll} is much higher than the symmetry of H_{coll}, thus the Schrödinger equation for H°_{coll} is simpler. This equation may be solved in various regions of the mass numbers. The distribution of the eigenvalues of H°_{coll} demonstrates the features of the spectrum of the collective excitations, we have already described.

The second question to be discussed concerns the terms of H contained in $H_{coll\text{-}intern}$. What is the role of the interaction between the collective and the internal degrees of freedom?. Undoubtedly this type of interaction is very important, so how is it possible to extract from $H_{coll\text{-}intern}$ some terms, both not too complicated (in order to be able to solve the problem) and not too trivial (in order to be able to take into account the features, existing in the reality). As the first step towards this goal, one may extract from $H_{coll\text{-}intern}$ some "anticollective" term, i.e. a term obtained, roughly speaking by averageing H with respect to the Cartesian indexes. More in details this question has been discussed in[6].

REFERENCES

1. V. Bargman, M. Moshinsky, Nucl. Phys. 18 697 (1960)
2. V. Bargman, M. Moshinsky, Nucl. Phys. 23 177 (1961)
3. M. Moshinsky, The Harmonic Oscillator in Modern Physics (Gordon and Breach, New York, 1969)
4. P. Kramer, M. Moshinsky, Nucl. Phys. A107 481 (1968)
5. R. Kalinauskas, V. Vanagas *), Liet. fiz. rink. 15 341 (1975)
6. V. Vanagas, The Microscopic Nuclear Theory within the Framework of the Restricted Dynamics, Lecture Notes (University of Toronto, Toronto, 1977)
7. J. Rainwater, Phys. Rev. 79 432 (1950)
8. A. Bohr, B. Mottelson, Dan Mat. Fys. Medd. 27, No.16 1 (1953)
9. A. Bohr, Rotational States of Atomic Nuclei (Ejnar Munksguards Forlage, Copenhagen, 1954)
10. J. M. Eisenberg, W. Greiner, Nuclear Models (North-Holland, Amsterdam, 1970)
11. A. Ya. Dzyublik, Preprint ITF-71-122R, Inst. Theoret. Phys., Kiev.
12. A. Ya. Dzyublick, V.I. Ovcharenko, A.I. Steshenko, G.F. Fillippov, Preprint ITF-71-134R, Inst. Theoret. Phys., Kiev.
13. W. Zickendraht, J. Math, Phys. 12 1663 (1971)
14. V. Vanagas, R. Kalinauskas, Yad. Fiz. 18 768 (1973) [Sov. J. Nucl. Phys. 18 395 (1974)]
15. V. Vanagas, Metody Teorii Predstavlenii Grupp i Vydelenie Kollektivnykh Stepenei Svobody Yadra. Konspekt Lektsii Shkoly MIFI [Methods of the Theory of Group Representations and Separation of Collective Degrees of Freedom of the Nucleus. Lecture Notes at the Moscow Engineering Physics Institute School. Izd. MIFI, Moscow, 1974]
16. V. Vanagas, Fiz. Élem. Chastits At. Yadra. 7 309 (1976) [Sov. J. Part. Nucl. 7 118 (1976)]
17. V. Vanagas, Fiz. Élem. Chastits At. Yadra. 11 454 (1980) [Sov. J. Part. Nucl. 11 (1980)]
18. R. Kalinauskas, M. Taurinskas, V. Vanagas, Liet. fiz. rink. 16 177 (1976)

19. D. E. Littlewood, The Theory of Group Characters and Matrix Representations of Groups (Clarendon Press, Oxford, 1950)

20. M. Kretzchmar, Z.f. Phys. 157 433 (1960); 158 284 (1960)

21. V. Vanagas, Algebraicheskie Metody v Teorii Yadra [Algebraic Methods in Nuclear Theory] (Izd. Mintis, Vilnius, 1971)

22. V. Vanagas, R. Kalinauskas, Liet. fiz. rink. 12 217 (1972)

23. V. Vanagas. Yad. Fiz. 23 950 (1976) [Sov. J. Nucl. Phys. 23 500 (1976)]

24. V. Vanagas, The Microscopical Theory of the Nuclear Collective Motion. Proc. International Symposium on Nuclear Structure Balatonfüred, Hungary, 1975 Vol. 1 (Budapest, 1976)

25. J. von Neumann, Mathematical Foundations of Quantum Mechanics (Univ. Press, Princeton, 1955)

26. V. Vanagas, M. Taurinskas *), Liet. fiz. rink. 17 717 (1977)

27. V. Vanagas *), Liet. fiz. rink. 17 409 (1977)

28. V. Vanagas *), Liet. fiz. rink. 17 421 (1977)

29. O. Katkevičius, V. Vanagas, Liet.fiz.rink. 21 No. 2, (1981) to be published

30. V. Vanagas, R. Kalinauskas, Liet. fiz. rink. 14 549 (1974)

31. M. Moshinsky, Rev. Mod. Phys. 34 813 (1962)

32. R. M. Asherova, Yu. F. Smirnov, Nucl. Phys. A144 116 (1970)

33. R. Kalinauskas, V. Vanagas, Liet.fiz.rink. 14 491 (1974)

34. V. Vanagas, E. Nadjakov, P. Raychev, Preprint IC/75/40, Trieste

35. M. Abramowitz, I. A. Stegun, Handbook of Mathematical Functions (National Bureau of Standards, 1964)

36. E. Chacón, M. Moshinsky, R.T. Sharp, J. Math. Phys. 17 668 (1976)

37. E. Chacón, M. Moshinsky, J. Math. Phys. 18 870 (1977)

38. O. Castaños, E. Chacón, A. Frank, M. Moshinsky, J. Math. Phys. 20 35 (1979)

39. V. Vanagas, S. Ališauskas, R. Kalinauskas, E.Nadjakov Bulg. J. Phys. 6 (1979), to be published

40. R. V. Dzholos, F. Dönau, D. Janssen, Yad. Fiz. 22 965 (1975) [Sov. J. Nucl. Phys. 22 503 (1976)]

41. A. Arima, F. Jachello, Ann Phys. (N.Y.) 99 253 (1976)

42. A. Arima, F. Jachello, Ann. Phys. (N.Y.) 111 201 (1978)

43. J. P. Elliott, Proc. Roy. Soc. A245 128 (1958)

44. J. P. Elliott, Proc. Roy. Soc. A245 562 (1958)

45. L. Sabaliauskas *), Liet.fiz.rink. 17 549 (1977)

46. L. Sabaliauskas, to be published

47. V. Vanagas, The Restricted Dynamic Nuclear Model. Proc. International Conference on Selected Topics in Nuclear Structure, Dubna, Vol. 1 46 (1976)

---

*) English translation in "Soviet Physics Collection" (Litovskij fizicheskij sbornik), Allerton Press, N.Y.

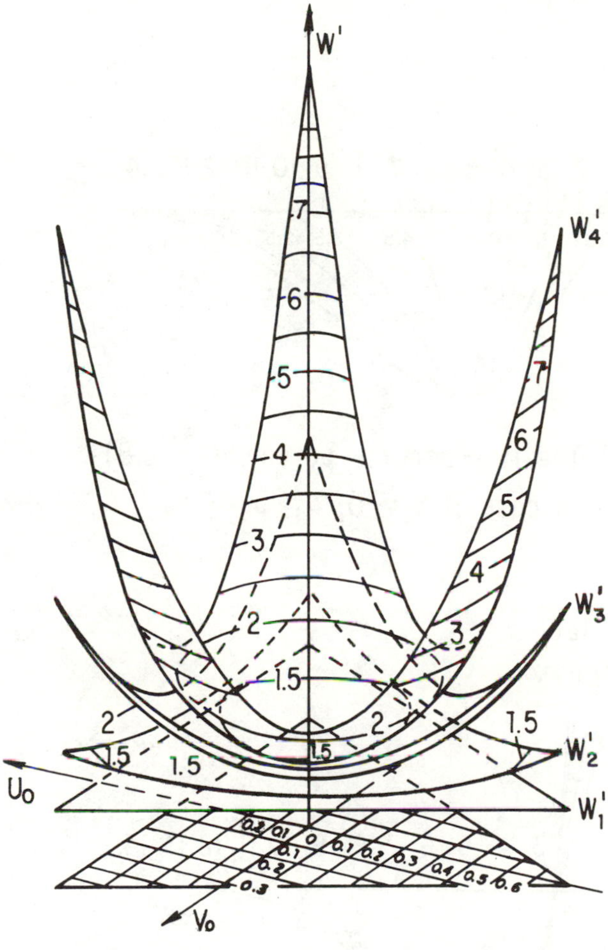

Fig. 1. The collective potentials (12.10) for a=1,2,3,4.

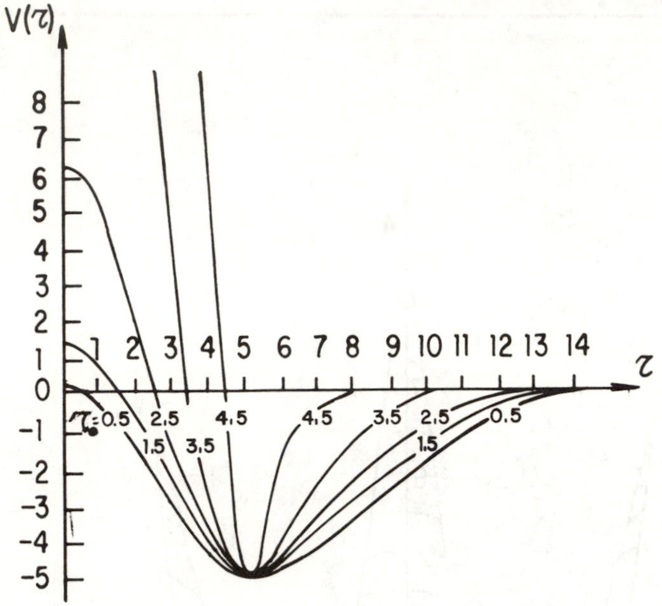

Fig. 2. The two-particle potentials (12.8) in case of $V_1 = -5$; $\tau_1 = 5$; $\varkappa = 1$

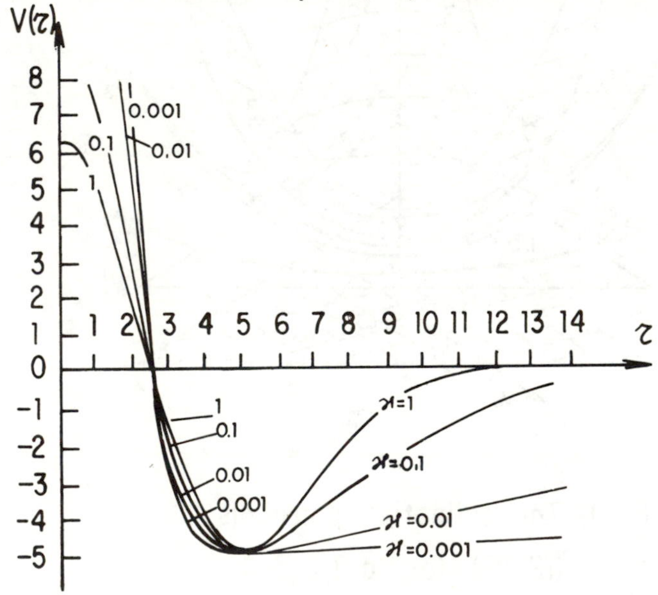

Fig. 3. The two-particle potentials (12.8) for $V_1 = -5$; $\tau_1 = 5$; $\tau_0 = 2.5$

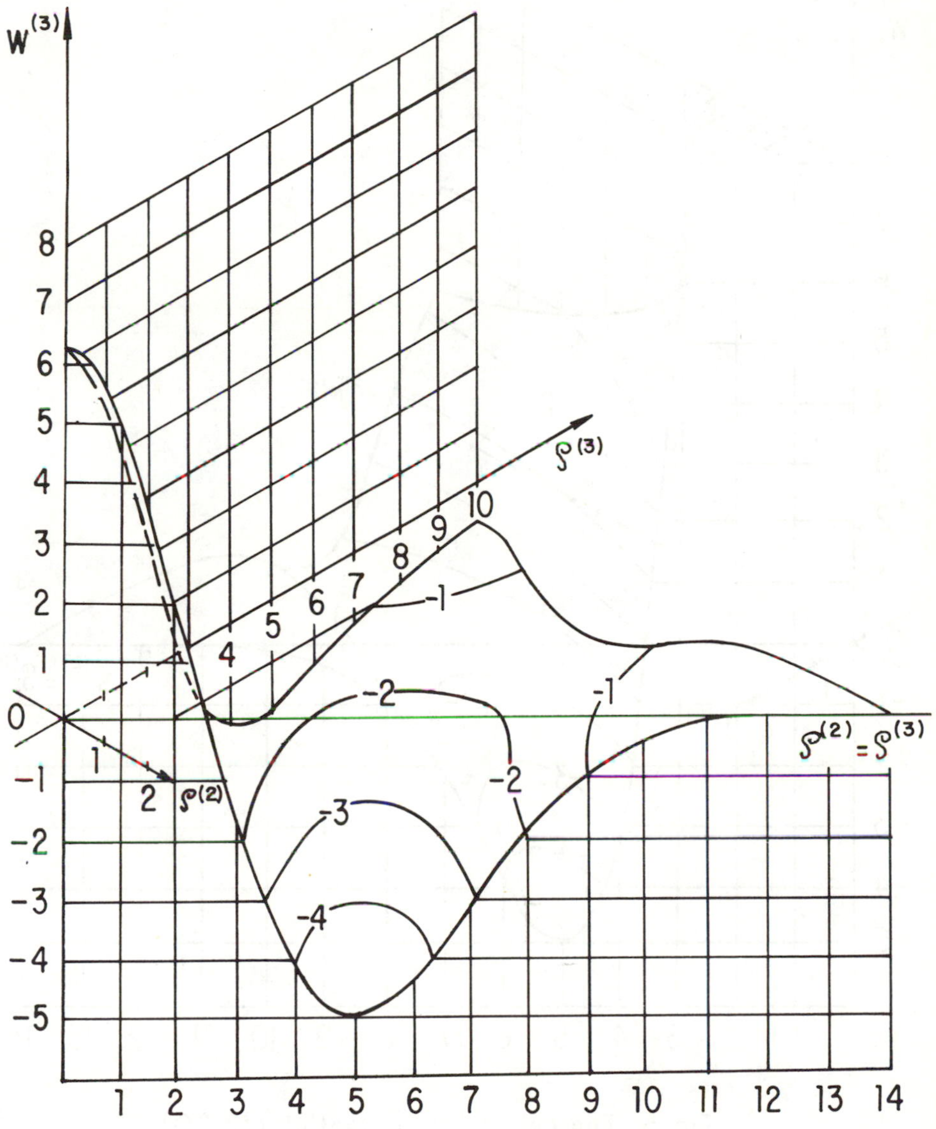

Fig. 4. The collective potential (12.20)
for $V_1 = -5$; $\tau_1 = 5$; $\tau_0 = 2,5$; $\mathcal{H} = 1$

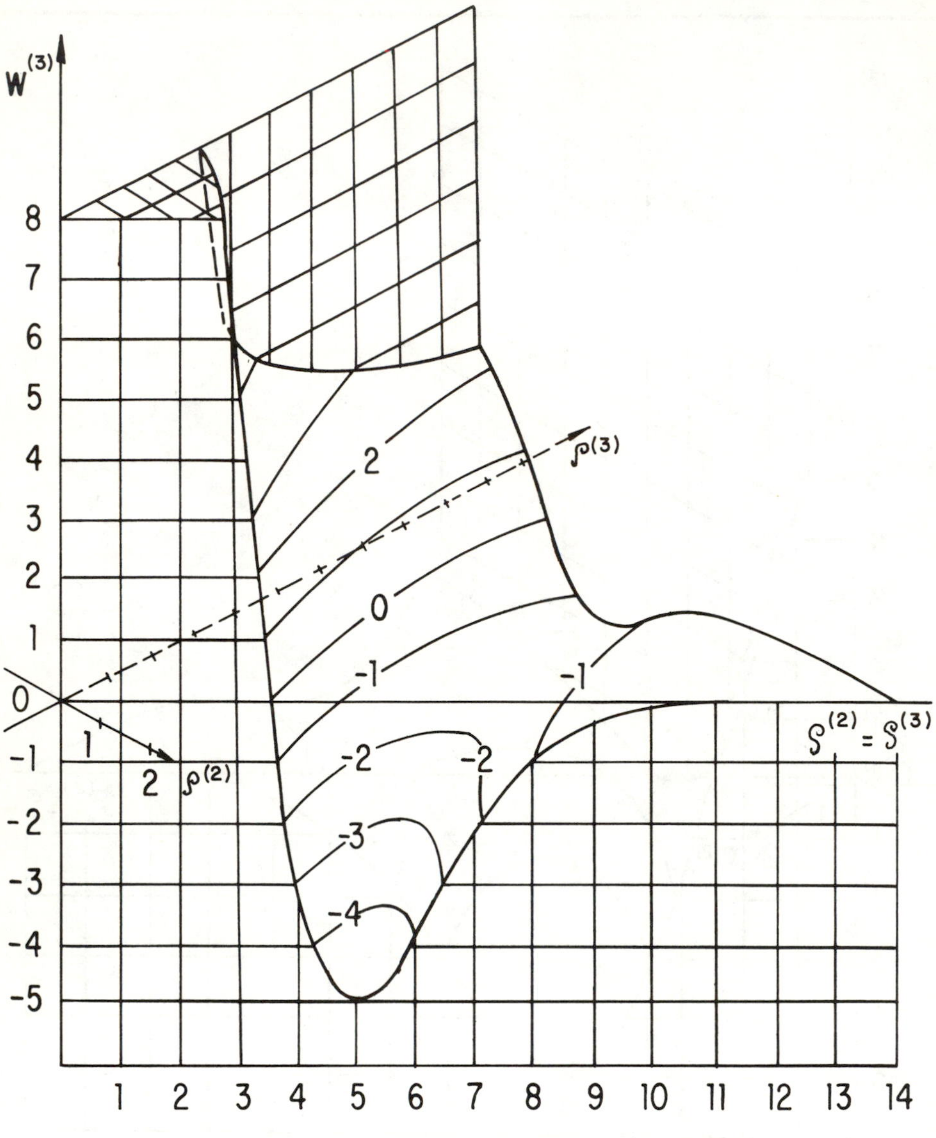

Fig. 5. The collective potential (12.20) for $V_1 = -5; \tau_1 = 5; \tau_0 = 3{,}5; \mathcal{H} = 1$

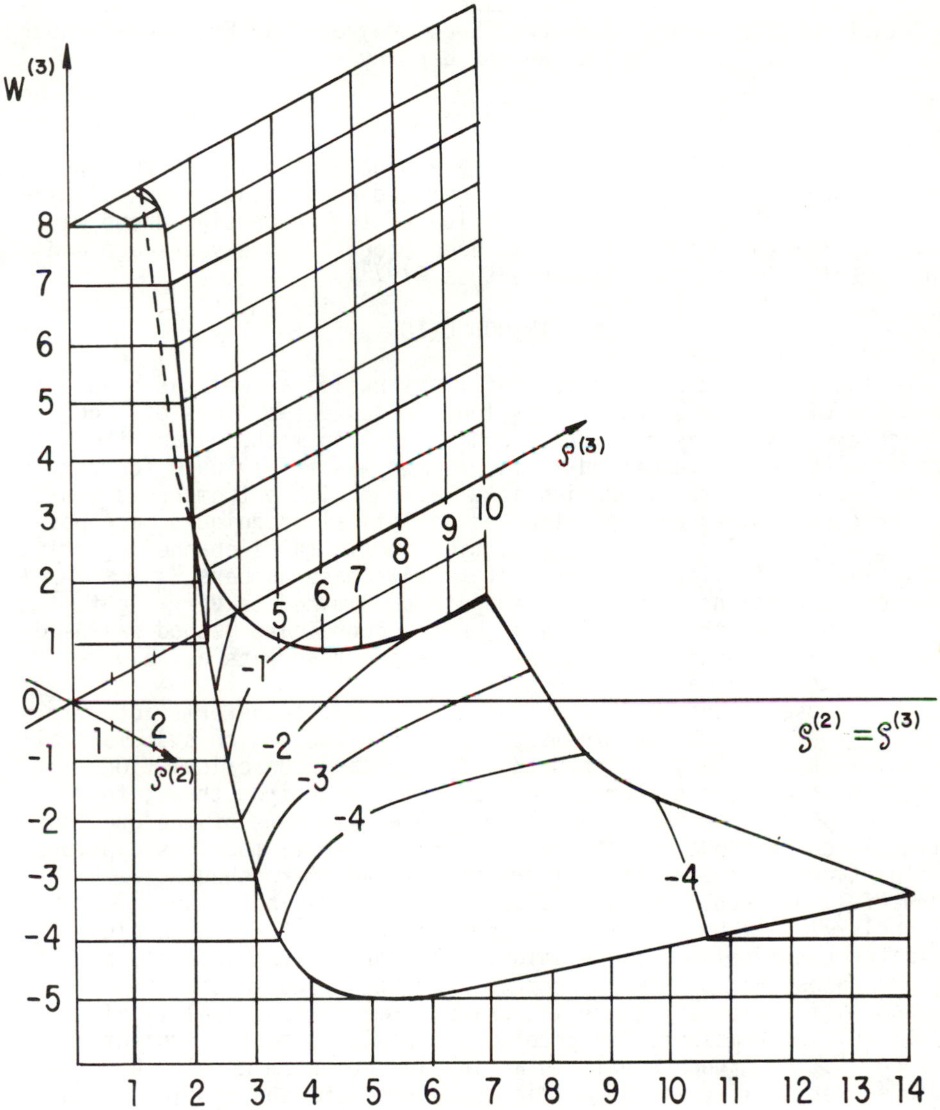

Fig. 6. The collective potential (12.20) for $V_1 = -5$; $\tau_1 = 5$; $\tau_0 = 2,5$; $\varkappa = 0,01$

INTEGRAL TRANSFORMS AND THE SYMMETRY GROUP OF THE HARMONIC OSCILLATOR, SU_3, AS A TOOL FOR RESONATING GROUP CALCULATIONS

W. Zahn
Institut für Theoretische Physik der Universität Erlangen-Nürnberg
Erlangen, Germany

ABSTRACT

Harmonic oscillator functions are widely used in modern low-energy physics, in general, and in refined resonating group calculations, in particular[1]. Classification of the resonating group wave functions with respect to SU_3 has been proposed some years ago and implemented in gret detail since then.[2][3][4].

INTRODUCTION

The treatment of reactions of light nuclei as well as clustering aspects of nuclear structure has long followed two lines of thought that may be summarized as the resonating group method[5] and the generator coordinate method[6]. The former was mainly used for very light nuclear systems; particularly Hackenbroich's tremendous computer code[1] based on sophisticated analytical techniques made possible very detailed and extensive microscopic multi-channel reaction calculations in light nuclear systems with mass numbers $4 \leq A \leq 12$ [7]. These calculations, however, seemed to be limited to very light nuclei. On the other hand, the generator coordinate method was used with particular success to somewhat heavier fragments as, e.g., in ^{16}O-^{16}O and α-^{40}Ca calculations[8].

The use of integral transforms - integral transforms actually relate the two methods[9]-seems to furnish a tool which systematically simplifies the analytical procedures necessary for calculations in light multicluster systems, i.e., systems built from three, four, five clusters, and thus gives hope to overcome some of the previous limitations without making the most drastic shell model assumptions. In particular, these integral transforms have furnished us with a powerful, new tool for the evaluation of resonating group kernels[8][10][11] by reducing the computational effort for nuclear cluster problems to a calculation of integrals over intercluster coordinates. Although much progress had been made in the technical problem of evaluating these integrals, some serious analytical difficulties remained. The problem of angular momentum projection of resonating group kernels in an angular momentum coupled basis[12] proved very difficult for multicluster problems and problems with heavy fragments other than closed-shell nuclei.

A practical solution to both of these problems has been effected by exploiting the SU_3 symmetry properties of the relative motion and internal harmonic oscillator functions[2]. In an SU_3-coupled cluster function basis SU_3-reduced matrix elements can be evaluated by integral transform techniques where any convenient basis can be used to effect the SU_3 coupling in the transform since the SU_3-reduced matrix elements are independent of SU_3 sub-

ISSN:0094-243X/81/710294-18$1.50

group labels. The technique can be applied to multicluster systems as readily as to simple cluster systems made up of only two fragments. The complexity of the cluster system is reflected only in the number of SU_3 recoupling coefficients needed for the evaluation of the reduced matrix elements.

An extensive compilation of the methods sketched in these short notes is found in refs 2) and 3).

THE NUCLEAR-CLUSTER MODEL ANSATZ

One of the basic assumptions in a cluster model ansatz is that cluster model wave functions may be expanded in terms of harmonic oscillator functions. A resonating group or cluster model wave function for an n-particle system made up from $\varkappa$ symmetric clusters may be written schematically as

$$\psi = \mathcal{A}\left(\prod_{i=1}^{\varkappa} \varphi_i\right)\left(\prod_{j=1}^{\varkappa-1} \chi_j\right) \xi^{\sigma,\tau} . \tag{2.1}$$

Here $\mathcal{A}$ denotes total antisymmetrization, the functions φ_i are cluster functions depending on the internal variables of the ith cluster, supposed to be square integrable and usually be chosen as oscillator functions of arbitrary, independent widths, the functions χ are representative for the collective modes of relative motion of the clusters and $\xi^{\sigma,\tau}$ stands for an n-particle spin-isospin function. The resonating group kernel $\mathcal{K}(\bar{\underset{\sim}{R}},\underset{\sim}{R})$ for an operator, H, acting on the above ψ, will contain Dirac δ's which stem, e.g., from the direct term of the antisymmetrizer. It is defined by

$$\int d\bar{\underset{\sim}{R}} \int d\underset{\sim}{R} \int d\bar{\underset{\sim}{\xi}} \int d\underset{\sim}{\xi} \; \prod_{i=1}^{\varkappa} \bar{\varphi}_i \prod_{j=1}^{\varkappa-1} \bar{\chi}_j(\bar{\underset{\sim}{R}})$$

$$H \mathcal{A} \prod_{k=1}^{\varkappa} \varphi_k \prod_{l=1}^{\varkappa-1} \chi_l(\underset{\sim}{R}) =$$

$$\int d\bar{\underset{\sim}{R}} \int d\underset{\sim}{R} \prod_{j=1}^{\varkappa-1} \bar{\chi}_j(\bar{\underset{\sim}{R}}) \mathcal{K}(\bar{\underset{\sim}{R}},\underset{\sim}{R}) \prod_{l=1}^{\varkappa-1} \chi_l(\underset{\sim}{R}) . \tag{2.2}$$

Here, ξ denote cluster-internal coordinates. How to technically implement the integration over these variables will be shown in the following using integral transforms with generating function properties. Once we have performed the integrations over the internal variables ζ the detailed solution of the nuclear cluster problem by a refined resonating group method is reduced to manageable proportions with the development of efficient techniques for the evaluation of integrals of the type

$$I = \int d\bar{\underset{\sim}{R}} \int d\underset{\sim}{R} \prod_{j=1}^{\varkappa-1} \bar{X}_j(\underset{\sim}{R}) K(\bar{\underset{\sim}{R}},\underset{\sim}{R}) \prod_{\ell=1}^{\varkappa-1} X_\ell(\underset{\sim}{R}). \tag{2.3}$$

Here, the use of integral transforms will yield a form easily accessible to the exploitation of SU_3 coupling techniques utilizing the fact that integral transforms may have very simple SU_3 properties. The expansion of (2.3) in the terms of SU_3-coupled polynomials will be demonstrated subsequently.

INTEGRAL TRANSFORMS

The (approximate) integration of the n-particle Schrödinger equation is a problem which is essentially twofold provided total antisymmetrization, necessary in any n-fermion system, is properly carried out. Firstly, the reduction of the n-particle system to a $\varkappa$-cluster system by explicitly integrating over the cluster-internal variables has to be performed. In order to analytically implement this task we utilize integral transforms which will be the subject of this chapter. Secondly, there is the problem of solving the coupled integro-differential equations in the remaining intercluster variables. This quite serious problem is the subject of the following chapter and its solution will be approximated by integrations given in closed form utilizing sophisticated SU_3 techniques for the harmonic oscillator functions.

Two of the most important integral transforms used in nuclear cluster calculations are the Hackenbroich transform (hereafter called H-transform)[10] and the Bargmann-Segal transform (hereafter called BS-transform)[11]. Both transforms, though having quite difficult transformation properties, are generating functions for harmonic oscillator functions.

The H-transform is defined via

$$f_\vartheta(x,k) = \pi^{-1/2} \exp\left\{-\frac{1}{2}\vartheta x^2 + 2ikx + \frac{1}{\vartheta}k^2\right\} \tag{3.1}$$

as a function of the variables x and k both ranging over $\mathbb{R}$. Hackenbroich originally considered $f_\vartheta(x,k)$ as a generating function for harmonic oscillator functions,

$$\varphi_n(\sqrt{\vartheta}x) = \left(\frac{\vartheta^{1/2}}{\pi^{1/2}2^n n!}\right)^{1/2} H_n(\sqrt{\vartheta}x) \exp\left\{-\frac{1}{2}\vartheta x^2\right\} \tag{3.2}$$

only, where H_n are Hermite polynomials. Then we have

$$f_{\vartheta}(x,k) = \pi^{-1/4} \sum_{n} k^{n} (-2)^{n/2} (\vartheta^{n-1} n!)^{-1/2} \varphi_{n}(\sqrt{\vartheta} x) . \quad (3.3)$$

Straightforward applications to complex problems using arbitrary oscillator frequencies in the variables, however, requires reinterpretations of $f_{\vartheta}(x,k)$ as an integral transform such that

$$\psi(x) = \int f_{\vartheta}(x,k)\, g(k)\, dk \quad (3.4)$$

holds. The mathematical problems involved, e.g., the mapping of dense subspaces of $\mathcal{L}^2(\mathbb{R})$ into a dense subspace of $\mathcal{L}^2(\mathbb{R})$, convergence and integrability questions, are extensively discussed elsewhere[10]. Suffice it to say that on the image space we may also define an inverse transform

$$\int f_{\vartheta}^{-1}(x,k)\, \psi(x)\, dx = g(k) \quad (3.5)$$

with

$$f_{\vartheta}^{-1}(x,k) = \pi^{-1/2} \exp\left\{\frac{1}{2}\vartheta x^2 - 2ikx - \frac{1}{\vartheta} k^2\right\} . \quad (3.6)$$

The relations

$$\int f_{\vartheta}^{-1}(\bar{k},x)\, f_{\vartheta}(k,x)\, dx = \delta(\bar{k}-k) \quad (3.7)$$

and

$$\int f_{\vartheta}^{-1}(k,\bar{x})\, f_{\vartheta}(k,x)\, dk = \delta(\bar{x}-x) \quad (3.8)$$

must hold, thus defining the kernel associated with the identity as

$$\int f^{*}_{\bar{\vartheta}}(\bar{k},x)\, f_{\vartheta}(k,x)\, dx = \pi^{1/2} \left(\frac{\bar{\vartheta}+\vartheta}{2}\right)^{-1/2}$$

$$\exp\left\{\left(\frac{1}{\bar{\vartheta}} - \frac{2}{\bar{\vartheta}+\vartheta}\right)\bar{k}^2 + \left(\frac{4}{\bar{\vartheta}+\vartheta}\right)\bar{k}\,k\right.$$

$$\left. + \left(\frac{1}{\vartheta} - \frac{2}{\bar{\vartheta}+\vartheta}\right)k^2\right\} \tag{3.9}$$

and avoiding Dirac δ's for any choice of $\bar{\vartheta}$ and ϑ in the kernel corresponding to the unit operator altogether. Extension to m dimensions of these transforms is trivially performed via

$$F(\underset{\sim}{k}, \underset{\sim}{x}) = \prod_{i=1}^{m} f_{\vartheta_i}(k_i, x_i)\,. \tag{3.10}$$

The BS-transform is widely used in connection with SU_3 symmetry. Its transformation properties are characterized by the unitary transformation

$$\beta(z) = \int_{-\infty}^{+\infty} A(x,z)\, f(x)\, dx \tag{3.11}$$

with

$$A(x,z) = \pi^{-1/4} \exp\left\{-\frac{1}{2}x^2 - \frac{1}{2}z^2 + \sqrt{2}\, x z\right\} \tag{3.12}$$

from $\mathcal{L}^2(\mathbb{R})$ to a space of square integrable analytical functions with the measure

$$d\mu(z) = \pi^{-1} e^{-z^* \cdot z} \, d\,\mathrm{Re}(z) \, d\,\mathrm{Im}(z) \tag{3.13}$$

where integration ranges over the complex plane. This unitary transformation may be inverted as

$$f(x) = \int A^*(x,z) \, \beta(z) \, d\mu(z) \tag{3.14}$$

where the asterisk denotes complex conjugation. We further have the important property that in BS space any linear operator is a bona fide integral operator and in particular the reproducing kernel is given as $\exp\{z^* \cdot z\}$. Finally we must recall the most significant fact that the BS transform kernel is a generating function for oscillator functions, i.e.,

$$A(x,z) = \sum_{n=0}^{\infty} \psi_n(x) \frac{z^n}{\sqrt{n!}} \tag{3.15}$$

where $\psi_n(x)$ is the nth oscillator eigenfunction in BS space. This fact implies that any integration over oscillator functions may be performed trivially in BS space by differentiations and limits $z \to 0$.

If we deal with n-dimensional problems we clearly take product spaces and the BS transform reads

$$A(\underset{\sim}{x}, \underset{\sim}{z}) = \prod_{i=1}^{n} A(x_i, z_i) \tag{3.16}$$

where $\underset{\sim}{z}$ and $\underset{\sim}{x}$ indicate n-component vectors. Note that the form of A is such that an orthogonal transformation in one set of variables implies the same in the other. Instead of evaluating resonating group kernels as defined in eq. (2.2) we shall now proceed to evaluate integral kernels

$$\mathcal{H}(\bar{\underset{\sim}{K}},\underset{\sim}{K}) = \iint A(\bar{\underset{\sim}{K}},\bar{\underset{\sim}{R}})\, \mathcal{K}(\bar{\underset{\sim}{R}},\underset{\sim}{R})\, A^{*}(\underset{\sim}{K},\underset{\sim}{R})\, d\bar{\underset{\sim}{R}}\, d\underset{\sim}{R} \tag{3.17}$$

under the assumption that the internal wave functions of the 0s clusters are oscillator functions of arbitrary frequency.

SU_3-SYMMETRY FOR CLUSTER PROBLEMS

The technique to be illustrated in this chapter can be exploited for the evaluation of kernels for both norms and two-body operators, and for cluster functions with fragments of equal and unequal oscillator size parameters. The special case of the norm kernel for a cluster system with fragments described by oscillator functions of equal size, however, forms a particularly simple example since such a norm kernel is an SU_3 scalar. Although the restriction to this simple problem can already be overcome right in the end of this chapter, a cluster function built from cluster components with oscillator functions of equal size may nevertheless be very useful for p - and sd-shell nuclei since a physically realistic description of such nuclei may involve the combination of a fairly rich shell model (valence) basis with core excitations described in terms of such cluster wave functions. Norm kernels for cluster wave functions built from oscillator functions of equal size are also useful for multi-nucleon transfer spectroscopy if norms can be calculated for relative motion functions of an excitation high enough to construct radial functions of realistic shape. For realistic applications the relative motion functions in a cluster basis built from oscillator functions of equal size must thus be able to carry excitations up to a high number of oscillator quanta. Since the BS-transform of such a kernel contains relative motion functions of arbitrarily high excitation an expansion of such a kernel in terms of SU_3 representations will lead directly to the closed evaluation of the needed matrix elements.

The BS-transform of the normalized 1-dimensional harmonic oscillator function $\psi_n(x)$, viz. $K_x^n/\sqrt{n!}$, is the normalized 1-dimensional oscillator function in Bargmann space. The fact that the oscillator ground state function is the simple number 1 leads to much of the simplicity of Bargmann space. The Bargmann space function

$$P(\underset{\sim}{K})^{(Q0)} = \frac{K_x^{\,n_x}}{\sqrt{n_x!}}\;\frac{K_y^{\,n_y}}{\sqrt{n_y!}}\;\frac{K_z^{\,n_z}}{\sqrt{n_z!}} \tag{4.1}$$

given here in a Cartesian oscillator basis, with $Q = n_x + n_y + n_z$ has SU_3 irreducible tensor character (Q0). The corresponding polynomial in $\underset{\sim}{K}^*$ has SU_3 irreducible tensor character (0Q). An SU_3-coupled Bargmann space polynomial can be defined by

$$[P(\underset{\sim}{K}_1)^{(Q_1 0)} \times P(\underset{\sim}{K}_2)^{(Q_2 0)}]^{(\lambda\mu)}_{\alpha} = \sum_{\alpha_1 \alpha_2} \langle (Q_1 0)\alpha_1, (Q_2 0)\alpha_2 | (\lambda\mu)\alpha \rangle \, P(\underset{\sim}{K}_1)^{(Q_1 0)}_{\alpha_1} \, P(\underset{\sim}{K}_2)^{(Q_2 0)}_{\alpha_2} \, , \tag{4.2}$$

where the SU_3 subgroup labels, α, can be chosen in a convenient basis; including the angular momentum basis $\alpha = \kappa L M$. For many steps in the calculation a detailed knowledge of the SU_3 Wigner coefficients will not be necessary.

For a simple 3-cluster nucleus with two relative motion functions, the generating function can be expanded in terms of SU_3-coupled functions

$$A(\underset{\sim}{K}_1, \underset{\sim}{R}_1)\, A(\underset{\sim}{K}_2, \underset{\sim}{R}_2) = \sum_{Q_1 Q_2} \sum_{(\lambda\mu)\alpha} [P(\underset{\sim}{K}_1)^{(Q_1 0)} \times P(\underset{\sim}{K}_2)^{(Q_2 0)}]^{(\lambda\mu)}_{\alpha} \left\{ [\chi(\underset{\sim}{R}_1)^{(Q_1 0)} \times \chi(\underset{\sim}{R}_2)^{(Q_2 0)}]^{(\lambda\mu)}_{\alpha} \right\}^* , \tag{4.3}$$

with obvious generalizations to multi-cluster systems. The B-S-transform of the kernel $\mathcal{K}$ can then be expanded in terms of SU_3-coupled K -space polynomials

$$H(\bar{\underset{\sim}{K}}, \underset{\sim}{K}) = \sum [P(\bar{\underset{\sim}{K}}_1)^{(\bar{Q}_1 0)} \times P(\bar{\underset{\sim}{K}}_2)^{(\bar{Q}_2 0)}]^{(\bar{\lambda}\bar{\mu})}_{\bar{\alpha}} [P(\underset{\sim}{K}_1^*)^{(0 Q_1)} \times P(\underset{\sim}{K}_2^*)^{(0 Q_2)}]^{(\mu\lambda)}_{\alpha} \iint d\bar{\underset{\sim}{R}}\, d\underset{\sim}{R}\, [\chi(\bar{\underset{\sim}{R}}_1)^{(\bar{Q}_1 0)} \times \chi(\bar{\underset{\sim}{R}}_2)^{(\bar{Q}_2 0)}]^{(\bar{\lambda}\bar{\mu})*}_{\bar{\alpha}} \, \mathcal{K} \, [\chi(\underset{\sim}{R}_1)^{(Q_1 0)} \times \chi(\underset{\sim}{R}_2)^{(Q_2 0)}]^{(\lambda\mu)}_{\alpha} . \tag{4.4}$$

If we now imagine that the kernel (for a two-body interaction) is expanded in terms of SU_3 irreducible tensor components

$$\mathcal{K} = \sum_{(\lambda_0\mu_0)} \mathcal{K}^{(\lambda_0\mu_0)}_{L_0=0} , \tag{4.5}$$

we can use the Wigner-Eckart theorem for the $(\lambda_0\mu_0)$ th component of the $\bar{\underset{\sim}{R}}$, $\underset{\sim}{R}$ -space integrals to express these integrals in terms of SU_3-reduced (double-barred) matrix elements and SU_3 Wigner coefficients by

$$\sum_{\rho_0} \langle \bar{Q}_1 \bar{Q}_2 (\bar{\lambda}\bar{\mu}) \| \mathcal{K}^{(\lambda_0\mu_0)} \| Q_1 Q_2 (\lambda\mu) \rangle_{\rho_0}$$

$$\langle (\bar{\lambda}\bar{\mu}) \bar{\alpha} (\mu\lambda) \alpha^* | (\lambda_0\mu_0) L_0 = 0 \rangle_{\rho_0} . \tag{4.6}$$

(The outer multiplicity label ρ_0 is needed when the Kronecker product $(\bar{\lambda}\bar{\mu}) \times (\mu\lambda)$ contains $(\lambda_0\mu_0)$ with a d-fold multiplicity, with $d > 1$.) The SU_3 Wigner coefficients carry all dependence on SU_3 subgroup labels α . Using this form for the $\bar{\underset{\sim}{R}}$, $\underset{\sim}{R}$ -space integrals, the $\bar{\alpha}$, α sums can be carried out; and the B-S transform of the kernel is given by the expansion

$$H(\bar{\underset{\sim}{K}}, \underset{\sim}{K}) = \sum_{\bar{Q}_1 \bar{Q}_2} \sum_{Q_1 Q_2} \sum_{(\bar{\lambda}\bar{\mu})(\lambda\mu)} \sum_{(\lambda_0\mu_0)\rho_0}$$

$$\langle \bar{Q}_1 \bar{Q}_2 (\bar{\lambda}\bar{\mu}) \| \mathcal{K}^{(\lambda_0\mu_0)} \| Q_1 Q_2 (\lambda\mu) \rangle_{\rho_0}$$

$$\Big[[P(\bar{\underset{\sim}{K}}_1)^{(\bar{Q}_1 0)} \times P(\bar{\underset{\sim}{K}}_2)^{(\bar{Q}_2 0)}]^{(\bar{\lambda}\bar{\mu})} \times$$

$$[P(\underset{\sim}{K}_1^*)^{(0 Q_1)} \times P(\underset{\sim}{K}_2^*)^{(0 Q_2)}]^{(\mu\lambda)} \Big]^{(\lambda_0\mu_0)\rho_0}_{L=0} . \tag{4.7}$$

By using the orthonormality of the

$$[P(\underset{\sim}{K}_1)^{(Q_1 0)} \times P(\underset{\sim}{K}_2)^{(Q_2 0)}]^{(\lambda\mu)}_{\alpha}$$

in K-space it is, in principle, possible to integrate over the complex variables $\bar{\underset{\sim}{K}}_1, \cdots, \underset{\sim}{K}_2^*$ with the Bargmann K-space measure[11]) to select a specific SU_3-reduced matrix element. In practice, a paritcular SU_3-reduced matrix element can be read directly from an expansion of $\mathcal{H}(\bar{\underset{\sim}{K}}, \underset{\sim}{K})$ in terms of the SU_3 irreducible tensor K-space polynomials, SU_3-coupled successively to resultant $(\lambda_0\mu_0)$, in the form illustrated above. Hence no laborious irreducible tensor decomposition of the kernel $\mathcal{K}(\bar{\underset{\sim}{R}}, \underset{\sim}{R})$ needs to be carried out in R-space. The SU_3-reduced matrix element will fall out of the B-S transform of this kernel quite naturally, if all factors in this transform are expressed in SU_3-coupled form.

The kernel of the unit operator is particularly simple since this is an SU_3-scalar, $(\lambda\mu)=(00)$-operator. The R-space integrals give the norm, N, of the fully antisymmetrized cluster wave function. In this case

$$\mathcal{H}(\bar{\underset{\sim}{K}}, \underset{\sim}{K}) = \sum_{\bar{Q}_1 \bar{Q}_2} \sum_{Q_1 Q_2} \sum_{(\lambda\mu)} \frac{1}{N^2} [\dim(\lambda\mu)]^{1/2}$$

$$[[P(\bar{\underset{\sim}{K}}_1)^{(\bar{Q}_1 0)} \times P(\bar{\underset{\sim}{K}}_2)^{(\bar{Q}_2 0)}]^{(\lambda\mu)} \times$$

$$[P(\underset{\sim}{K}_1^*)^{(0 Q_1)} \times P(\underset{\sim}{K}_2^*)^{(0 Q_2)}]^{(\mu\lambda)}]^{(00)} , \tag{4.8}$$

with straightforward generalizations to more complicated cluster systems: If the oscillator excitations, $Q_1, \cdots, Q_{n-1}$ of the first $(n-1)$ relative motion variables $\underset{\sim}{K}_1, \cdots, \underset{\sim}{K}_{n-1}$ are restricted to their lowest Pauli-allowed values and SU_3-coupled to resultant $(\lambda_c\mu_c)$, the

$$[\ldots[P(\underset{\sim}{K}_1)^{(Q_1 0)} \times P(\underset{\sim}{K}_2)^{(Q_2 0)}]^{(\lambda_{12}\mu_{12})} \times \ldots$$

$$\times P(\underset{\sim}{K}_{n-1})^{(Q_{n-1}, 0)}]^{(\lambda_c\mu_c)}$$

$$\equiv P(\underset{\sim}{K}_1, \cdots, \underset{\sim}{K}_{n-1})^{(\lambda_c\mu_c)} \tag{4.9}$$

become the B-S transforms of a heavy-fragment internal wave function, and, for fixed $\bar{Q}_1 = Q_1, \ldots, \bar{Q}_{n-1} = Q_{n-1}$, the generalization of the above equation gives the norm, $\mathcal{N}$, of a [heavy-fragment + $(n+1)^{st}$ cluster component] -wave function with quantum numbers $(\lambda_c\mu_c), (Q_n 0), (\lambda\mu)$. For the 3-cluster system with arbitrary excitations $\bar{Q}_1, \bar{Q}_2, Q_1, Q_2$, the above $1/\mathcal{N}^2$ is shorthand notation for a matrix which factors into submatrices for each $(\lambda\mu)$, with $\bar{Q}_1 + \bar{Q}_2 = Q_1 + Q_2 = \lambda + 2\mu$. The eigenvectors for non-zero eigenvalues of $1/\mathcal{N}^2$ give the Pauli-allowed states of this cluster system. Pauli-forbidden states have zero eigenvalues for $1/\mathcal{N}^2$.

To show how the above program can be implemented in a practical cluster model calculation let us consider the simple 3-cluster system, $^{12}C = \alpha + \alpha + \alpha$. For such a cluster system the B-S transform of the kernel for the full interaction Hamiltonian is of simple Gaussian form if the two-body interaction can be expanded in terms of Gaussians. That is, in this case[11)]

$$H(\bar{\underset{\sim}{K}}, \underset{\sim}{K}) = \sum_{\beta} a_{\beta} \exp\{\varrho(\beta)\} \exp\{\sigma(\beta)\} \exp\{\tau(\beta)\}. \tag{4.10}$$

The β-sum arises from the antisymmetrizer, which can be expressed in terms of a double coset decomposition. The β -sum runs over the full set of double coset generators. The weighting factors a_β are calculated by the techniques of Kramer and Seligman.[13)] The exponentials are abbreviations for

$$\exp\{\sigma(\beta)\} \equiv \exp\left\{ \sum_{i,j=1}^{2} \sigma_{ij}(\beta) (\bar{\underset{\sim}{K}}_i \cdot \underset{\sim}{K}_j^*) \right\}, \tag{4.11a}$$

$$\exp\{\varrho(\beta)\} \equiv \exp\left\{ \sum_{i,j=1}^{2} \varrho_{ij}(\beta) (\bar{\underset{\sim}{K}}_i \cdot \bar{\underset{\sim}{K}}_j) \right\}, \tag{4.11b}$$

$$\exp\{\tau(\beta)\} \equiv \exp\left\{ \sum_{i,j=1}^{2} \tau_{ij}(\beta) (\underset{\sim}{K}_i^* \cdot \underset{\sim}{K}_j^*) \right\}. \tag{4.11c}$$

Anexpansion of these exponentials followed by a succession of SU_3-recoupling transformations of the κ-space polynomials will lead to the needed final form in which $\#(\bar{\kappa},\kappa)$ is expanded in terms of the SU_3-coupled tensors

$$[[P(\bar{\kappa}_1)^{(\bar{Q}_1 0)} \times P(\bar{\kappa}_2)^{(\bar{Q}_2 0)}]^{(\bar{\lambda}\bar{\mu})} \times [P(\kappa_1^*)^{(0 Q_1)} \times P(\kappa_2^*)^{(0 Q_2)}]^{(\mu\lambda)}]^{(\lambda_0\mu_0)\rho_0}_{0} . \tag{4.12}$$

The basic building blocks for this expansion are

$$\frac{(\bar{\kappa}_i\cdot\kappa_j^*)^n}{n!} = [P(\bar{\kappa}_i)^{(n0)} \times P(\kappa_j^*)^{(0n)}]^{(00)}_{L=0}\,[\dim(n0)]^{1/2} \tag{4.13}$$

and, with $i=j$

$$\frac{(\bar{\kappa}_i\cdot\bar{\kappa}_i)^n}{n!} = \frac{[(2n+1)!]^{1/2}}{n!}\,P(\bar{\kappa}_i)^{(2n,0)}_{L=0} \tag{4.14}$$

while, with $i\neq j$

$$\frac{(\bar{\kappa}_i\cdot\bar{\kappa}_j)^n}{n!} = \sum_{(\lambda\mu)} [\lambda+1]^{1/2}(-1)^{\mu/2}\,[P(\bar{\kappa}_i)^{(n0)} \times P(\bar{\kappa}_j)^{(n0)}]^{(\lambda\mu)}_{L=0} , \tag{4.15}$$

where $(\lambda\mu)$ runs over the values $(2n,0)$, $(2n-4,2), \cdots, (0n)$ for n=even, $\cdots (2, n-1)$ for n =odd.

With SU_3-recoupling transformations which take us from

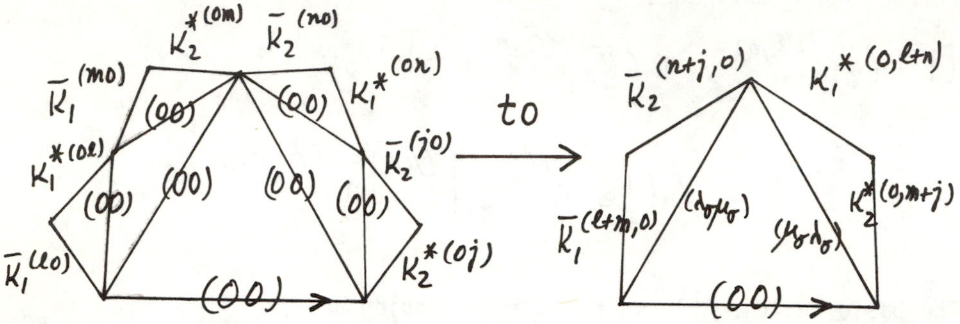

the exp $\{\sigma(\beta)\}$ factor can be expanded by

$$\exp\{\sigma\} = \sum_{(\lambda_\sigma\mu_\sigma)} \sum_{\substack{l+m,\, n+j \\ l+n,\, m+j}} C_\sigma(l+m, n+j, l+n, m+j; (\lambda_\sigma\mu_\sigma))$$

$$\Big[\big[P(\underset{\sim}{\bar{K}}_1)^{(l+m,0)} \times P(\underset{\sim}{\bar{K}}_2)^{(n+j,0)}\big]^{(\lambda_\sigma\mu_\sigma)} \times \qquad (4.16)$$

$$\big[P(\underset{\sim}{K}_1^*)^{(0,l+n)} \times P(\underset{\sim}{K}_2^*)^{(0,m+j)}\big]^{(\mu_\sigma\lambda_\sigma)}\Big]^{(00)}_{0}$$

with

$$C_\sigma = \sum_{\substack{l \\ (m,n,j)}} \sigma_{11}^{\,l}\, \sigma_{12}^{\,m}\, \sigma_{21}^{\,n}\, \sigma_{22}^{\,j} \left[\dim(\lambda_\sigma\mu_\sigma)\right]^{1/2}$$

$$\left[\binom{l+m}{l}\binom{n+j}{j}\binom{l+n}{l}\binom{m+j}{j}\right]^{1/2}$$

$$\begin{bmatrix} (l0) & (m0) & (l+m,0) \\ (n0) & (j0) & (n+j,0) \\ (l+n,0) & (m+j,0) & (\lambda_\sigma\mu_\sigma) \end{bmatrix} \qquad (4.17)$$

where the U_3 9-($\lambda\mu$) coefficient contains at most 2-rowed representations so that it is equivalent to a very simple SU_2 9-j coefficient with 4 "stretched" couplings.

The expansion of the exp $\{\varrho\}$ factor in terms of SU_3-coupled κ-space polynomials is given by

$$\exp\{\varrho\} = \sum_{\bar{q}_1,\bar{q}_2} \sum_{(\lambda_\varrho\mu_\varrho)} B_\varrho(\bar{q}_1\,\bar{q}_2\,;(\lambda_\varrho\mu_\varrho)) \left[P(\bar{K}_1)^{(\bar{q}_1 0)} \times P(\bar{K}_2)^{(\bar{q}_2 0)}\right]^{(\lambda_\varrho\mu_\varrho)}_{L_\varrho=0} \tag{4.18}$$

where

$$B_\varrho(\bar{q}_1\,\bar{q}_2\,;(\lambda_\varrho\mu_\varrho)) = \sum_{\substack{n,m \\ n+m=\frac{1}{2}(\bar{q}_1+\bar{q}_2)}} \varrho_{aa}^{\,n}\,\varrho_{bb}^{\,m} \frac{[(2n+1)!\,(2m+1)!]^{1/2}}{n!\,m!} \langle(2n,0)0;(2m,0)0\,||\,(\lambda_\varrho\mu_\varrho)0\rangle\, d^{\frac{1}{2}\lambda_\varrho}_{\frac{1}{2}(\bar{q}_1-\bar{q}_2),(n-m)}(\beta)\,. \tag{4.19}$$

The quadratic form $\{\sum \varrho_{ij}\bar{K}_i\cdot\bar{K}_j\}$ has first been put into diagonal form: $\{\varrho_{aa}(\bar{K}_a\cdot\bar{K}_a) + \varrho_{bb}(\bar{K}_b\cdot\bar{K}_b)\}$.

The SU_2 d-coefficient, with β given by $\tan\beta = 2\varrho_{12}/(\varrho_{22}-\varrho_{11})$ is the generalized Moshinsky bracket for SU_3-coupled oscillator functions which takes us from the $[P(\bar{K}_a)\times P(\bar{K}_b)]^{(\lambda\mu)}$ to the $[P(\bar{K}_1)\times P(\bar{K}_2)]^{(\lambda\mu)}$ basis. The double-barred coefficient is a very simple $SU_3 \supset R_3$ Wigner coefficient with all L's equal to zero. A similar expansion holds for exp $\{\tau\}$.

By coupling the expansions of the ϱ, σ and τ factors and combining these by SU_3 recoupling transformations which take us from

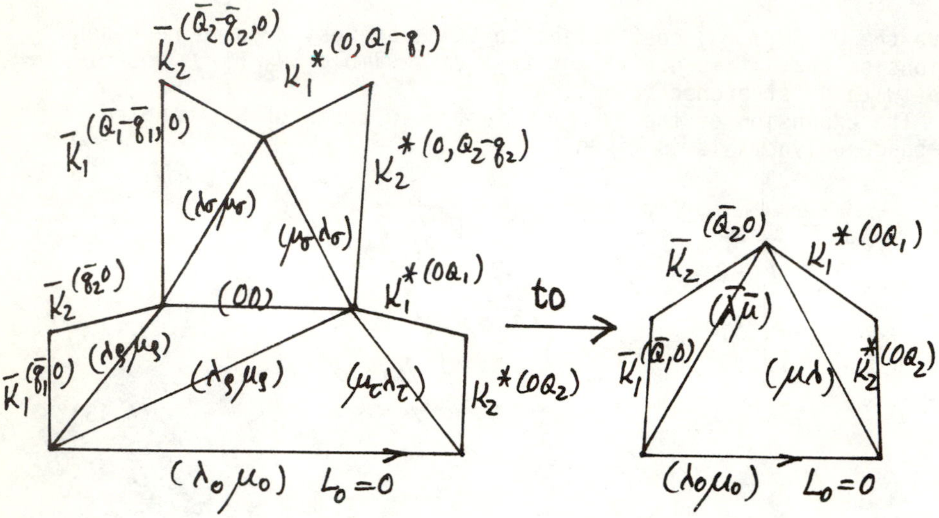

the resultant is expanded in a form needed to pick off the desired SU_3-reduced matrix elements

$$\exp\{\rho\}\,\exp\{\sigma\}\,\exp\{\tau\} =$$

$$\sum c_\sigma\left(\bar{Q}_1-\bar{q}_1,\ \bar{Q}_2-\bar{q}_2,\ Q_1-q_1,\ Q_2-q_2\ ;(\lambda_\sigma\mu_\sigma)\right)$$

$$B_\rho(\bar{q}_1,\bar{q}_2;(\lambda_\rho\mu_\rho))\ B_\tau(q_1 q_2;(\lambda_\tau\mu_\tau))$$

$$Z(\bar{Q}_1\bar{Q}_2\,Q_1Q_2\,\bar{q}_1\bar{q}_2\,q_1q_2\,(\lambda_\rho\mu_\rho)(\lambda_\sigma\mu_\sigma)(\lambda_\tau\mu_\tau)(\bar{\lambda}\bar{\mu})$$

$$(\lambda\mu)(\lambda_0\mu_0)\,\rho_0)$$

$$[[P(\underset{\sim}{\bar{K}}_1)^{(\bar{Q}_1 0)} \times P(\underset{\sim}{\bar{K}}_2)^{(\bar{Q}_2 0)}]^{(\bar{\lambda}\bar{\mu})} \quad \times$$

$$\left[P(\underset{\sim}{K}_1^*)^{(0Q_1)} \times P(\underset{\sim}{K}_2^*)^{(0Q_2)} \right]^{(\mu\lambda)} \Big]^{(\lambda_0\mu_0)\rho_0}_{L_0=0} \tag{4.20}$$

,

where

$$I = U((\lambda_\rho\mu_\rho)(\lambda_\sigma\mu_\sigma)(\lambda_\rho\mu_\rho)(\mu_\sigma\lambda_\sigma);(\bar{\lambda}\bar{\mu}),(00))$$

$$\Big\{ \sum_{\bar{\rho}_0} \langle (\lambda_\rho\mu_\rho) L_\rho=0;(\mu_\tau\lambda_\tau) L_\tau=0 \| (\lambda_0\mu_0) L_0=0 \rangle_{\bar{\rho}_0}$$

$$U((\bar{\lambda}\bar{\mu})(\mu_\sigma\lambda_\sigma)(\lambda_0\mu_0)(\mu_\tau\lambda_\tau);(\lambda_\rho\mu_\rho)1\bar{\rho}_0,(\mu\lambda)1\rho_0)\Big\}$$

$$\left[\binom{\bar{Q}_1}{\bar{q}_1}\binom{\bar{Q}_2}{\bar{q}_2}\binom{Q_1}{q_1}\binom{Q_2}{q_2} \right]^{1/2}$$

$$\begin{bmatrix} (\bar{q}_1 0) & (\bar{q}_2 0) & (\lambda_\rho\mu_\rho) \\ (\bar{Q}_1-\bar{q}_1,0) & (\bar{Q}_2-\bar{q}_2,0) & (\lambda_\sigma\mu_\sigma) \\ (\bar{Q}_1 0) & (\bar{Q}_2 0) & (\bar{\lambda}\bar{\mu}) \end{bmatrix}$$

$$\begin{bmatrix} (Q_1-q_1,0) & (Q_2-q_2,0) & (\lambda_\sigma\mu_\sigma) \\ (q_1 0) & (q_2 0) & (\lambda_\tau\mu_\tau) \\ (Q_1 0) & (Q_2 0) & (\lambda\mu) \end{bmatrix} \tag{4.21}$$

The U-coefficients are SU_3 Racah coefficients in unitary form.[14] The 9-($\lambda\mu$) coefficients contain at most 2-rowed U_3 representations so that they are again equivalent to simple SU_2 coefficients.

OUTLOOK

The basic achievements of the approach sketched in these short notes are

1). Arbitrary two-fragment (multicluster) problems in the p- and sd-shell can be treated. Analytic expressions for the resonating group matrix elements can be given in closed form.

2). Total antisymmetrization is taken into account. In refs.[2] we show, by using combined methods for permutation and SU_3 symmetry, how the computational effort can be kept at a feasible level.

3). The problem of angular momentum projection in the multi-center case is reduced to purely algebraic techniques using SU_3 coupling and recoupling coefficients which are readily available[14].

It is our aim to show how such techniques presented here and elsewhere[2] can be used to throw some light on an old and essentially unsolved problem in nuclear reaction theory, the problem of the so-called molecular resonances in the scattering of ^{12}C on ^{12}C and similar heavy ions. First results from such a microscopic theory of $^{12}C + ^{12}C$ molecular resonances[3] will be presented at the Group Theory Conference.

ACKNOWLEDGEMENTS

This work has been supported by the Deutsche Forschungsgemeinschaft. It is my sincere pleasure to acknowledge innumerable discussion with my friends and teachers K.T. Hecht and T.H. Seligman.

REFERENCES

1) H.H. Hackenbroich, in Symposium on Present Status and Novel Developments in the Nuclear Many-Body-Problem, Rome, Editrice Compositori, Bologna, 1973

2) K.T. Hecht and W. Zahn, Nucl. Phys. A313 (1979) 77;
Nucl. Phys. A318 (1979) 1;
Lecture Notes in Physics, Springer, 94 (1979) 408;
Nukleonika (1980) in press;
K.T. Hecht, E.J. Reske, T.H. Seligman, and W. Zahn, submitted for publication

3) K.T. Hecht and W. Zahn,
Lectures Notes in Physics, Springer, Group Theory Conference, Mexico, 1980

4) K.T. Hecht and W. Zahn, submitted for publication

5) J.A. Wheeler, Phys. Rev. 52 (1937) 1083

6) D.L. Hill and J.A. Wheeler, Phys. Rev. 89 (1953) 1102

7) H. Stöwe and W. Zahn, Nucl. Phys. A269 (1976) 138, A286 (1977)317, Z. Phys. A286 (1978) 173, J. Phys. G4 (1978) 1423

8) H. Friedrich, Nucl. Phys. A224 (1974) 537, H. Friedrich and K. Langanke, Nucl.Phys. A252 (1975) 47

9) D.M. Brink, International School of Physics Enrico Fermi, course 36, 1965

10) W. Zahn, Burg Monographs in Science, Vol. 2, Burg Verlag, Basel, 1975
H.H. Hackenbroich, T.H. Seligman, and W. Zahn, Helv. Phys. Acta 50 (1977) 723
L. Martignon and T.H. Seligman, Nucl. Phys. A286 (1977) 177

11) T.H.Seligman and W. Zahn, J. Phys. G2 (1976) 79

12) T.H. Seligman and W. Zahn, Helv. Phys. Acta 49 (1976) 217

13) P. Kramer and T.H. Seligman, Nucl. Phys. A136 (1969) 545, A186 (1972) 49

14) Y. Akiyama and J.P. Draayer, J. Math. Phys. 14 (1973) 1904, Comp. Phys. Commun. 5 (1973) 405.

CANONICAL TRANSFORMATIONS AND THEIR REPRESENTATION IN QUANTUM MECHANICS

M. Moshinsky*

Instituto de Física, UNAM

Apdo. Postal 20-364, México 20, D.F.

Table of Contents

* Member of the Instituto Nacional de Investigaciones Nucleares y el Colegio Nacional

ISSN:0094-243X/81/710312-38$1.50

INTRODUCTION AND SUMMARY

Canonical transformation in classical mechanics were, in the twenties, one of the cornerstones on which quantum mechanics was built. It is sufficient to mention the representation of Poisson brackets by commutators, or the unitary representation, i.e. Fourier kernel, that takes the states from configuration to momentum space, to realize how fundamental was the subject at the time.

After the initial importance, the field of canonical transformations in quantum mechanics went into a period of long hibernation. It was mainly in the decade of the sixties when it again became fashionable, mainly through the effort of those working in geometrical quantization.

In the seventies the author and a number of collaborators became interested in the subject, at first because the representation of linear canonical transformations were needed in a number of problems of nuclear physics. Later because of the intrinsic interest in the representation of non-linear and non-bijective (i.e. not one to one onto) canonical transformations.

The author has discussed this subject in a number of monographs with the same title, though in each case he has added some of the material developed in the intervening time. Thus sections 1 to 3 correspond to older and well known parts of the subject while the rest are related with articles recently published or in press.

We briefly summarize the contents of the paper: In section 1 we briefly review some standard concepts for the representation of canonical transformations in quantum mechanics. In section 2 we discuss some properties and examples of classical canonical transformations.

In section 3 we derive equations that allow us to determine the representation of canonical transformations if these involve functions $f(x,p)$ of the original coordinates and momenta, which can be written unambiguously as operators $f(x',-i\partial/\partial x')$, where we use Dirac's notation[1] for the c-number index associated with the coordinate x and take $\hbar=1$. We apply then the analysis to the representation of linear canonical transformations.

In section 4 we discuss the case of non-linear but bijective canonical transformations. An example of this occurs when we have the canonical transformation from the original x,p to $\bar{x}=H(x,p)$, $\bar{p}=-T(x,p)$, where H is a Hamiltonian and $-T$ its canonically conjugate variable. In H we have a potential $V(x)$ that monotonically decreases from $V=+\infty$ at $x=-\infty$ to $V=-\infty$ at $x=+\infty$. A representation is obtained explicitly and shown with the help of the WKB approximation, to have the appropriate classical limit when $\hbar \to 0$.

In all the previous cases we dealt with bijective canonical transformations. When they are non-bijective there is a problem even in classical mechanics which we discuss in section 5 under the name of structure of phase space.

In section 6 we turn our attention to the non-linear and non-bijective case which we illustrate through the example of canonical transformations to action and angle variables.

In section 7 we discuss the representation of canonical transformations in Wigner distribution function space, where the connection with the classical limit is more direct than in standard representation theory in Hilbert space.

Finally in the concluding section we stress some of the more important features of the problem and indicate possible applications.

1. CONCEPTS OF REPRESENTATION THEORY

Classically a canonical transformation is a change of variables which leaves the Poisson brackets invariant

$$\bar{x} = \bar{x}(x,p) \qquad \bar{p} = \bar{p}(x,p) \tag{1.1a}$$

$$\{\bar{x},\bar{p}\} = \{x,p\} = 1 \tag{1.1b}$$

We can ask ourselves what is its representation in quantum mechanics. Following the notation used by Dirac[1], if we have two different bases,

$$x|x'\rangle = x'|x'\rangle$$

$$\bar{x}|\bar{x}') = \bar{x}'|\bar{x}') \tag{1.2}$$

the quantum mechanical representation of the canonical transformation (1.1a) is given by the matrix elements of the operator U which lets us pass from one base to the other

$$U|\bar{x}'\rangle = |\bar{x}'), \quad \langle x'|\bar{x}') = \langle x'|U|\bar{x}'\rangle \tag{1.3a}$$

$$\bar{x} = UxU^{-1}, \qquad \bar{p} = UpU^{-1} \tag{1.3b}$$

We distinguish here by the angular brackets the states $|x'\rangle$ in which the original coordinate x is diagonal, and by the round ones those in which the new coordinate $\bar{x}$ is diagonal.

We now give some examples. Let us consider the fol

lowing transformation, which we find already in Dirac's book[1];

$$\bar{x} = p \qquad \bar{p} = -x \tag{1.4}$$

This is the canonical transformation which makes us pass from coordinate to momentum space; using units such that $\hbar=1$, we know that its representation is given by the Fourier transform

$$\langle x'|U|\bar{x}'\rangle = e^{ix'\bar{x}'}/\sqrt{2\pi} \tag{1.5}$$

We now consider the simplest generalization of transformation (1.4), that is,

$$\bar{x} = ax + bp, \quad \bar{p} = cx + dp, \quad b \neq 0, \quad a,b,c,d \in R \tag{1.6}$$

This canonical transformation was studied some ten years ago by Itzykson[2], Bargmann[3] and a few years later, but independently, by Moshinsky and Quesne[4,5]. If (1.6) is to be a canonical transformation, it must satisfy the following condition:

$$ad - bc = 1 \tag{1.7a}$$

that is

$$\det \begin{vmatrix} a & b \\ c & d \end{vmatrix} = 1 \tag{1.7b}$$

which is equivalent to say that the Poisson bracket relation $\{x,p\} = 1$ is invariant under the group SL(2,R) or equivalently, Sp(2,R).

The quantum mechanical representation of this transformation, which will be deduced afterwards (see section 3), is

$$\langle x'|U|\bar{x}'\rangle = e^{-\frac{i}{2b}(ax'^2 - 2x'\bar{x}' + d\bar{x}'^2)}/\sqrt{2\pi|b|} \tag{1.8}$$

We must point our that some 30 years passed between Dirac's and Bargmann's work on quantum mechanical representation of linear canonical transformations, though from the mathematical point of view all the necessary ideas are contained in Dirac's book[1]. So why did the physicists not interest themselves in the representation in quantum mechanics of canonical transformations for so long a time? Bargmann answered this through the following analogy: suppose you want to rob a house; you have to learn before how to open locks; when you have managed to get into the house you do not interest yourself anymore in locks but in the valuables you find in the house. If you find another locked door you will interest yourself again in locks, if you want to go on stealing . So considering quantum mechanics as a house, the canonical transformations were studied as a means for understanding it i.e., getting into the house. When physicists found their way into it, that is, when quantum mechanics was formalized, people went on working in this field without worrying about understanding its complete meaning. Now, when some things seem not to work so well, people are again interested in the structure of quantum mechanics i.e. there is a new interest in trying to open doors as some rooms are still locked.

2. CANONICAL TRANSFORMATIONS

We consider now canonical transformations in an n-dimensional space

$$\bar{x}_i = \bar{x}_i(x,p) \quad \bar{p}_i = \bar{p}_i(x,p) \qquad i = 1,\dots,n \tag{2.1}$$

$$\{\bar{x}_i,\bar{x}_j\} = \{\bar{p}_i,\bar{p}_j\} = 0 \quad \{\bar{x}_i,\bar{p}_j\} = \delta_{ij} \tag{2.2}$$

One can discuss them more generally by the use of a 2n-dimensional phase space

$$z_\alpha \quad , \alpha = 1,\dots,2n \tag{2.3a}$$

defined in the following way:

$$z_i = x_i \quad , \quad z_{i+n} = p_i \tag{2.3b}$$

If one considers the 2n x 2n matrix

$$K = \| K_{\alpha\beta} \| = \begin{bmatrix} 0 & I \\ -I & 0 \end{bmatrix} \tag{2.4}$$

where I is the identity matrix in n-dimensional space, the Poisson bracket can be written generally as:

$$\{F,G\} = \sum_{\alpha,\beta} \frac{\partial F}{\partial z_\alpha} K_{\alpha\beta} \frac{\partial G}{\partial z_\beta} \tag{2.5}$$

Thus to satisfy eqs. (2.2), we must have

$$\{\bar{z}_\alpha , \bar{z}_\beta\} = \sum_{\gamma,\delta} \frac{\partial \bar{z}_\alpha}{\partial z_\gamma} K_{\gamma\delta} \frac{\partial \bar{z}_\beta}{\partial z_\delta} = K_{\alpha\beta} \tag{2.6}$$

which is the general equation which must be satisfied by canonically conjugate variables; this is equivalent to saying that a canonical transformation leaves the matrix $K_{\alpha\beta}$ invariant. Thus they are elements of the general symplectic group. In the particular case of linear canonical transformations

$$\bar{z}_\alpha = S_{\alpha\beta} z_\beta \qquad (2.7)$$

(where we use Einstein's convention on summation over repeated indices), we must have:

$$S_{\alpha\gamma}K_{\gamma\delta}S_{\delta\beta} = K_{\alpha\beta} \qquad S_{\delta\beta} = S_{\beta\delta} \qquad (2.8)$$

If we decompose S in n n submatrices in the following way

$$S = \begin{bmatrix} A & B \\ C & D \end{bmatrix} \qquad (2.9)$$

S satisfies eq. (2.8) if the following relations are fulfilled

$$A\tilde{B} = B\tilde{A} \qquad D\tilde{C} = C\tilde{D} \qquad A\tilde{D} - B\tilde{C} = 1 \qquad (2.10)$$

If we restrict ourselves to the case that det B $\neq$ 0 we can determine C

$$C = (D\tilde{A} - I)\tilde{B}^{-1} \qquad (2.11a)$$

and thus obtain the following restrictions for the other matrices

$$A\tilde{B} = B\tilde{A} \qquad B\tilde{D} = D\tilde{B} \qquad (2.11b)$$

Examples of linear canonical transformations

Let us consider the one-dimensional harmonic oscillator with Hamiltonian $H = \frac{1}{2}(p^2 + x^2)$, and let us ask ourselves which is the most general linear canonical transformation which leaves H invariant. It is easy to find that

$$\bar{x} = x \cos\alpha + p \sin\alpha$$

$$\bar{p} = -x \sin\alpha + p \cos\alpha \qquad (2.12)$$

is a canonical transformation and leaves H invariant. This is a canonical transformation which is not limited to the coordinate space, and thus is more general than the rotations, translations or permutations which are all point transformations. As the one-dimensional system considered has no degeneracy because it is described completely by the energy eigenvalue, we get no further information from the group (2.12).

We consider now the three-dimensional case, that is

$$H = \frac{1}{2}(p^2 + r^2) \tag{2.13}$$

where p and r are respectively, the moment and coordinate vectors in three-dimensional space. If H is to be invariant,the transformation matrix S given by eq. (2.9), which must satisfy formula (2.8), verifies also the condition

$$S\tilde{S} = I \tag{2.14}$$

Writing

$$U = A + iB$$

$$V = C - iD \tag{2.15}$$

where A, B, C, D are real matrices, eq. (2.8) and eq. (2.14) give:

$$V = U \qquad U^{+}U = 1 \tag{2.16}$$

Thus the transformation which leaves H invariant corresponds to a representation of U(3) such that

$$S = \frac{1}{2}\begin{bmatrix} U + U^* & -i(U - U^*) \\ i(U - U^*) & U + U^* \end{bmatrix} \tag{2.17}$$

The U(3) in its representation (2.17) is the symmetry group of the three-dimensional harmonic oscillator.

We can also apply canonical transformations to the development in time of a given system. If x,p are the coordinates and momenta of a system at time t_o, we can ask which is the transformation which supplies us with the coordinates and momenta $\bar{x},\bar{p}$ at time t:

$$\bar{x} = \bar{x}(x,p,t)$$

$$\bar{p} = \bar{p}(x,p,t) \tag{2.18}$$

For a Hamiltonian system the equations are

$$\frac{dx}{dt} = \frac{\partial H}{\partial p}$$

$$\frac{dp}{dt} = -\frac{\partial H}{\partial x} \tag{2.19}$$

Thus, if $\{x,p\} = 1$, then

$$\begin{aligned}\{x + dx, p + dp\} &\simeq \{x,p\} + \{x,dp\} + \{dx,p\} = \\ &= 1 + \{x, -\frac{\partial H}{\partial x}dt\} + \{\frac{\partial H}{\partial p}dt, p\} = \\ &= 1 - \frac{\partial^2 H}{\partial x \partial p}dt + \frac{\partial^2 H}{\partial x \partial p}dt = 1\end{aligned} \tag{2.20}$$

up to first order in dx and dp. Applying eqs. (2.19), the Poisson bracket given by formula (2.20) is also satisfied, so that for a Hamiltonian system the development in time is a canonical transformation.

For a free particle we have (in units such that $h = m = 1$)

$$\bar{x} = x + pt$$

$$\bar{p} = p \tag{2.21}$$

and its quantum mechanical representation is (using eq. (1.8) with $a = d = 1$, $b = t$)

$$\langle x'|U|\bar{x}'\rangle = \frac{e^{i\alpha}}{\sqrt{2\pi t}} e^{-i(x'-\bar{x}')^2/2t} \tag{2.22}$$

which is exactly the Green function of the problem (α is an arbitrary phase).

In many problems is not easy to find the representation in quantum mechanics of a classical canonical transformation. This is the case in particular when the canonical transformation is not linear, as in the following example.

We consider the transformation between two systems which have centrifugal potentials of different strength

$$H = \frac{1}{2}p^2 + \frac{\lambda^2}{2x^2} \quad , \quad \bar{H} = \frac{1}{2}\bar{p}^2 + \frac{\bar{\lambda}^2}{2\bar{x}^2} \tag{2.23a}$$

$$T = (2H)^{-1}\, xp \quad , \quad \bar{T} = (2\bar{H})^{-1}\, \overline{xp}$$

(2.23b)

where T is the conjugate variable to the Hamiltonian and has been obtained by the following equation

$$T = \int_a^x \frac{dy}{\sqrt{2H(x,p) - 2V(y)}} \tag{2.24}$$

with $V(y) = \lambda^2/2y^2$. Considering the transformation

$$H = \bar{H} \quad , \quad T = \bar{T} \tag{2.25}$$

or equivalently

$$H = \bar{H} \quad , \quad 2HT = 2\bar{H}\bar{T} \tag{2.26}$$

we get for the new coordinates and momenta

$$\bar{x} = x \left[\frac{x^2p^2 + \bar{\lambda}^2}{x^2p^2 + \lambda^2} \right]^{1/2} \quad , \quad \bar{p} = p \left[\frac{x^2p^2 + \lambda^2}{x^2p^2 + \bar{\lambda}^2} \right]^{1/2} \tag{2.27}$$

It is simple to show that $\{\bar{x},\bar{p}\} = 1$; but it could be very difficult to obtain the representation in quantum mechanics from (2.27), though it is simple from (2.26). We want to prove that it is possible to define a classical canonical transformation in an implicit way through the use of some functional relations which connect the canonical variables.

If we have 2n functional relations between the two sets of 2n variables z_α and $\bar{z}_\alpha$, that is

$$f_\alpha(z) = \bar{f}_\alpha(\bar{z}), \quad \alpha = 1,2,\ldots,2n \tag{2.28}$$

then by differentiating eq. (2.28) with respect to z_β, we get:

$$\frac{\partial f_\alpha(z)}{\partial z_\beta} = \frac{\partial \bar{f}_\alpha(\bar{z})}{\partial z_\beta} = \frac{\partial \bar{f}_\alpha(\bar{z})}{\partial \bar{z}_\gamma} \frac{\partial \bar{z}_\gamma}{\partial z_\beta} \tag{2.29}$$

from which

$$\frac{\partial \bar{z}_\gamma}{\partial z_\beta} = \frac{\partial \bar{z}_\gamma}{\partial \bar{f}_\alpha(\bar{z})} \frac{\partial f_\alpha(z)}{\partial z_\beta} \tag{2.30}$$

Thus eq. (2.6) becomes

$$\frac{\partial \bar{z}_\alpha}{\partial \bar{f}_\gamma(\bar{z})} \frac{\partial f_\gamma(z)}{\partial z_\delta} K_{\delta\mu} \frac{\partial \bar{z}_\beta}{\partial \bar{f}_\lambda(\bar{z})} \frac{\partial f_\lambda(z)}{\partial z_\mu} = K_{\alpha\beta} \tag{2.31}$$

Multiplying eq. (2.31), on the left by $\partial \bar{f}_\gamma(\bar{z})/\partial \bar{z}_\alpha$ and on the right by $\partial \bar{f}_\lambda(\bar{z})/\partial \bar{z}_\beta$ eq. (2.31) can be cast in the following form

$$\frac{\partial f_\gamma(z)}{\partial z_\delta} K_{\delta\mu} \frac{\partial f_\lambda(z)}{\partial z_\mu} = \frac{\partial \bar{f}_\gamma(\bar{z})}{\partial \bar{z}_\alpha} K_{\alpha\beta} \frac{\partial \bar{f}_\lambda(\bar{z})}{\partial \bar{z}_\beta} \tag{2.32}$$

or equivalently

$$\{f_\gamma, f_\lambda\}_z = \{\bar{f}_\gamma, \bar{f}_\lambda\}_{\bar{z}} \tag{2.33}$$

Thus, if eq. (2.33) is satisfied, a canonical transformation $\bar{z} = \bar{z}(z)$ between the variables z_α and $\bar{z}_\alpha$ can be implemented, that is, the functional relations (2.28) implicitly define a canonical transformation.

In the example considered in (2.23) the relations (2.25) give us a canonical transformation as $\{H,T\}_{x,p} = \{\bar{H},\bar{T}\}_{\bar{x},\bar{p}} = 1$.

3. REPRESENTATION OF LINEAR CANONICAL TRANSFORMATIONS

Now we want to determine the representation in quantum mechanics of a canonical transformation given implicitly by eq. (2.28). We are thus faced with the problem of finding an operator U which relates the barred to the unbarred operators and which takes us from one base in which we can express our system to the other (see eqs. 1.3). In terms of the implicit definition of the canonical transformation given by (2.28), U must satisfy relations analogous to (1.3), that is

$$\bar{f}_\alpha(\bar{x},\bar{p}) = U\,\bar{f}_\alpha(x,p)U^{-1} = f_\alpha(x,p) \tag{3.1}$$

or, by multiplication on the right by U

$$f_\alpha(x,p)U = U\,\bar{f}_\alpha(x,p) \qquad (3.2)$$

Starting from (3.2) we are looking forward to find a set of differential equations which could define completely the transformation operator U. Multiplying (3.2) on the right by the vectot $|\bar{x}'>$ and on the left by $<x'|$ and, introducing the unit operator

$$\int |x''> dx'' <x''| \qquad (3.3)$$

we have

$$\int <x'|\, f_\alpha\,(x,p)|\, x''> dx''<x''|U|\,\bar{x}'> = \int <x'|\, U|x''> dx''<x''|\bar{f}_\alpha(x,p)|\, x'> \qquad (3.4)$$

If the functions f (x,p) can be written unambiguosly as operators in configuration space we have

$$<x'|f_\alpha(x,p)|x''> = f_\alpha(x',-i\,\frac{\partial}{\partial x'})\,\delta(x'-x'') \qquad (3.5)$$

and, by the use of formula (1.3a) the l.h.s. of eq. (3.4) reads

$$\int f_\alpha(x',-i\,\frac{\partial}{\partial x'})\;\delta(x'-x'')dx''<x''|\bar{x}')=f_\alpha(x',\,-i\,\frac{\partial}{\partial x'})<x'|\bar{x}') \qquad (3.6)$$

By the usual matrix calculus we deduce the following relations

$$U_{ij}F_{jk} = (UF)_{ik} = [(UF)^+_{ki}]^* = [(F^+U^+)_{ki}]^* = (F^+_{kj}U^+_{ji})^* \qquad (3.7)$$

and thus the r.h.s. of eq. (3.4) reads

$$\int <x'|U|x''>dx''<x''|\bar{f}_\alpha(x,p)|\bar{x}'> = \left[\int <\bar{x}'|\bar{f}^+_\alpha(x,p)|\; x''>dx''<x''|U^+|x'>\right]^*. \qquad (3.8)$$

Then by using eq. (1.3a), eq. (3.8) becomes

$$\left[\int \bar{f}_{\alpha}^{+}(\bar{x}', -i \frac{\partial}{\partial \bar{x}'}) \delta(\bar{x}'-x'') dx'' (x''|x'> \right]^{*} =$$

$$= [\bar{f}^{+}(\bar{x}', -i \frac{\partial}{\partial \bar{x}'})]^{*} <x'|\bar{x}') \quad (3.9)$$

Thus eq. (3.4) can be written, by the use of eq. (3.6) and eq. (3.9) as:

$$f_{\alpha}(x', -i \frac{\partial}{\partial x'})<x'|\bar{x}') = [\bar{f}_{\alpha}^{+}(\bar{x}', -i \frac{\partial}{\partial \bar{x}'})]^{*} <x'|\bar{x}') ,$$

$$\alpha = 1,\ldots, 2n . \quad (3.10)$$

We have thus found a set of 2n coupled partial differential equations with respect to $\bar{x}_i'$ and x_i' , i = 1,...,n, which allows us to obtain the transformation bracket $<x'|\bar{x}')$ i.e. the unitary representation $<x'|U|\bar{x}'>$ in configuration space of the canonical transformations.

Sometimes the transformation between the barred and unbarred variables is particularly simple, as in the case of the linear transformations, which we will consider below, and thus we can work directly with the variables themselves, but, if that is not the case, as, for example for the transformation given by formula (2.27), then it might prove advantageous to use some function f_{α} of the variables x,p to get a set of partial differential equations which we know how to solve. If, as in the example of formula (2.27), we have a square root of an operator, we do not know how to implement equations (3.10).

We work out first the case of linear canonical transformations in two-dimensional phase space, for which we shall prove eq. (1.8). In this case the canonical transformation is given by (1.6), and we have now the identification

$$f_1 = ax + bp , \quad \bar{f}_1 = \bar{x} \quad (3.11a)$$

$$f_2 = cx + dp , \quad \bar{f}_2 = \bar{p} \quad (3.11b)$$

Then eqs. (3.10) become:

$$(ax' - ib \frac{\partial}{\partial x'})<x'|\bar{x}') = \bar{x}'<x'|\bar{x}') \quad (3.12a)$$

$$(cx' - id \frac{\partial}{\partial x'})<x'|\bar{x}') = i \frac{\partial}{\partial \bar{x}'} <x'|\bar{x}') \quad (3.12b)$$

where, to obtain the r.h.s. of eq. (3.12b) we had to take the Hermitian conjugate and conjugate of $\bar{p}$.

We propose a solution for this set of two differential equations

$$\langle x'|\bar{x}') = G\, e^{i(\lambda x'^2+\mu x'\bar{x}'+\nu\bar{x}'^2)} \tag{3.13}$$

which is an exponential function of the most general quadratic form in x' and $\bar{x}'$. On applying (3.12) to (3.13) we get the two algebraic equations

$$ax' + 2b\lambda x' + b\mu\bar{x}' = \bar{x}' \tag{3.14a}$$

$$cx' + 2d\lambda x' + d\mu\bar{x}' = -\mu x' - 2\nu\bar{x}' \tag{3.14b}$$

whose solution is obtained by requiring that eqs. (3.14) be satisfied for any choice of the values of x' and $\bar{x}'$. So, from eq. (3.14a) we get:

$$b\mu= 1, \quad a + 2b\lambda= 0, \quad \mu = \frac{1}{b}, \quad \lambda = -\frac{a}{2b} \tag{3.15a}$$

and from (3.14b)

$$c + 2d\lambda+ \mu = 0, \quad d\mu + 2\nu = 0, \quad \nu = -\frac{d\mu}{2}, \quad \mu = -c -2d\lambda \tag{3.15b}$$

If we substitute the value of μ and λ obtained in (3.15a) into (3.15b), we get a consistent set of values of λ,μ,ν only when ad-bc=1 (i.e. when the transformation is canonical) and then

$$\lambda = -\frac{a}{2b}, \quad \mu = \frac{1}{b}, \quad \nu = -\frac{d}{2b} \tag{3.16}$$

The coefficient G is obtained by requiring that the transformation bracket be normalized

$$\int_{-\infty}^{+\infty} (\bar{x}'|x'\rangle dx'\langle x'|\bar{x}'') = \delta\,(\bar{x}' - \bar{x}'') \tag{3.17}$$

and thus the solution of eqs. (3.12) is, up to a phase,

$$\langle x'|\bar{x}') = \langle x'|U|\bar{x}'\rangle = e^{-\frac{i}{2b}(ax'^2-2x'\bar{x}'+d\bar{x}'^2)}/\sqrt{2\pi|b|} \tag{3.18}$$

It is easy to see that (3.18) form a representation of Sp(2,R) or, equivalently, of SL(2,R) in configuration space as, up to a phase factor ϕ, we have[4,5]

$$\int \langle x'|U\begin{pmatrix} a & b \\ c & d \end{pmatrix}|\bar{x}'\rangle d\bar{x}' \langle \bar{x}'|U\begin{pmatrix} \bar{a} & \bar{b} \\ \bar{c} & \bar{d} \end{pmatrix}|x''\rangle =$$

$$e^{i\phi}\langle x'|U\begin{pmatrix} \alpha & \beta \\ \gamma & \delta \end{pmatrix}|x''\rangle \qquad (3.19)$$

where

$$\begin{pmatrix} \alpha & \beta \\ \gamma & \delta \end{pmatrix} = \begin{pmatrix} a & b \\ c & d \end{pmatrix}\begin{pmatrix} \bar{a} & \bar{b} \\ \bar{c} & \bar{d} \end{pmatrix} \qquad (3.20)$$

Formula (3.18) is valid only for $b \neq 0$; we therefore look for a formula which gives the transformation bracket $\langle x'|\bar{x}')$ in the case $b = 0$. The simplest way of doing this is by passage to the limit in eq. (3.18) as $b = \varepsilon$ goes to zero. Taking the transformation matrix

$$\begin{bmatrix} a+\delta & \varepsilon \\ c & a^{-1}+\delta \end{bmatrix} \qquad \varepsilon, \delta > 0 \qquad (3.21)$$

where ε [illegible] are small quantities, condition (1.7a) implies that

$$\delta = \varepsilon c \left[a + a^{-1}\right]^{-1} \qquad (3.22)$$

Thus, in the limit $\varepsilon \to 0$, the transformation matrix (3.21) becomes $\begin{bmatrix} a & 0 \\ c & a^{-1} \end{bmatrix}$ and its unitary representation is given by:

$$\langle x'|\bar{x}') = \exp\left\{-\frac{i}{2}\frac{c(x'^2+\bar{x}'^2)}{(a+a^{-1})}\right\}$$

$$\lim_{\varepsilon\to 0}\left[2\pi\varepsilon\right]^{-1/2}\exp\left\{-\frac{i\pi}{4}+\frac{i}{2\varepsilon}\left(a^{-1/2}x' - a^{1/2}\bar{x}'\right)\right\} =$$

$$= \frac{1}{\sqrt{a}}\exp\left\{-\frac{ic}{2a}x'^2\right\}\delta\left(\frac{x'}{a}-\bar{x}'\right) \qquad (3.23)$$

where the phase factor mentioned in (3.18) is taken as $e^{-i\pi/4}$. There are two interesting cases to be considered if we set $a=a^{-1}=1$, then the representation bracket becomes

$$\langle x'|U\begin{pmatrix} 1 & 0 \\ c & 1 \end{pmatrix}|\bar{x}'\rangle = \exp\left(-\frac{ic}{2}x'^2\right)\delta(x'-\bar{x}') \qquad (3.24)$$

which, for $c = 0$, corresponding to the identity transformation matrix, gives:

$$\langle x'|U\begin{pmatrix} 1 & 0 \\ 0 & 1 \end{pmatrix}|\bar{x}'\rangle = \delta(x'-\bar{x}') \qquad (3.25)$$

We notice that, for a complex extension of the canonical transformation, corresponding to c imaginary, the expression (3.24) would be the matrix representation of the Gaussian interaction[6]:

$$\langle x'|U\begin{pmatrix}1 & 0\\ -i\gamma & 1\end{pmatrix}|\bar{x}'\rangle = e^{-\gamma x'^2/2}\,\delta(x'-\bar{x}') \qquad (3.26)$$

It is easy to generalize the results thus far obtained to the 2n-dimensional phase space, where the transformation is given by formula (2.7) and where the transformation matrix (2.9) must satisfy conditions (2.10) if the transformation is to be canonical. In this case the unitary representation of the Sp(2n,R) group is[4]:

$$\langle \underset{\sim}{x}'|U\begin{pmatrix}A & B\\ C & D\end{pmatrix}|\underset{\sim}{x}''\rangle = \left[(2\pi)^n|\det B|\right]^{-1/2}\cdot$$

$$\exp\left\{-\frac{i}{2}\left(\tilde{\underset{\sim}{x}}'B^{-1}A\underset{\sim}{x}' - 2\tilde{\underset{\sim}{x}}'B^{-1}\underset{\sim}{x}'' + \tilde{\underset{\sim}{x}}''DB^{-1}\underset{\sim}{x}''\right)\right\} \qquad (3.27)$$

where by $\tilde{\underset{\sim}{x}}$ we denote the transpose of a vector $\underset{\sim}{x}$.

We notice that this is a representation of [illegible],R) and is equivalent to a representation of SL(2n,R) only in the case of n=1.

4. REPRESENTATIONS OF NON-LINEAR BUT BIJECTIVE CANONICAL TRANSFORMATIONS

In the previous section we showed how to obtain the representations of canonical transformations when they are written implicitly in terms of the equations (2.33) i.e.

$$f_\alpha(x,p) = \bar{f}_\alpha(\bar{x},\bar{p}); \quad \alpha = 1,2,\dots 2n \qquad (4.1)$$

and the f_α, $\bar{f}_\alpha$ can be expressed unambiguously as quantum mechanical operators. This procedure was applied to linear canonical transformations but it also can be used for some non-linear ones as, for example, (2.23) where we write

$$f_1(x,p) = H = \frac{1}{2}p^2 + \frac{\lambda^2}{2x^2} \quad ; \quad \bar{f}_1(\bar{x},\bar{p}) = \bar{H} = \frac{1}{2}\bar{p}^2 + \frac{\bar{\lambda}^2}{2\bar{x}^2}$$

$$f_2(x,p) = 2HT = \frac{1}{2}(xp+px) \qquad \bar{f}_2(\bar{x},\bar{p}) = 2\bar{H}\bar{T} = \frac{1}{2}(\bar{x}\bar{p}+\bar{p}\bar{x}) \qquad (4.2)$$

The unitary representation $\langle x'|\bar{x}') = \langle x'|U|\bar{x}'\rangle$ for (4.2) was obtained explicitly in reference[7].

More generally, if we are interested in the representation of the canonical transformation relating two arbitrary Hamiltonians, we can obtain it from the product of two representations. The first one takes uf from the coordinates x,p (where we restrict ourselves to problems in one degree of freedom) to the $\bar{x},\bar{p}$ associated with the initial Hamiltonian H and its canonically conjugate variable i.e. the time T of (2.24), so we can write

$$\bar{x} = H(x,p) = (2m)^{-1}\ p^2 + V\ (x), \tag{4.3a}$$

$$\bar{p} = -T(x,p) = -\int_a^x (m/2)^{1/2}\left[H(x,p)-V(y)\right]^{-1/2}dy, \tag{4.3b}$$

where in this section we shall use normal units so that m is the mass of the particle. The second representation is the inverse of the first but now associated with the final Hamiltonian.

It is thus quite interesting to consider the representation of canonical transformations of the type (4.3). This is, in general, a difficult and subtle problem[8-10] and we shall only consider in this section the case when the potential V(x) is a monotonically decreasing function of x with the properties

$$V\ (x) = \begin{cases} +\infty & \text{for } x = -\infty \\ -\infty & \text{for } x = +\infty \end{cases} \tag{4.4}$$

As an example we have linear potential in which $V(x) = -F_o x$ where F_o is a positive constant with the dimension of force. In this case we get from (4.3)

$$\bar{x} = (2m)^{-1}\ p^2 - F_o x \qquad\qquad x = (F_o/2m)\ \bar{p}^2 + F_o^{-1}\bar{x}$$

or

$$\bar{p} = -\ F_o^{-1}\ p \qquad\qquad p = -\ F_o\bar{p} \tag{4.5}$$

and thus the canonical transformation is clearly bijective as is also true for the general case[10]. In what follows we shall consider that the "a" appearing in the expression (4.3b) for the time T is the turning point of the potential i.e.

$$a = x_-\ (H), \qquad V\left[x_-\ (H)\right] = H. \tag{4.6}$$

We wish now to discuss the representation of the canonical transformation (4.3) with the help of the equations (3.4), (3.8) i.e.

$$\int_{-\infty}^{\infty} <x'|H|x''>dx''<x''|U|\bar{x}'>=$$

$$\left(- \frac{h^2}{2m} \frac{\partial^2}{\partial x'^2} + V\,(x')\right)<x'|U|\bar{x}'> = \bar{x}'<x'|U|\bar{x}'> \qquad (4.7)$$

$$\int_{-\infty}^{\infty} <x'|T|x''>dx''<x''|U|\bar{x}'> = - \frac{h}{i}\frac{\partial}{\partial \bar{x}'} <x'|U|\bar{x}'> \ . \qquad (4.8)$$

The expression (4.7) is a standard differential equation and if we denote by $\psi_\nu\,(x')$ the solution of

$$\left[- \frac{h^2}{2m} \frac{\partial^2}{\partial x'^2} + V\,(x')\right]\psi_\nu(x') = \nu\psi_\nu(x'), \quad -\infty<\nu<\infty \qquad (4.9)$$

where ψ_ν is normalized in the sense of $\delta(\nu'-\nu)$ as its spectrum is continuous in the range indicated, we immediately conclude that

$$<x'|U|\bar{x}'>= \psi_{\bar{x}'}(x')\ \exp[\,i\ \phi(\bar{x}')\,] \qquad (4.10)$$

where the phase $\phi(\bar{x}')$ is, so far, an arbitrary real function of $\bar{x}'$ which we expect to determine[7-10] with the help of eq. (4.8). The phase of $\psi_{\bar{x}'}(x')$ will be fixed by the requirement that for values of x' far to the right of the turning point $x_-\,(\bar{x}')$, $\psi_{\bar{x}'}\,(x')$ takes the form of the WKB approximation[11])

$$\psi_{\bar{x}'}(x') \simeq (2m)^{1/2}(\pi\hbar)^{-1/2}\ \{2m[\,\bar{x}'- V(x')\,]\}^{-1/4}$$

$$\sin\left[\hbar^{-1}\int_{x_-(\bar{x}')}^{x'}\{2m\,[\bar{x}'- V(y)]\}^{1/2}dy+ \frac{\pi}{4}\right] \qquad (4.11)$$

Equation (4.8) for determining $\phi(\bar{x}')$ has a number of problems. To begin with we may inquire about the matrix form $<x'|T|x''>$ of the operator associated with the T of (4.3b). In the case of the linear potential (4.5) this operator can be written down immediately as

$$<x'|T|x''> = - F_o^{-1}\ (\hbar/i)\ (\partial/\partial x')\ \delta(x'-x'') \qquad (4.12)$$

and thus -as discussed in reference 9- the phase vanishes i.e. $\phi(\bar{x}') = 0$.

For the general problem it is clearly impossible to have a well defined and unambiguous procedure to determine $<x'|T|x''>$but, as we proceed to whow, this is not really necessary. For U to be a representation it is enough that (4.8) be satisfied when we are close to the classical limit i.e. when we replace in (4.10) the $\psi_{\bar{x}'}(x')$ of (4.9) by it WKB approximation (4.11). Furthermore we do not need the equation (4.8) to come from $\bar{p}=-T(x,p)$ but we can equally well derive it from $F(\bar{p})=F[-T(x,p)]$ where F is some appropriate function. This point is important because in references 8-10 we found out that to satisfy equations (4.7), (4.8) in a direct fashion it was necessary to consider functions of x,p invariant under the symmetries of the problem. In the Hamiltonian whose potential has the form (4.4), the trajectories are invariant under time reflection and this suggests that, in classical mechanics, rather than the relation $\bar{p}=-T(x,p)$ we consider the squared one

$$\bar{p}^2 = T^2(x,p) \tag{4.13}$$

in which both sides are invariant under time reversal. The corresponding euation, which could also be derived by applying $(h/i)\partial/\partial\bar{x}'$ to both sides of (4.8), becomes then

$$\int \langle x'|T^2|x''\rangle\, dx''\langle x''|U|\bar{x}'\rangle = -\hbar^2 \frac{\partial^2}{\partial\bar{x}'^2}\langle x'|U|\bar{x}'\rangle \tag{4.14}$$

We now apply $-h^2\partial^2/\partial\bar{x}'^2$ to $\psi_{\bar{x}'}(x')$, considering its WKB form (4.11) and immediately see, disregarding terms containing positive powers of $\hbar$, that we obtain

$$-\hbar^2 \frac{\partial^2}{\partial\bar{x}'^2}\psi_{\bar{x}'}(x') \simeq$$

$$\int_{x_-(\bar{x}')}^{x'} (m/2)^{1/2}\,[\bar{x}'-V(y)]^{-1/2}dy\;\; \psi_{\bar{x}'}(x') \tag{4.15}$$

which would be precisely the effect that we would expect in the classical limit for the quantum mechanical operator associated with $T^2(x,p)$ of (4.3b). Thus, we can consider that in (4.10) the phase $\phi(\bar{x}')$ vanishes if (4.8) is satisfied and so $\langle x'|U|\bar{x}'\rangle = \psi_{\bar{x}'}(x')$ is the unitary representation of the canonical transformation (4.3) when "a" is selected as in (4.6)

Inversely, once $\phi(\bar{x}')=0$ is established from the semiclassical limit, $\langle x'|T|x''\rangle$ is completely defined and we can use (1.3b), (4.8) and (4.10) to write

$$\langle x'|T|x''\rangle = -\int \langle x'|U|\bar{x}'\rangle d\bar{x}'\langle\bar{x}'|\bar{p}|\bar{x}''\rangle\, d\bar{x}''\langle\bar{x}''|U^{+}|x''\rangle$$

$$= \left[\int \psi^{*}_{\bar{x}'}(x')\left[\frac{\hbar}{i}\frac{\partial}{\partial\bar{x}'}\psi_{\bar{x}'}(x'')\right]d\bar{x}'\right]^{*} \tag{4.16}$$

thus having a definite matrix element for the operator associated with the classical observable $T(x,p)$ of (4.3b)

In the canonical transformation (4.3) to Hamiltonian and time, the potential (4.4) implied that the motion was unbounded in the direction $x \to +\infty$. If the potential had a minimum, then there is a region of energies in which the motion is bounded and thus periodic. The canonical transformation is then nonbijective[8-10] and for its representation we require first to discuss the structure of the classical problem, which we proceed to do in the next section.

5. THE STRUCTURE OF PHASE SPACE

The aim of this section is to analyze the structure of phase space as a carrier of a canonical transformation in analogy with the way that the complex plane, or more correctly the complex manifold, is analyzed as the carrier of conformal transformations. In the latter case when the transformation is non-bijective as, for example, when $\bar{z} = z^\kappa$ with κ integer, we are led to the concept of a κ sheeted Riemann surface[12] for the $\bar{z}$ plane. An analogous concept can be introduced for the phase plane, i.e. the phase space for a problem with one degree of freedom[13]

It turns out thought that, even for conformal transformations, there is an alternative to the Riemann sufface analysis. This alternative uses the concepts of ambiguity group and ambiguity spin[13] and with their help we can carry out the conformal transformations keeping the picture of single sheeted complex planes for both z and $\bar{z}$.

We shall review this alternative procedure for the conformal transformation $\bar{z}=z^k$ and then show that a similar method can be used for canonical transformations relating oscillator Hamiltonians, the frequency of one of which is k times the frequency of the other. From these examples it will be possible to see how we can implement the program for a wide class of non-bijective conformal or canonical transformations, with the help of the fundamental concepts of ambiguity group and ambiguity spin.

a) The conformal transformation $\bar{z} = z^\kappa$

If $\bar{z} = z^\kappa$, κ integer, it is clear that the transformation is non-bijective as there are κ points z mapped on a single point $\bar{z}$, where the former are connected by the operations of the cyclic group C_κ i.e.

$$C_\kappa : z \to z \exp(i\, 2\pi\, r/\kappa),\ r=0,1,\ldots,\kappa-1 \quad (5.1)$$

We shall denote groups of this type by the name of ambiguity groups as they take into account the multiplicity of the points in one complex plane that are mapped on a single one in the other.

We can now ask about the unambiguous mapping of entire functions of z onto functions of $\bar{z}$. For this purpose it is convenient to note that entire functions of z (the only ones of interest to us here) can be interpreted in two ways: As basis functions, in which case they will be denoted by greek letters $\psi(z), \chi(z)$, and as operators acting by multiplication on the basis functions in which case they will be noted by latin letters f(z), g(z). Thus when f(z) acts on $\psi(z)$ by multiplication we get the new basis function $\chi(z)$ i.e.

$$\chi(z) = f(z)\, \psi(z) \quad (5.2)$$

The basis functions

$$\psi(z) = \sum_{\nu}^{\infty} \alpha_\nu z^\nu \quad (5.3)$$

can be decomposed into their irreducible parts $\psi^\lambda(z)$, $\lambda = 0,1,\ldots k-1$ with respect to the ambiguity group C_κ of (5.1) i.e.

$$\psi(z) = \sum_{r=0}^{\kappa-1} \psi^\lambda(z) \quad (5.4a)$$

$$\psi^\lambda(z) = \kappa^{-1} \sum_{r=0}^{\kappa-1} \exp(i2\pi\lambda r/\kappa)\, \psi[z\exp(-i2\pi r/\kappa)] \quad (5.4b)$$

where $\exp(i\, 2\pi\lambda\, r/\kappa)$, $r = 0, 1, \ldots \kappa-1$ are the irreducible representations, characterized by $\lambda = 0, 1, \ldots \kappa-1$, of the operations in the abelina cyclic group C_κ.

Clearly then if we write $\nu = n\kappa + \lambda$, $\lambda = 0, 1, \ldots \kappa-1$; $n = 0, 1, 2, \ldots$, the $\psi^\lambda(z)$ become

$$\psi^\lambda(z) = z^\lambda \Phi^\lambda(z^\kappa), \quad (5.5a)$$

$$\Phi^\lambda(z^\kappa) = \sum_{n=0}^{\infty} \alpha_{n\kappa+\lambda} (z^\kappa)^n. \quad (5.5b)$$

and thus to the basis function (z) we can associate in the $\bar{z}$ plane a κ dimensional vector of <u>entire</u> functions in $\bar{z}$, i.e.

$$\psi(z) \rightarrow \begin{bmatrix} \Phi^0(\bar{z}) \\ \Phi^1(\bar{z}) \\ \cdots\cdots \\ \Phi^{\kappa-1}(\bar{z}) \end{bmatrix} \qquad (5.6)$$

This vector has an index $\lambda = 0, 1, \ldots \kappa-1$ which we can denote as ambiguity spin as it represents the alternative to introducing in $\bar{z}$ a κ -sheeted Riemann surface.

If the basis functions $\phi(z)$ in the z plane transform into the vectors (5.6) it is clear that the operators f(z) that act by multiplication will transform into matrices. To obtain them is enough to find the matrix representation of z as any f(z) is a linear combination of powers of z. We can then write

$$\chi(z) = z\phi(z) = \bar{z}^{1/\kappa} \sum_{\lambda=0}^{\kappa-1} \bar{z}^{\lambda/\kappa} \Phi^{\lambda}(\bar{z})$$

$$= \sum_{\lambda=0}^{\kappa-2} \bar{z}^{\frac{\lambda+1}{\kappa}} \Phi^{\lambda}(\bar{z}) + \bar{z}\Phi^{\kappa-1}(\bar{z})$$

$$= \bar{z}\,\Phi^{\kappa-1}(\bar{z}) + \sum_{\lambda=1}^{\kappa-1} \bar{z}^{\lambda/\kappa} \Phi^{\lambda-1}(\bar{z}) = \sum_{\lambda=0}^{\kappa-1} \bar{z}^{\lambda/\kappa} X^{\lambda}(\bar{z})$$

where $\{ X^{\lambda}(\bar{z}), \lambda = 0,1,\ldots \kappa-1 \}$ is the vector in the $\bar{z}$ plane associated with $\chi(z)$.

From (5.7) we have the correspondance

$$z \quad \leftrightarrow \quad \begin{bmatrix} 0 & 0 & \cdots\cdots & 0 & \bar{z} \\ 1 & 0 & \cdots\cdots & 0 & 0 \\ \cdots & \cdots & \cdots\cdots & \cdots & \cdots \\ 0 & 0 & \cdots\cdots & 1 & 0 \end{bmatrix} \qquad (5.8)$$

and thus the matrix representation in the $\bar{z}$ plane of the operator z.

Clearly the matrices obey the same algebra as the original functions of z they come from and, in particular, we notice that to z^{κ} corresponds $\bar{z}$ multiplied by the $\kappa\times\kappa$ unit matrix.

b) Canonical transformations relating oscillator Hamiltonians of different frequencies

Taking units in which the mass of the particle, and frequency of one of the oscillators are 1, we have the two Hamiltonians

$$\frac{1}{2}\,(\bar{p}^2 + \bar{x}^2)\ ,\quad \frac{1}{2}\,(p^2 + \kappa^{-2}\ x^2) \tag{5.9}$$

where we assume that κ is integer. The frequencies of the oscillators in $(\bar{x},\bar{p})$, (x, p) are then related by the factor κ . We can carry the point transformation $x \rightarrow \kappa x$ $p \rightarrow p/\kappa$ and thus our canonical transformation can be defined by the implicti equations

$$\frac{1}{2}\,(\bar{p}^2 + \bar{x}^2) \;=\; \frac{1}{2\kappa}\,(p^2 + x^2), \tag{5.10a}$$

$$\text{angtan}\;(\bar{p}/\bar{x}) = \kappa\ \ \text{angtan}\;(p/x) \tag{5.10b}$$

Introducing the observables η , ξ by the definition

$$\eta = (1/\sqrt{2})\;(x-ip),\ \xi = (1/\sqrt{2})\;(x+ ip) \tag{5.11a,b}$$

and a similar expression for $\bar{\eta}$, $\bar{\xi}$, we see that (5.10) implies

$$\bar{\eta} = \frac{\eta^{\kappa}}{\kappa^{1/2}\ (\eta\xi)^{(\kappa-1)/2}},\quad \bar{\xi} = \frac{\xi^{\kappa}}{\kappa^{1/2}\ (\eta\xi)^{(\kappa-1)/2}} \tag{5.12a,b}$$

which are alternative implicit equations defining the canonical transformation. From them we conclude that (5.10a) can also be written as

$$\eta\xi \;=\; \kappa\bar{\eta}\ \bar{\xi} \tag{5.13}$$

We now notice from (5.12), (5.13) that the transformation with the C_κ group of operations

$$C_\kappa\ :\ \ \eta \rightarrow \eta\,\exp\,(i2\Pi r/\kappa),\ \xi \rightarrow \xi\,\exp(-i2\Pi r/\kappa),$$

$$r = \ 0,\ 1,\ \ldots\ \kappa-1 \tag{5.14}$$

does not affect the values of $\bar{\eta},\bar{\xi}$. Thus we can speak of the C_κ defined by the operations (5.14) as the ambiguity group of the canonical transformation (5.10) or, equivalently, (5.12).

We now wish to investigate how can we transform entire functions of x, p or, equivalently, of η, ξ into functions of $\bar{x}$, $\bar{p}$ or $\bar{\eta}$, $\bar{\xi}$. Denoting them as $\psi(\eta,\xi)$ we have that

$$\psi(\eta,\xi) = \sum_{\nu,\nu'=0}^{\infty} \alpha_{\nu\nu'}\eta^{\nu}\xi^{\nu'}$$

$$= \sum_{\Lambda,\Lambda'=0}^{\kappa-1} \left(\frac{\eta}{\xi}\right)^{(\Lambda-\Lambda')/2} (\eta\xi)^{(\Lambda+\Lambda')/2} \left[\sum_{n,n'=0}^{\infty} \alpha_{n\kappa+\Lambda,n'\kappa+\Lambda'} (\eta^{\kappa})^{n} (\xi^{\kappa})^{n'}\right] = \sum_{\lambda=0}^{\kappa-1} \left[\left(\frac{\eta}{\xi}\right)^{1/2}\right]^{\lambda} \Phi^{\lambda}(\bar{\eta},\bar{\xi}) \tag{5.15}$$

The r.h.s. of (5.15) follows if we replace $\Lambda-\Lambda'$ by λ when $\Lambda-\Lambda'>0$ or by $\lambda-\kappa$ if $\Lambda-\Lambda'<0$. It is clear then that the remaining function Φ^{λ} is invariant under the group C_{κ} and thus can be written unambiguously as a function of $\bar{\eta}$, $\bar{\xi}$.

The observable $\eta\xi$ is invariant under C_{κ} and thus we expect, by an analysis similar to the one at the end of subsection a) that it corresponds to $\kappa\bar{\eta}\bar{\xi}$ multiplied by the $\kappa\times\kappa$ unit matrix. We postulate a similar relation for any function of $\eta\xi$ and thus, for example, we have

$$(\eta\xi)^{1/2} \leftrightarrow (\kappa\bar{\eta}\bar{\xi})^{1/2} \begin{bmatrix} 1 & 0 & 0 & \cdots & 0 \\ 0 & 1 & 0 & \cdots & 0 \\ 0 & 0 & 1 & \cdots & 0 \\ \cdots & \cdots & \cdots & \cdots & \cdots \\ 0 & 0 & 0 & \cdots & 1 \end{bmatrix} \tag{5.16}$$

On the other hand for $(\eta/\xi)^{1/2}$ we can consider the function $(\eta/\xi)^{1/2}\psi(\eta,\xi)$ and from (5.8), (5.15) we immediately conclude that

$$(\eta / \xi)^{1/2} \leftrightarrow \begin{bmatrix} 0 & 0 & \dots\dots\dots & 0 & (\bar{\eta} / \bar{\xi})^{1/2} \\ 1 & 0 & \dots\dots\dots & 0 & 0 \\ 0 & 1 & \dots\dots\dots & 0 & 0 \\ & & \dots\dots\dots & & \\ 0 & 0 & \dots\dots\dots & 1 & 0 \end{bmatrix} \qquad (5.17)$$

Thus by multiplication of (5.16) by (5.17) or its inverse we have the correspondance

$$\eta \leftrightarrow \begin{bmatrix} 0 & 0 & \dots\dots\dots & 0 & \kappa^{1/2}\,\bar{\eta} \\ (\kappa\bar{\eta}\bar{\xi})^{1/2} & 0 & \dots\dots\dots & 0 & 0 \\ 0 & (\kappa\bar{\eta}\bar{\xi})^{1/2} & \dots\dots & 0 & 0 \\ \dots & \dots & \dots\dots\dots & \dots & \\ 0 & 0 & \dots\dots & (\kappa\bar{\eta}\bar{\xi})^{1/2} & 0 \end{bmatrix}$$

$$\xi \leftrightarrow \begin{bmatrix} 0 & (\kappa\bar{\eta}\bar{\xi})1/2 & \dots\dots\dots & 0 \\ \dots & \dots & \dots\dots\dots & \dots \\ 0 & 0 & \dots\dots\dots & (\kappa\bar{\eta}\bar{\xi})^{1/2} \\ \kappa^{1/2}\xi & 0 & \dots\dots\dots & 0 \end{bmatrix} \qquad (5.18a,b)$$

We note that the r.h.s. of (5.18a,b) are hermitian conjugate to one another as we would like to have in view of the fact that in the (x,p) phase space $\xi^{*}=\eta$. Furthermore we obtain from (5.16), (5.18) the correspondances

$$\kappa^{-1/2}\,(\eta\xi)^{\frac{1-\kappa}{2}}\,\eta^{\kappa} \leftrightarrow \bar{\eta}\ I\ ,\quad \kappa^{-1/2}\,(\eta\xi)^{\frac{1-\kappa}{2}}\,\xi^{\kappa} \leftrightarrow \bar{\xi}\ I \qquad (5.19)$$

(where I is the $\kappa\times\kappa$ unit matrix) which reflects the defining relations (5.12).

As shown in reference[13] the quantum mechanical correspondances require matrix representations in the $\bar{\eta},\bar{\xi}$ creation and annihilation operators of a form similar to those appearing in (5.18). The ambiguity spin plays then also a fundamental role in the representation in quantum mechanics of non-bijective classical canonical transformations.

c) Correspondances between functions of x,p and $\bar{x},\bar{p}$ for general non-bijective canonical transformations.

We have illustrated the procedure to be followed in the specific canonical transformation (5.10) for the mapping of functions of x, p into matrix functions of $\bar{x},\bar{p}$. For the general problem it is clear that the first requirement is to find the ambiguity group of the problem i.e. the groups of transformations that connect all points x,p mapped on a single point $\bar{x},\bar{p}$ and viceversa. The second step is to find the irreducible representations λ of this group and decompose the functions $\psi(x,p)$ into their irreducible parts so that $\psi(x,p)$ can be correlated with a vector function in the $\bar{x},\bar{p}$ observables of components $\Phi^{\lambda}(\bar{x},\bar{p})$, as in (5.15). Note that in the problem discussed in subsection b), $\lambda = 0,1, \ldots \kappa-1$ takes a finite number of values. For other groups, like for example the translation one[8], λ can take a continuous range of variables in the interval $0 \leq \lambda < 1$. Thus the matrices in $\bar{x},\bar{p}$ associated with x and p may also have continuous indices or a mixture of continuous and discrete.

The essential features of the correspondance are already clear from the example discussed in subsection b). In the next section we shall consider canonical transformation to action and angle variables which are non-bijective but still have simple ambiguity groups. To obtain their representation in quantum mechanics we shall follow much of the analysis given in this and in the previous sections.

6. CANONICAL TRANSFORMATIONS TO ACTION AND ANGLE VARIABLES AND THEIR REPRESENTATION IN QUANTUM MECHANICS

If we have a Hamiltonian whose potential V(x) decreases monotonically, from $+\infty$ at $x = -\infty$, to a single minimum 0 at $x=0$ and then increases monotonically to $+\infty$ when $x = +\infty$, the motion of the particle for any energy will be bounded and periodic.

It is then possible to define the action for this type of Hamiltonian for all values of H as

$$J(H) = (2\pi)^{-1} \oint p\,dx$$

$$= \pi^{-1} \int_{x_-(H)}^{x_+(H)} \{ 2 [H-V(y)] \}^{1/2} dy \qquad (6.1)$$

where again we use dimensionaless units and $x_\pm(H)$ are the right and left turning points for which $V[x_\pm(H)]=H$. From the properties of the potential described above J is a monotonically increasing function of H starting with J=0 for H=0 and ending with J=∞ when H=∞ .

The angle ϕ canonically conjugate to J will satisfy the Poisson bracket relation

$$\{ J,\phi\} = \left(- \frac{\partial J}{\partial H}\frac{\partial \phi}{\partial T} + \frac{\partial J}{\partial T}\frac{\partial \phi}{\partial H}\right) \{ H, -T \} = 1 \qquad (6.2)$$

where H,T are given in (4.3). But as the last Poisson bracket is 1 and J is a function of H only, we obtain

$$\phi(x,p) = - T(x,p) \left(\frac{dH}{dJ}\right) \qquad (6.3)$$

where, from the above discussion, H(J) is a monotonically increasing function of J.

Now, we propose the canonical transformation

$$|\bar{x}| = J[H(x,p)] \qquad (6.4a)$$

$$\bar{x}\,\bar{p}/|\bar{x}| = \phi(x,p) \qquad (6.4b)$$

where we take absolute value $|\bar{x}|$ as J is always positive and the canonical conjugate variable[8] to $|\bar{x}|$ is not $\bar{p}$ but $(\bar{x}/|\bar{x}|)\,\bar{p}$.

We proceed to analyze some of the properties of the canonical transformation (4.4). The periodic nature of the motion[14] indicates that any observable f(x,p), considered as a function of J, ϕ can be expanded in the Fourier series

$$f(x,p) = \sum_{s=-\infty}^{\infty} b_s(J) \exp(is\phi) \qquad (6.5)$$

Therefore, in particular, both x and p are amenable to this type of expansion and from (6.4) we conclude that they are functions of $|\bar{x}|$, $\exp(i\,\bar{x}\bar{p}/|\bar{x}|)$ only. As the latter are invariant under the transformation

$$\bar{x} \to \bar{x}, \quad \bar{p} \to \bar{p} + 2m\pi, \quad m \text{ integer};$$

$$\bar{x} \to -\bar{x}, \quad \bar{p} \to -\bar{p} \tag{6.6}$$

The mapping between phase spaces x,p and $\bar{x},\bar{p}$ is nonbijective[8-10] as is illustrated in Fig. 1. From (6.5) and (6.6) it is clear that the strip $0 \leq \bar{x} \leq \infty$, $0 \leq \bar{p} \leq 2\pi$, in the new phase space $(\bar{x},\bar{p})$ is mapped on the full original (x,p) plane with dotted and dashed lines in both corresponding to each other. The heavy line $\bar{p}=0$ bordering the strip in the $(\bar{x},\bar{p})$ plane from below corresponds to the line p=0 in (x,p) phase space. This is due to the fact that from (4.3b), (4.6) $T(x,0)=0$ as $x_-(H)=V^{-1}[(p^2/2m) + V(x)] = x$ for $p = 0$ and thus $\phi(x,0)$ of (6.3) also vanishes. Thus we can make a cut, marked by a heavy line, along the x axis from x = 0 to ∞ along which we enter new sheets in the (x,p) phase space when we cross the lines $\bar{p}=0$ or 2π in the $(\bar{x},\bar{p})$ plane. The right and left half planes in $(\bar{x},\bar{p})$ correspond to two independent denumerable infinite of sheets in the (x,p) phase space which are connected at the point (x,p) = (0,0), marked with a dot in the figure, that corresponds to the line $\bar{x}=0$ in the $(\bar{x},\bar{p})$ plane. Thus, by transforming the (x,p) plane into a kind of Riemann surface, we can recover the bijectiveness in the correspondance of the points (x,p) with those of $(\bar{x},\bar{p})$

We showed though in the previous section another way of recovering bijectiveness which is more useful when passing to the representation in quantum mechanics of canonical transformations of the type (6.4). We note that (6.6) defines a group of canonical transformations in the new phase space that connects all points $(\bar{x},\bar{p})$ that are mapped on a single point (x,p) in the original plane. This is the ambiguity group mentioned in previous section and it is clear from (6.6) that it is $T \wedge I$ the semidirect product[8] of translations T_m by $2m\pi$ in the momentum variable $\bar{p}$ and inversions I i.e. the operation $(\bar{x},\bar{p}) \to (-\bar{x},-\bar{p})$. Note incidentally that now the ambiguity group (6.6) appears in the new $(\bar{x},\bar{p})$ phase space and thus the Riemann surface is in the old one (x,p). This is opposite to what we had in the example discussed in the previous section. Thus here we expect the ambiguity spin to be associated with functions in the old (x,p) rather than the new one as happened in (5.6) and (5.15).

As indicated in the previous section we need to find the irreducible representations (irreps) of this ambiguity group. To achieve this we note that if we have a variable $-\infty < y < \infty$, the plane waves

$$(2\pi)^{-1/2} \exp(i\nu y), \quad -\infty < \nu < \infty \tag{6.7}$$

are basis of irreducible representations of the translation group $y \to y+a$ where a is an arbitrary real constant. The irreps themselves are the $\exp(i\nu a)$ and they are characterized by ν in the interval indicated in (6.7). If we have though instead of $y \to y+a$, the operations (6.6) i.e.

$$T_m : \quad y \to y + 2m\pi \; ; \quad I : \quad y \to -y, \qquad (6.8)$$

then it is convenient to write the parameter in the form

$$\nu = \sigma(n+\lambda), \; \sigma = \pm 1, \; n=0,1,2,\ldots; \quad 0 \leq \lambda < I \qquad (6.9)$$

and consider instead of (6.7) the two component basis

$$(2\pi)^{-1/2} \begin{bmatrix} \exp\{i\ (n+\lambda\ )\ y\} \\ \exp\{-1\ (n+\ \lambda)\ y\} \end{bmatrix} \qquad (6.10)$$

Clearly then to operations (6.8) there correspond the matrices

$$T_m \to \begin{bmatrix} \exp\ (i2\pi\lambda m) & 0 \\ 0 & \exp\ (-i2\pi\lambda m) \end{bmatrix}, \quad m=0, \pm 1, \pm 2\ldots, \quad I \to \begin{bmatrix} 0 & 1 \\ 1 & 0 \end{bmatrix} \qquad (6.11)$$

which are the irreps[8)] of the $T \wedge I$ group. Thus the irreps are characterized by the real number λ in the interval $0 \leq \lambda < 1$ and, as they are two dimensional, they require an extra index $\sigma = \pm 1$.

We may wish to map, as in the previous section, any function $\bar{\phi}\ (\bar{x},\bar{p})$ in the new phase space into those of the old one (x,p). We have then to decompose $\bar{\phi}\ (\bar{x},\bar{p})$ into its irreducible parts with respect to the group $T \wedge I$ and, as they will be characterized by the irrep λ, $0 \leq \lambda < 1$ and row $\sigma = \pm 1$, we see that there will be a correspondance between a scalar $\bar{\phi}$ and a "spinor" ϕ i.e.

$$\bar{\phi}\ (\bar{x},\bar{p}) \leftrightarrow \{\phi_{\lambda\sigma}(x,p);\ 0 \leq \lambda < 1, \quad \sigma = \pm 1\} \qquad (6.12)$$

The explicit form of the $\phi_{\lambda\sigma}(x,p)$ is discussed in reference[15)].

In this section we wish rather to go directly into the quantum mechanical problem of representing the canonical transformation (6.4). To achieve our purpose we first return to the analysis of the representation of the canonical transformation that takes us from x and p to the Hamiltonian (H) and time (T) variables as discussed in section 4. In there the potential was the monotonically decreasing one of (4.4) and the unitary representation (4.10) (where we proved that the phase vanishes) can also be written as

$$<x'|U|\bar{x}'>=\psi_{\bar{x}'}(x') = \int_{-\infty}^{\infty} \psi_{\nu}(x')\ \delta(\bar{x}'-\nu)\ d\nu \qquad (6.13)$$

The right hand side of (6.13) has a clear physical meaning as $\psi_{\nu}(x')$, $\delta(\bar{x}'-\nu)$ are respectively the eigenfunctions of $H(x', -i\,\partial/\partial x')$, $\bar{x}'$ with the same eigenvalue ν and thus $<x'|U|\bar{x}'>$ transforms one set into the other as required by (4.7).

At first one is tempted to follow the same procedure (6.13) for the potential discussed at the beginning of this section but an immediate difficulty arises: The spectra of $H(x',-i\partial/\partial x')$, and thus also of $J(H)$, is now discrete while that of $\bar{x}'$ is continuous. The concept of ambiguity spin appearing in (6.12) for the classical picture comes now to our rescue. We note that if we had plane waves in the new momentum space $(2\pi)^{-1/2}\exp(i\nu\bar{p}')$, $-\infty\leq\nu,\bar{p}'\leq\infty$, they are basis for irreducible representation of the translation group $\bar{p}'\to\bar{p}'+a$. If we have though the operations of the ambiguity group (6.6) i.e. $\bar{p}'\to\bar{p}'+2m\pi$, $\bar{p}'\to-\bar{p}'$, then the discussion carried above indicates that we have to write $\nu=\sigma'(n+\lambda')$, $\sigma'=\pm 1$, $n=0,1,2,\ldots$, $0\leq\lambda'<1$ to get basis for irreducible representations of the group $T\wedge I$. The passage from momentum to configuration space implies that $(2\pi)^{-1/2}\exp(i\nu\bar{p}')$ goes into $\delta(\bar{x}'-\nu)$ and thus if we want to decompose the latter into its irreducible parts with respect to $T\wedge I$ we have

$$\underset{-\infty\leq\nu\leq\infty}{\delta(\bar{x}'-\nu)} \to \underset{\sigma'=\pm1;\ 0\leq\lambda'<1;\ n=0,1,2\ldots}{\delta\left[\bar{x}'-\sigma'(n+\lambda')\right]} \qquad (6.14)$$

where underneath each of the δ functions we have indicated the ranges of the parameters appearing in it.

We can now introduce in the quantum mechanical representation the spinor indices in a similar way as they appear in (6.12) for the classical case. We still follow the procedure indicated in (6.13), but now the mapping of the discrete eigenstates $\psi_n(x')$ of the Hamiltonian

$$\left[-\frac{1}{2}\frac{\partial^2}{\partial x'^2} + V(x')\right]\psi_n(x')=E_n\psi_n(x'), \quad n=0,1,2\ldots \qquad (6.15)$$

(which are also eigenstates of $J(H)$ with eigenvalues $J(E_n)$) is not done on the full set of states $\delta(\bar{x}'-\nu)$ but onto a subset of them $\delta\left[\bar{x}'-\sigma'(n+\lambda')\right]$ characterized by[8-10] the irrep $0\leq\lambda'<1$ of $T\wedge I$ and the row $\sigma'=\pm1$ i.e.

$$\langle x'\lambda'\sigma'|U|\bar{x}'\rangle = \sum_{n=0}^{\infty} \psi_n(x')\delta[\bar{x}'-\sigma'(n+\lambda')] \tag{6.16}$$

Again we can use, as in section 4, the WKB approximation to show that the application to the representation (6.16) of operators such as $|\bar{x}|$ and $\cos[(\bar{x}/|\bar{x}|)\bar{p}]$ (which are invariant under the ambiguity group) lead in the classical limit to the appearance of the action J and the cosine of the angle ϕ. We prefer though, for didactical reasons, to carry this analysis only for a particular case of potentials of the type mentioned at the beginning of this section, that of the harmonic oscillator. We have then that J=H, $\phi = -T$ and the canonical transformation becomes

$$|\bar{x}| = \frac{1}{2}(p^2 + x^2), \tag{6.17a}$$

$$\bar{x}\,\bar{p}\,/|\bar{x}| = \text{ang tan}\,(p/x) \tag{6.17b}$$

which we can invert to have

$$x = (2|\bar{x}|)^{1/2}\cos(\bar{x}\,\bar{p}\,/|\bar{x}|) \tag{6.18a}$$

$$p = (2\,|\bar{x}|)^{1/2}\sin(\bar{x}\,\bar{p}\,/\,|\bar{x}|) \tag{6.18b}$$

Let us now apply $|\bar{x}|$ as an operator to the representation (6.16) which implies that we must multiply the latter by the eigenvalue $|\bar{x}'|$ of its ket to get

$$|\bar{x}'|\langle x'\lambda'\sigma'|U|\bar{x}'\rangle = \sum_{n=0}^{\infty} (n+\lambda)\,\psi_n(x')\delta[\bar{x}'-\sigma'(n+\lambda')]$$

$$= \sum_{n=0}^{\infty} [(\eta\,\xi+\lambda)\,\psi_n(x')]\,\delta[\bar{x}'-\sigma'(n+\lambda')] \tag{6.19}$$

where η, ξ are now the creation and annihilation operators

$$\eta = \frac{1}{\sqrt{2}}(x' - \frac{\partial}{\partial x'}), \qquad \xi = \frac{1}{\sqrt{2}}(x' + \frac{\partial}{\partial x'}) \tag{6.20}$$

Thus we have the following correspondance between operators in the new and old Hilbert spaces

$$|\bar{x}| \leftrightarrow \left\| \left[\frac{1}{2}(p^2+x^2) + \lambda' - \frac{1}{2}\right] \delta(\lambda' - \lambda'')\delta_{\sigma'\sigma''} \right\| \tag{6.21}$$

where, as discussed already for the classical case in the previous section, the observables in the old space become matrices associated with the ambiguity spin which takes the continuous set of values $0 \leq \lambda', \lambda'' < 1$ and discrete ones $\sigma', \sigma'' = \pm 1$.

Turning now our attention to the operator $\cos\left[(\bar{x}/|\bar{x}|)\,\bar{p}\right]$ we note that when acting on the ket $|\bar{x}'>$ it becomes

$$\cos\left(\frac{1}{i}\frac{\bar{x}'}{|\bar{x}'|}\frac{\partial}{\partial\bar{x}'}\right) = \frac{1}{2}\left[\exp\left(\frac{\bar{x}'}{|\bar{x}'|}\frac{\partial}{\partial\bar{x}'}\right) + \exp\left(\frac{-\bar{x}'}{|\bar{x}'|}\frac{\partial}{\partial\bar{x}'}\right)\right] \tag{6.22}$$

and as $\exp(ad/dx)\,f(x) = f(x+a)$, we see from the δ function in (6.16) that $(x'/|\bar{x}'|) = \sigma'$ and

$$\cos\left(\frac{1}{i}\frac{x'}{|\bar{x}'|}\frac{\partial}{\partial x'}\right) < x'\lambda'\sigma'|U|\bar{x}'>$$

$$= \frac{1}{2}\sum_n\left[\psi_{n-1}(x') + \psi_{n+1}(x')\right]\delta\left[\bar{x}' - \sigma'(n+\lambda')\right]$$

$$= \left[\xi(\eta\xi)^{-1/2} + (\eta\xi)^{-1/2}\eta\right] < x'\lambda'\sigma'|U|\bar{x}'> \tag{6.23}$$

where in the r.h.s. we have used well known properties of the creation and annihilation operators of the harmonic oscillator. We can then establish the following correspondance between operators

$$\cos\left[(\bar{x}/|\bar{x}|)\,\bar{p}\right]$$

$$\left\{\frac{1}{\sqrt{2}}(x+ip)\left[\frac{1}{2}(p^2+x^2)-\frac{1}{2}\right]^{-1/2} + \left[\frac{1}{2}(p^2+x^2)-\frac{1}{2}\right]^{-1/2}\frac{1}{\sqrt{2}}(x-ip)\right\}$$

$$\|\delta(\lambda'-\lambda'')\delta_{\sigma'\sigma''}\| \tag{6.24}$$

where the appearance of operators such as $\left[\frac{1}{2}(p^2+x^2)-\frac{1}{2}\right]^{-1/2}$ cause us no trouble if we think of them as acting on a complete set of harmonic oscillator functions $\psi_n(x)$ that are eigenstates of $1/2(p^2+x^2)$. Again we have a matrix form on the r.h.s. of (6.24) depending on the ambiguity spin index $0 \leq \lambda', \lambda'' < 1$, $\sigma', \sigma'' = \pm 1$, though here only the unit matrix in these indices appears.

In the classical limit, $\lambda' - 1/2$ which is in the interval $-\frac{1}{2} \leq \lambda' - \frac{1}{2} < \frac{1}{2}$, as well as $\frac{1}{2}$ itself, can be disregard ed as compared with the eigenvalues $n+\frac{1}{2}$ of the Hamiltonian and thus we recover the classical relations

$$|\bar{x}| = \frac{1}{2}(p^2+x^2), \quad \cos(\bar{x}\,\bar{p}/|\bar{x}|) = x(p^2+x^2)^{-1/2}, \tag{6.25}$$

except for the unit matrix $\|\delta(\lambda'-\lambda'')\delta\sigma'\sigma''\|$ which in any case is required in the classical picture as discussed in the previous section. The expression (6.16) is then the representation in quantum mechanics of the non-bijective canonical transformation (6.4) to action and angle variables.

7. WIGNER DISTRIBUTION FUNCTIONS AND THE REPRESENTATION OF CANONICAL TRANSFORMATIONS IN QUANTUM MECHANICS

In the previous sections we discussed the representation of canonical transformations (which are described in phase space) in quantum mechanics, which operates in Hilbert space. Thus the relation between the transformations and their representations —discussed in detail in the previous sections— is not immediately obvious.

It is known though that there are several ways of describing quantum mechanics in phase space and one —of particular interest to us— is with the help of Wigner distribution functions[16]. In this section we shall represent the canonical transformation on the basis of these distribution functions getting then instead of the representation $\langle x'|U|\bar{x}'\rangle$ in Hilbert space a kernel $\langle xp|K|\bar{x}\bar{p}\rangle$ in phase space, where the latter has a much more direct connection with the canonical transformation than the former.

We begin by recalling the definition of Wigner distribution function $\phi(x,p)$ for a given wave function $\psi(x)$ i.e.

$$\phi(x,p) = (\pi\hbar)^{-1}\int_{-\infty}^{\infty}\langle\psi|x+y\rangle\langle x-y|\psi\rangle\exp(i2py/h)\,dy \tag{7.1}$$

where we use Dirac notation (without primes) for the wavefunctions i.e. $\langle x|\psi\rangle = \psi(x)$, $\langle\psi|x\rangle = \psi^*(x)$, and restrict ourselves to a single degree of freedom. As is well known[16] the integration of $\phi(x,p)$ with respect to p or x gives respectively the probability density for the state $|\psi\rangle$ in configuration or momentum space.

We consider now the canonical transformation

$$\bar{x} = \bar{x}(x,p), \quad \bar{p} = \bar{p}(x,p), \quad \{\bar{x},\bar{p}\}_{x,p} = 1, \tag{7.2}$$

under which a classical distribution function $\phi(x,p)$ would of course transform into $\bar{\phi}(x,p)$ given by

$$\bar{\phi}(x,p) = \phi[\bar{x}(x,p), \bar{p}(x,p)] . \tag{7.3}$$

In quantum mechanics, as we showed in the previous sections, a state $|\psi>$ transforms into

$$|\psi> \to |\bar{\psi}> = U|\psi> , \tag{7.4}$$

where U is the unitary representation of the canonical transformation. Thus we have then

$$\bar{\phi}(x,p) = (\pi\hbar)^{-1}\int_{-\infty}^{\infty}<\bar{\psi}|x+y><x-y|\bar{\psi}>\exp(i2py/\hbar)dy$$

$$= (\pi\hbar)^{-1}\int\int_{-\infty}^{\infty}\int dz_+dydz_-<\psi|z_+><z_+|U^+|x+y><x-y|U|z_->$$

$$<z_-|\psi>\exp(i2py/\hbar) . \tag{7.5}$$

Writing $z_{\pm} = \bar{\bar{x}} \pm y'$ when it is associated with ψ and $z_{\pm} = \bar{x} \pm \bar{y}$ when it is associated with U and integrating over $\bar{\bar{x}},y',\bar{y},y$ with the extra factor

$$\delta(y'-\bar{y}) = (\pi\hbar)^{-1}\int_{-\infty}^{\infty}\exp[i2\bar{p}(y'-\bar{y})/\hbar]d\bar{p} , \tag{7.6}$$

we immediately arrive at the relation

$$\bar{\phi}(x,p) = \int\int_{-\infty}^{\infty}<xp|K|\bar{x}\bar{p}>\phi(\bar{x},\bar{p})d\bar{x}d\bar{p} , \tag{7.7}$$

in which the kernel K is given by

$$<xp|K|\bar{x}\bar{p}> = (2\hbar/\pi)\int\int dyd\bar{y}<\bar{x}+\bar{y}|U^+|x+y><x-y|U|\bar{x}-\bar{y}>$$

$$\exp[i(2py-2\bar{p}\bar{y})/\hbar] \tag{7.8}$$

We expect from (7.3) that in the classical limit, i.e. when $\hbar \to 0$, the kernel becomes

$$<xp|K|\bar{x}\bar{p}> = \delta[\bar{x}-\bar{x}(x,p)]\delta[\bar{p}-\bar{p}(x,p)] , \qquad (7.9)$$

where $\bar{x}(x,p)$, $\bar{p}(x,p)$ are the explicit functions of x,p appearing in (7.2).

In the kernel (7.8) the matrix elements of the representation U are precisely the $<x'|U|\bar{x}'>$ and $<\bar{x}'|U^+|x'> = <x'|U|\bar{x}'>^*$ that we have discussed in the previous sections, but where now we replace

$$x' = x \pm y, \quad \bar{x}' = \bar{x} \pm \bar{y} , \qquad (7.10)$$

and in which the $\pm$ signs are associated with U^+, U respectively.

We are then in the position of obtaining the kernel $<xp|K|\bar{x}\bar{p}>$ for any canonical transformation in which we know $<x'|U|\bar{x}'>$. Thus, for example, for the linear canonical transformation

$$\bar{x} = ax+bp, \quad \bar{p} = cx+dp, \quad ad-bc = 1, \quad b \neq 0 , \qquad (7.11)$$

we have the representation (1.8) and from (7.8) we immediately conclude that the kernel K becomes

$$<xp|K|\bar{x}\bar{p}> = \delta[\bar{x}-(ax+bp)]\delta[\bar{p}-(cx+dp)] . \qquad (7.12)$$

Thus for the linear canonical transformation the kernel coincides with its classical limit (7.9) in agreement with the fact that for this type of transformation the Poisson and Moyal[17] brackets coincide.

We consider now the less simple example of the canonical transformation to Hamiltonian (H) and time (T) for the linear potential mentioned in (4.5) i.e.

$$\bar{x} = (p^2/2m)-F_o x , \quad \bar{p} = - p/F_o , \qquad (7.13)$$

where m is the mass of the particle, F_o is a constant of dimensions of force and it is obvious that $\{\bar{x},\bar{p}\} = 1$. The equations that can be derived from (3.10) for this case are then

$$\left[-\frac{\hbar^2}{2m}\frac{\partial^2}{\partial x'^2} - F_o x'\right] <x'|U|\bar{x}'> = \bar{x}'<x'|U|\bar{x}'> , \qquad (7.14a)$$

$$-\frac{1}{F_o}\frac{\hbar}{i}\frac{\partial}{\partial x'}\langle x'|U|\bar{x}'\rangle = -\frac{\hbar}{i}\frac{\partial}{\partial \bar{x}'}\langle x'|U|\bar{x}'\rangle \ . \tag{7.14b}$$

The normalized —in the sense of $\delta(\bar{x}-\bar{x}'')$— solution of the first equation[11] is the Airy function $A\Phi(-\xi)$

$$\xi = [x'+(\bar{x}'/F_o)](2mF_o/\hbar^2)^{1/2} \ , \tag{7.15a}$$

$$A = (2m)^{1/3}\,\pi^{-1/2}\,F_o^{-1/6}\,\hbar^{-2/3} \ , \tag{7.15b}$$

$$\Phi(\xi) = (4\pi)^{-1/2}\int_{-\infty}^{\infty} \exp i[(u^3/3) + u\xi]du \ . \tag{7.15c}$$

As the second equation (7.14b) is automatically satisfied because of the form (7.15a) of the ξ variable, we conclude that[18]

$$\langle x'|U|\bar{x}'\rangle = A\Phi(-\xi) \ . \tag{7.16}$$

To get the kernel K for the canonical transformation (7.13) we just have to substitute (7.16) in (7.8), thus obtaining[18]

$$\langle xp|K|\bar{x}\bar{p}\rangle = \{2\left(\frac{m}{\hbar^2F_o^2}\right)^{1/2}\pi^{-1/2}$$

$$\Phi\left[2\left(\frac{m}{\hbar^2F_o^2}\right)^{1/3}\left(\frac{p^2}{2m} - F_o x-\bar{x}\right)\right]\}\ \delta[\bar{p}+(p/F_o)]. \tag{7.17}$$

We proceed to analyze the behaviour of this kernel. We note first when $\hbar \to 0$ the function Φ becomes[11] either very small or very rapidly oscillating except when $\bar{x} \simeq (p^2/2m)-F_o x$. Furthermore, with the help of (7.15c), we easily see that

$$\pi^{-1/2}\int_{-\infty}^{\infty} \Phi(\xi)d\xi = 1 \tag{7.18}$$

Thus the curly bracket in (7.17) tends to a δ function in the limit $\hbar \to 0$ so that the kernel K goes into its classical limit (7.9) where $\bar{x}(x,p)$, $\bar{p}(x,p)$ are given by the r.h.s. of (7.13).

To see what are the quantum corrections it is best to apply K of (7.17) to a smooth distribution function $\phi(x,p)$ rather than study it directly. We choose

$$\phi(x,p) = (\pi ab)^{-1} \exp\left[-(x^2/a^2)-(p^2/b^2)\right] , \qquad (7.19)$$

where from (7.1) we would have the relation $b = (\hbar/a)$ if ϕ is obtained from a Gaussian state in configuration space. Again using (7.15c) we obtain for the new distribution function $\bar{\phi}(x,p)$ the expression

$$\bar{\phi}(x,p) = \phi(\bar{x},\bar{p}) \sum_{k=0}^{\infty} \frac{\hbar^{2k} F_o^{2k} m^{-k}}{(2a)^{3k}} \times \left[\sum_{\substack{t=0 \\ (3k-t)\,\mathrm{even}}}^{3k} \frac{(-1)^{(3k-t)/2} (2\bar{x})^{t} (3k)!}{a^{t} t! \, [(3k-t)/2]! \, k! \, 3^{k}} \right] \qquad (7.20)$$

where $\bar{x},\bar{p}$ are given by r.h.s. of (7.13). As indicated in (7.3) $\phi(\bar{x},\bar{p})$ is the classical change in the distribution function due to the canonical transformation and it will be the only one remaining in (7.20) when $\hbar \to 0$. Thus the terms associated with higher powers of $\hbar^2$ indicate the successive quantum corrections to the distribution function when we perform the canonical transformation.

The examples discussed in this section are very special, but they clearly indicate the procedure for the general derivation of the kernel K. Even when the canonical transformation is non-bijective we can use (7.8) to obtain K though in this case we note that in the unitary representation U we have ambiguity spin indices as for $\langle x'\lambda'\sigma'|U|\bar{x}'\rangle$ in the canonical transformation to action and angle discussed in the previous section. The kernel for non-bijective transformations besides being a function of x,p; $\bar{x},\bar{p}$ is also a function of the indices that characterize ambiguity spin space.

CONCLUSION

We have tried in this paper to give a systematic outline of the work of the author and his collaborators in relation with the representation in quantum mechanics of classical canonical transformations. Initially —and in particular in relation with lineal canonical transformations— it seemed that the procedures for finding the representation were already implicit in the standard formalism of quantum mechanics as presented in Dirac's book. Very quickly though we found out that for non-linear and, particularly, for non-bijective canonical transformation, the representation problem becomes very subtle. In the non-bijective case even the canonical transformations themselves have to be reformulated, with the help of the concepts of ambiguity group and ambiguity spin, so as to recover a bijectiveness fundamental for the representation problem in quantum mechanics. Once the obstacle was understood we were able to find representations in quantum mechanics for such non-bijective canonical transformations as the ones that lead to action and angle variables.

While the framework developed here seems interesting from a conceptual standpoint, one would like to ask whether it has physical applications. For the linear canonical transformations they are well known[5,6] but until recently the author was not aware of applications of the representations of non-linear and non-bijective canonical transformations. Recent work on the microscopic derivation of nuclear collective models[19] has shown though that there may be interesting applications in this field[20].

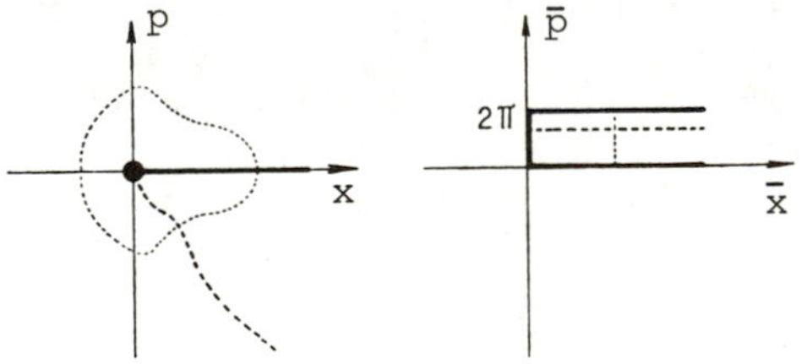

Fig. 1. Mapping of phase spaces by the canonical transformation leading to action and angle variables for a potential taking the value ∞ at $x = \pm\infty$ with a single minimum 0 at $x = 0$. Heavy, dotted and dashed lines correspond to each other.

REFERENCES

1. P.A.M. Dirac, Quantum Mechanics, Clarendon Press, Oxford 1947
2. C. Itzykson, Comun. Math. Phys. 4, 92 (1967)
3. V. Bargmann, "Group Representations on Hilbert Spaces of Analytic Functions in Analytic Methods of Mathematical Physics", edited by P.P. Gilbert and R.G. Newton (Gordon & Breach, New York, 1968)
4. M. Moshinsky and C. Quesne, J. Math. Phys. 12, 1772 (1971)
5. M. Moshinsky and C. Quesne, "Oscillator Systems" in Symmetry Properties of Nuclei, Proceedings of the 15th Solvay Conference on Physics (Gordon & Breach, London 1974)
6. P. Kramer, M. Moshinsky and T.H. Seligman, "Complex Extensions of Canonical Transformations and their Representation in Quantum Mechanics" in Group Theory and its Applications, Vol. III, edited by E.M. Loebl, (Academic Press, New York 1975)
7. P.A. Mello and M. Moshinsky, J. Math. Phys. 16, 2017 (1975)
8. M. Moshinsky and T.H. Seligman, Ann. Phys. (N.Y.) 114, 243 (1978)
9. M. Moshinsky and T.H. Seligman, Ann. Phys. (N.Y.) 120, 402 (1979)
10. M. Moshinsky and T.H. Seligman, Ann. Phys. (N.Y.) 121, 458 (1980)
11. L.D. Landau and E.M. Lifshitz, Quantum Mechanics, pp. 73-74, 158-165, Oxford Univ. Press (London, New York 1959)
12. L.V. Ahlfors, Complex Analysis, pp. 79-81, (McGraw-Hill, New York, 1953)
13. P. Kramer, M. Moshinsky and T.H. Seligman, J. Math. Phys. 19, 683 (1978)
14. H. Goldstein, Classical Mechanics, (Addison-Wesley, Reading, Mass. 1957)
15. M. Moshinsky and T.H. Seligman, J. Math. Phys. (submitted for publication) 1980
16. E.P. Wigner, Phys. Rev. 40, 749 (1932)
17. J.E. Moyal, Proc. Cambridge Phil. Soc. 45, 99 (1949)
18. G. García-Calderón and M. Moshinsky, J. Phys. A Math. Gen. 13, L185, (1980)
19. V. Vanagas, The Microscopic Nuclear Theory within the Framework of the Restricted Dynamics, Lecture Notes in Physics, University of Toronto, 1977
20. E. Chacón, M. Moshinsky and V. Vanagas, J. Math. Phys. February or March (1981).

AIP Conference Proceedings

		L.C. Number	ISBN
No.1	Feedback and Dynamic Control of Plasmas	70-141596	0-88318-100-2
No.2	Particles and Fields - 1971 (Rochester)	71-184662	0-88318-101-0
No.3	Thermal Expansion - 1971 (Corning)	72-76970	0-88318-102-9
No.4	Superconductivity in d-and f-Band Metals (Rochester, 1971)	74-18879	0-88318-103-7
No.5	Magnetism and Magnetic Materials - 1971 (2 parts) (Chicago)	59-2468	0-88318-104-5
No.6	Particle Physics (Irvine, 1971)	72-81239	0-88318-105-3
No.7	Exploring the History of Nuclear Physics	72-81883	0-88318-106-1
No.8	Experimental Meson Spectroscopy - 1972	72-88226	0-88318-107-X
No.9	Cyclotrons - 1972 (Vancouver)	72-92798	0-88318-108-8
No.10	Magnetism and Magnetic Materials - 1972	72-623469	0-88318-109-6
No.11	Transport Phenomena - 1973 (Brown University Conference)	73-80682	0-88318-110-X
No.12	Experiments on High Energy Particle Collisions - 1973 (Vanderbilt Conference)	73-81705	0-88318-111-8
No.13	π-π Scattering - 1973 (Tallahassee Conference)	73-81704	0-88318-112-6
No.14	Particles and Fields - 1973 (APS/DPF Berkeley)	73-91923	0-88318-113-4
No.15	High Energy Collisions - 1973 (Stony Brook)	73-92324	0-88318-114-2
No.16	Causality and Physical Theories (Wayne State University, 1973)	73-93420	0-88318-115-0
No.17	Thermal Expansion - 1973 (lake of the Ozarks)	73-94415	0-88318-116-9
No.18	Magnetism and Magnetic Materials - 1973 (2 parts) (Boston)	59-2468	0-88318-117-7
No.19	Physics and the Energy Problem - 1974 (APS Chicago)	73-94416	0-88318-118-5
No.20	Tetrahedrally Bonded Amorphous Semiconductors (Yorktown Heights, 1974)	74-80145	0-88318-119-3
No.21	Experimental Meson Spectroscopy - 1974 (Boston)	74-82628	0-88318-120-7
No.22	Neutrinos - 1974 (Philadelphia)	74-82413	0-88318-121-5
No.23	Particles and Fields - 1974 (APS/DPF Williamsburg)	74-27575	0-88318-122-3
No.24	Magnetism and Magnetic Materials - 1974 (20th Annual Conference, San Francisco)	75-2647	0-88318-123-1
No.25	Efficient Use of Energy (The APS Studies on the Technical Aspects of the More Efficient Use of Energy)	75-18227	0-88318-124-X

No.26	High-Energy Physics and Nuclear Structure - 1975 (Santa Fe and Los Alamos)	75-26411	0-88318-125-8
No.27	Topics in Statistical Mechanics and Biophysics: A Memorial to Julius L. Jackson (Wayne State University, 1975)	75-36309	0-88318-126-6
No.28	Physics and Our World: A Symposium in Honor of Victor F. Weisskopf (M.I.T., 1974)	76-7207	0-88318-127-4
No.29	Magnetism and Magnetic Materials - 1975 (21st Annual Conference, Philadelphia)	76-10931	0-88318-128-2
No.30	Particle Searches and Discoveries - 1976 (Vanderbilt Conference)	76-19949	0-88318-129-0
No.31	Structure and Excitations of Amorphous Solids (Williamsburg, VA., 1976)	76-22279	0-88318-130-4
No.32	Materials Technology - 1975 (APS New York Meeting)	76-27967	0-88318-131-2
No.33	Meson-Nuclear Physics - 1976 (Carnegie-Mellon Conference)	76-26811	0-88318-132-0
No.34	Magnetism and Magnetic Materials - 1976 (Joint MMM-Intermag Conference, Pittsburgh)	76-47106	0-88318-133-9
No.35	High Energy Physics with Polarized Beams and Targets (Argonne, 1976)	76-50181	0-88318-134-7
No.36	Momentum Wave Functions - 1976 (Indiana University)	77-82145	0-88318-135-5
No.37	Weak Interaction Physics - 1977 (Indiana University)	77-83344	0-88318-136-3
No.38	Workshop on New Directions in Mossbauer Spectroscopy (Argonne, 1977)	77-90635	0-88318-137-1
No.39	Physics Careers, Employment and Education (Penn State, 1977)	77-94053	0-88318-138-X
No.40	Electrical Transport and Optical Properties of Inhomogeneous Media (Ohio State University, 1977)	78-54319	0-88318-139-8
No.41	Nucleon-Nucleon Interactions - 1977 (Vancouver)	78-54249	0-88318-140-1
No.42	Higher Energy Polarized Proton Beams (Ann Arbor, 1977)	78-55682	0-88318-141-X
No.43	Particles and Fields - 1977 (APS/DPF, Argonne)	78-55683	0-88318-142-8
No.44	Future Trends in Superconductive Electronics (Charlottesville, 1978)	77-9240	0-88318-143-6
No.45	New Results in High Energy Physics - 1978 (Vanderbilt Conference)	78-67196	0-88318-144-4
No.46	Topics in Nonlinear Dynamics (La Jolla Institute)	78-057870	0-88318-145-2
No.47	Clustering Aspects of Nuclear Structure and Nuclear Reactions (Winnepeg, 1978)	78-64942	0-88318-146-0
No.48	Current Trends in the Theory of Fields (Tallahassee, 1978)	78-72948	0-88318-147-9
No.49	Cosmic Rays and Particle Physics - 1978 (Bartol Conference)	79-50489	0-88318-148-7

No. 50	Laser-Solid Interactions and Laser Processing - 1978 (Boston)	79-51564	0-88318-149-5
No. 51	High Energy Physics with Polarized Beams and Polarized Targets (Argonne, 1978)	79-64565	0-88318-150-9
No. 52	Long-Distance Neutrino Detection - 1978 (C.L. Cowan Memorial Symposium)	79-52078	0-88318-151-7
No. 53	Modulated Structures - 1979 (Kailua Kona, Hawaii)	79-53846	0-88318-152-5
No. 54	Meson-Nuclear Physics - 1979 (Houston)	79-53978	0-88318-153-3
No. 55	Quantum Chromodynamics (La Jolla, 1978)	79-54969	0-88318-154-1
No. 56	Particle Acceleration Mechanisms in Astrophysics (La Jolla, 1979)	79-55844	0-88318-155-X
No. 57	Nonlinear Dynamics and the Beam-Beam Interaction (Brookhaven, 1979)	79-57341	0-88318-156-8
No. 58	Inhomogeneous Superconductors - 1979 (Berkeley Springs, W.V.)	79-57620	0-88318-157-6
No. 59	Particles and Fields - 1979 (APS/DPF Montreal)	80-66631	0-88318-158-4
No. 60	History of the ZGS (Argonne, 1979)	80-67694	0-88318-159-2
No. 61	Aspects of the Kinetics and Dynamics of Surface Reactions (La Jolla Institute, 1979)	80-68004	0-88318-160-6
No. 62	High Energy e^+e^- Interactions (Vanderbilt , 1980)	80-53377	0-88318-161-4
No. 63	Supernovae Spectra (La Jolla, 1980)	80-70019	0-88318-162-2
No. 64	Laboratory EXAFS Facilities - 1980 (Univ. of Washington)	80-70579	0-88318-163-0
No. 65	Optics in Four Dimensions - 1980 (ICO, Ensenada)	80-70771	0-88318-164-9
No. 66	Physics in the Automotive Industry - 1980 (APS/AAPT Topical Conference)	80-70987	0-88318-165-7
No. 67	Experimental Meson Spectroscopy - 1980 (Sixth International Conference , Brookhaven)	80-71123	0-88318-166-5
No. 68	High Energy Physics - 1980 (XX International Conference, Madison)	81-65032	0-88318-167-3
No. 69	Polarization Phenomena in Nuclear Physics - 1980 (Fifth International Symposium, Santa Fe)	81-65107	0-88318-168-1
No. 70	Chemistry and Physics of Coal Utilization - 1980 (APS, Morgantown)	81-65106	0-88318-169-X
No. 71	Group Theory and its Applications in Physics - 1980 (Latin American School of Physics, Mexico City)	81-66132	0-88318-170-3